The Adventures of an Itinerant Engineer

The Adventures of an Itinerant Engineer

By

Douglas Young

Blackie & Co
Publishers Ltd

A BLACKIE & CO PUBLISHERS PAPERBACK

First published in 2003

A CIP catalogue record for this title is
available from the British Library

ISBN 1 84470 046 1

Blackie & Co Publishers Ltd
107-111 Fleet Street
LONDON EC4A 2AB

The World Mercator Projection

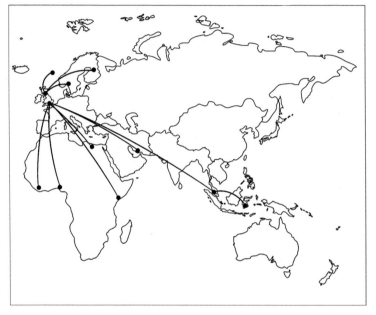

CONTENTS

PART ONE – EAST AFRICA, 1955-1957
General map of Kenya and Tanganyika, with routes and locations
marked

Section 1
NAIROBI AND MOMBASA

Section 2
THE GREAT WATERFALL SAFARI

Section 3
PANGANI FALLS

PART TWO – TIREE, 1958-1959
General map of the United Kingdom

PART THREE – THE PERSIAN PERIOD, 1959-1961
General map of Iran

PART FOUR – SIERRA LEONE, NIGERIA AND LIBYA, 1962-1970
General map of Sierra Leone

Section 1
GUMA DAM

Section 2
PEPEL PORT, NIGERIA AND LIBYA
General maps of Nigeria and Libya

Section 3
RUTILE

PART FIVE – ANGLESEY, INVERGORDON, INDONESIA AND THE OIL BUSINESS, 1970-1989

Section 1
ANGLESEY

Section 2
INVERGORDON

Section 3
INDONESIA
General map of Indonesia

Section 4
THE OIL BUSINESS

PART ONE –
EAST AFRICA, 1955-1957
General map of Kenya and Tanganyika

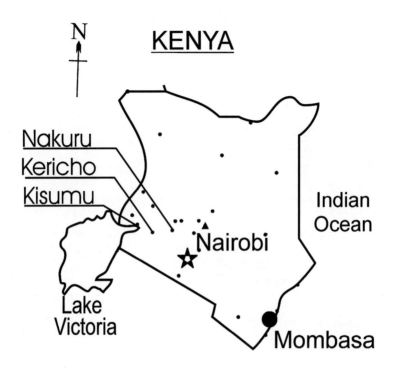

KENYA

N

Nakuru
Kericho
Kisumu

Nairobi

Indian
Ocean

Lake
Victoria

Mombasa

0   500 Miles

TANZANIA

Lake Victoria

Arusha

N

Moshi
Korogwe
Pangani

Falls
Kilosa

Dar es Salaam

Morogoro

Iringa

Lake Tanganyika

Mbeya

Songea

Lake Nyasa

0   500 Miles

3

SECTION 1 – NAIROBI AND MOMBASA

CHAPTER 1
The Kit List

It all began with a casual glance through the 'Situations Vacant' columns of the morning paper, as I was on my way into the Glasgow office one dreary November morning. 'Civil Engineer required for interesting work in East Africa', I read.

During the remainder of the day the thought lingered in my mind that a spell in Africa. might be the antidote for another winter in Scotland, and when I returned home in the evening, and heard the wireless forecasting more rain and fog for the weekend, I sat down and wrote off for some further information, enclosing an outline of my experience.

Rather to my surprise I had a letter back in a few days, informing me that the work was in connection with a new electric power line from the Owen Falls in Uganda to Nairobi in Kenya – a distance of about 300 miles – and asking me to telephone a London number to arrange an interview there. Next day I spoke to the partner in charge of my section of the office concerning my prospects, and he agreed that a period overseas would broaden my experience – and, of course, the pay would be better.

The meeting in London was arranged, and I spent the next few days in intensive research. This would be the first time I had attended a job interview, having gone straight from school to an articled apprenticeship, then to the army for two years' national service, followed by three years as a junior engineer with the same firm.

I had no idea of the questions, both technical and personal, which might be posed, and as my knowledge of Africa was strictly limited I visited the local library, looked up some particulars about the company in question, leafed through a number of travel and reference books, and consulted with two colleagues who had been in Kenya during their flying training with the RAF during the war.

Their lasting impressions were of heat, dust and insects. They both related the old story about the young airman who came into the squadron office complaining: "That bloody old Albert was a bit hostile this morning when I was shifting him off the end of the runway," and then collapsed in a heap when informed: "That wasn't the CO's lion you were shifting. Look, there he is over there. He has been sitting under that tree in the garden all morning!"

Somewhat confused by all this conflicting information, I decided that my best course would be to give the impression of quiet confidence and reasonable sobriety, and hope not to be rumbled.

I travelled down on the overnight train and arrived in the London office in good time, where I was fed tea and biscuits, and then wheeled in to meet the director in charge of East Africa operations. He explained the scope and details of the work, and we discussed the difficulties of surveying in wild and remote country, and then more general topics. I asked about the situation with the Mau Mau, which was then at its height, and was known to be fairly active in the general area of the work. He replied that the company had not suffered much interference with their work, and that, anyway, until I was fluent in Swahili, it would be most unlikely that I would be called up to the Kenya Police Reserve; and also that my recent experience in the military would stand me in good stead (what good my time spent trundling around Salisbury Plain in a tank, and three territorial army summer camps, would have done to improve my chances against a terrorist with a panga still left me in some doubt).

Anyway, he said he would be pleased to have me join, providing I could pass the medical examination, and had no criminal record, which might cause problems with the fidelity assurance and guarantee – whatever that might be. I reflected briefly on the number of times I had been drinking after hours, and on the sabbath, and how often I had exceeded the speed limit, or parked my car in forbidden places; but as I had never actually been apprehended, I was able to reassure him truthfully that, indeed, I was no criminal.

I was then directed to the company doctor, in Harley Street, who was a specialist in tropical medicine, and also an authority on wildfowl – a subject we discussed at some length as he subjected my person to prods, and gave hammer blows to my knees and ankles. He then asked me to jump up and down a few times, took my blood pressure, and, having discovered that I could read a notice on the far side of his consulting room, declared me fit for service in Africa – with the comment that, if I could stand the rain in winter and the midges in summer in Scotland, Africa would be like a long holiday.

So saying, he summoned his nurse, and – together – they filled my arms with vaccinations and inoculations against all known forms of disease, and gave me a bundle of certificates to prove it.

I returned home with very stiff arms, and about a week later received a letter confirming my state of good health, enclosing a cheque for my expenses, a draft contract of employment, some guff about banking and insurance, a paper on tropical hygiene, and some information concerning suitable clothing for work in the tropics. Studying the papers on health gave me to wonder if going to Africa was such a good idea after all, what with malaria, typhoid, sandfly fever, sleeping sickness, etc. – and the chances of being bitten by a snake, taken by lions, trampled by elephants, charged by rhinos, gored by a buffalo, not to mention being chopped up by the Mau Mau, or, much more likely, accidentally shot by the Home Guard. The last set of papers was about suggested clothing and equipment, and came with a note stating that the vast sum of £150 (well in excess of my monthly pay) would be dispatched when my letter of acceptance was received in London. This appeal to my mercenary instincts, and reflections concerning my current cash flow situation, did the trick. I signed up on the spot, and began my preparations for departure, discarding all my earlier misgivings and worries.

Armed with the kit list, I made an expedition into the interior of Glasgow, where I found a shop with the splendid title of Naval, Military and Tropical Outfitters emblazoned in gold lettering across the windows. The displays featured figures in naval uniforms, with gold sleeves, senior officers of line regiments in all their finery, and a tableau of some people dressed in safari suits, set against a jungle background, being served drinks by dark-skinned attendants wearing long white nightshirts, and red flowerpot hats. Undaunted by this display, I entered the shop and presented my kit list to an assistant.

Very quickly the counter was stacked up with an astonishing pile of equipment. For a start, there was a pith helmet, complete in its own carrying case (about one coolie-load already); jackets and shirts, with an amazing array of pockets and flaps; a curious garment designated 'back protector', to be worn round the waist and held in position with a harness of tapes; shorts (?), which came down to about mid-calf level, rather after the style of the 'New Look' recently adopted by ladies; underwear and stockings more appropriate to persons contemplating an Arctic expedition; and a pair of special mosquito boots, also considered proof against low-flying snakes, which would probably have given a small crocodile a bad case of indigestion.

As this pile grew, and I contemplated the rapid depletion of my £150, it occurred to me that this list must have been prepared by some humorist, probably with a financial interest in Boer War army surplus stores, and this suspicion was confirmed when the shop manager remarked in passing that Dr Livingstone had set off with much the same equipment. Recalling what had happened to him (as a result of my recent study of African affairs), I resolved to have none of it, and settled for a lightweight dinner suit of the most modern cut, a tin trunk with allegedly insect-proof seals, a pair of desert boots and a bush hat.

These purchases left a reasonable balance in my account, and this was allocated under two headings: one for riotous living before my departure, and the remainder for living expenses and such clothing as I might find available in Kenya when I arrived.

I had been reliably informed that clothing made locally was about half the UK price, and equally susceptible to the damage inflicted upon it by the local laundries, whose practice it was to beat the garments on stones in the river, and then burn small holes all over, as a result of sparks blowing out of the charcoal iron. I later found that there was more than a grain of truth in these allegations.

I was very busy during the next short period, attending numerous farewell parties, disengaging delicately from some social entanglements while continuing the round of seasonal festivities, closing my account with the Inland Revenue, and finding a good home for my old car.

I made a number of trial packings of my trunk to see how the weight allowance on the flight out would match up with my equipment, and was alarmed to note that by including my technical books I would be well over the limit. The only way round this dilemma was to buy a canvas holdall for my books, and declare them as reading matter for the voyage. My bush hat also gave me some concern; I could just imagine the interest it might cause among the street urchins if worn in Glasgow, and it would obviously be very difficult to pack without seriously impairing its shape.

The great day of departure arrived, and there I was in my best Sunday go-to-the-meeting suit, settled in a corner seat of the London train, my tin trunk in the guard's van, my books and my unworn hat on the luggage rack over my head. The seats opposite were taken by a lady and her young son of about six or seven. He sat staring at me for some time before asking, "Ma, is he a cowboy?"

This brought much mirth from the other occupants of the compartment, and having assured them that I was not a cowboy, and did not have a gun, I set out along the train to see if there was anyone of my acquaintance aboard. On returning to my seat I discovered that some oaf had dumped his suitcase on top of my fine new hat, and its beautiful pristine shape had been damaged beyond recovery. I saw at once, however, that I could now stow it in the holdall, and would no longer need to worry about looking a bit of a charlie walking about London in my best suit and wearing a bush hat.

I had been booked into one of the railway hotels for the night, and, as I was registering, I observed another young man who appeared destined for the same job. He must have followed the kit list to the letter, and was burdened with the pith helmet, complete with carrying case, and wearing the regulation mosquito boots. I introduced myself, exercising what I thought was commendable restraint by not presuming him to be Dr Livingstone, and we arranged to meet in the bar later, before having what we imagined would be our last civilised meal for a couple of years.

In honour of this occasion I dressed in my new dinner suit, and when I approached my new colleague he failed to recognise me. Obviously he thought I was one of the waiters. Once over this minor misunderstanding we had a most enjoyable dinner, and saw off the remainder of my allocated funds in a brandy and a fine cigar. Then upstairs for a swift repacking in readiness for an early start in the morning.

CHAPTER 2
The Voyage out

Next morning it was up with the band and, after an excellent breakfast, off by coach to the airfield at Heathrow. This was developing from a very basic pair of runways, taxi tracks and aircraft hardstandings, serviced by a number of wooden huts, and a few large hangars. We were soon through the formalities of customs and passport controls, and, after a short wait, being ushered out to our waiting aircraft, a Canadair Argonaut of BOAC.

Once aboard I was greatly impressed by the size of the machine, and the amount of space available to each passenger, my previous experience of passenger aircraft being limited to flights to the Channel Islands in a de Havilland Rapide. The free drinks service was an additional attraction. After a brief welcome by the crew, and a demonstration of how to put on a life jacket, the engines were fired up – and we were off. The weather was clear and frosty, and soon we were over the Channel, then across France, over the Alps – which looked magnificent in their deep covering of sparkling snow – down the coast of Italy, and then coming in over Rome for our first refuelling stop. The geometrical pattern of the streets and squares stood out in strong relief against their shadows in the glare of the midday sun.

The airport, which was under reconstruction, as were most airports at that time (and, it appears, ever since), was an extremely scruffy place, memorable only for the extortionate cost of a small glass of whisky – which I chose as the alternative to a cup of so-called tea – and a sticky bun, which had probably been rejected by one of Hannibal's elephants, to judge by the texture.

After a short break we were off again, across the spine of Italy, over the sea to Greece, and then Crete, then once more over the sea in the gathering dusk towards Egypt. We landed in Cairo in the dark, and trundled around the country for about ten minutes before coming to a halt outside yet another airport in the throes of reconstruction, where we were escorted off, had our passports confiscated while we remained in the transit lounge, and were proffered tea and wads. (A 'wad' was a form of stale, flat bun, previously experienced only in Salvation Army vans or NAAFI canteens.)

Our section of the building was partitioned off from the

domestic passengers by a low screen of plants and potted shrubs, and while we were relaxing some sort of commotion developed on the other side of the screen. Policemen and soldiers were to be seen running around, waving their guns in the air, in pursuit of a young man. None of the customers in the near vicinity of this pursuit seemed much interested, and we transit passengers, not to let the side down (being British), sat fascinated, outwardly calm but, as a man – I believe – ready to dive under the tables if any shooting started. The pursuit receded through the doors of the kitchen, and normal service was soon resumed. For all we knew, this was a performance for the benefit of visitors, or perhaps they were shooting a scene for a film. Whatever it was, it added a bit of local colour to our impressions of Cairo.

Soon we were off again in our comfortable seats, heading south towards Khartoum, for our overnight stop. We left the lights of Cairo behind, and continued over the desert, following the course of the Nile, which could be distinguished as a darker patch against the otherwise monotonous grey. Soon even this became indistinct, and we continued in bright moonlight until we could see the lights of the airfield at Khartoum, where we landed without incident, taxied to the stand and disembarked with our overnight bags to await the arrival of the bus that would take us into the terminal building and the hotel. As we stood waiting I became aware of the peculiar smell of Africa – a mixture of woodsmoke, old fur, decaying vegetation, sweat, goats and camels. The only buildings visible in the distance were the hangars and service facilities for the airport, and, in the silence, the ticking sound of the engines cooling signalled the end of this stage of our voyage.

After the very minimum of formalities we moved into the hotel nearby. This was a spacious low building, with a central complex of lounges, bars and dining rooms, and numerous small bungalows set in the grounds. The hotel had been an overnight stopping place for the old Imperial Airways, when they ran the flying boat service to the Cape. It was comfortably furnished and spotlessly clean, and the staff appeared to be helpful and friendly.

After a pleasant dinner, and a couple of drinks in the bar, I retired to my room in one of the bungalows. It was a three-sided affair, with one side opening onto a deep veranda, with a tiled floor, and solid Victorian furniture. If this was to be the state of accommodation available in Africa, I could see that I might learn to enjoy my tour.

This state of euphoria was soon to be disturbed.

I had switched on the light in the bedroom and gone into the bathroom for a few moments, all innocent. When I returned I was appalled to find the place full of the local insect population. Apart from things whizzing round the light, there appeared to be a full-scale invasion of beetles scuttling across the floor, pursued by a pack of small lizards, and – for all I knew – snakes and scorpions as well.

This situation caused me to pause and consider.

I needed a plan.

First, I had to clear the bed and set up the mosquito net, which was draped in an elegant knot from a hoop hanging from the ceiling.

Second, I had to get into bed and undress; there was no way I would consider leaving my clothes outside the net overnight.

Third, I had to extinguish the light. There was a secondary problem here, as this would entail a sortie outside the net, and God alone knows what I might tread on, or collect, during this performance.

Getting the net set up was obviously the first priority, but this turned out to be much more difficult than originally imagined. The knot had been tied by some fiendish dervish, and was unwilling to unravel, despite all my efforts. The attachments of the hoop from the ceiling appeared pretty frail, and looked unlikely to stand up to much strong-arm treatment from the floor, so I decided that I should stand upon the bed the better to reach it.

Surprise, surprise.

The bed was made up from a series of wooden slats, covered by a thin mattress, and as I stood up the slats parted, and I descended gracefully to the floor, up to my knees in mattress. The immediate reaction was to get it up off the floor, quickly, before it became infested with beetles etc., and then to try to unravel the net while balancing on the frame of the bed.

This strategy succeeded. After a bit of a struggle the knot unravelled, the net came down, and I was enveloped in a cloud of dust and long-dead, assorted winged insects, which had probably been lying there since the last flying boat had departed.

Eventually everything was sorted out, and just before falling asleep I recalled the list of instructions and advice attached to the kit list concerning insect bites, and the probable presence of scorpions in my shoes in the morning. Just as a matter of insurance, I took my shoes in with me under the net.

In the clear light of dawn everything seemed to have sorted itself out; all the carnivores had eaten one another, and there were no scorpions visible.

What a relief.

After a quick shower and breakfast we were off again, still flying south over the desert towards Entebbe, our next stop. We came in over the lake, the blue of the water and the green of the vegetation around the shores a welcome change in colour after the greys and browns of the desert. After touchdown we rolled up to the airport buildings. This was a much poorer establishment than Khartoum, with a standard RAF-type control tower, a few tin sheds, and a touch of the grass huts showing through around the edges. We only stopped briefly to refuel, and were on our way again, on the last leg towards Nairobi.

More brown desert passed beneath, with the odd huddle of huts around infrequent waterholes. The cabin crew came around with coffee, and a questionnaire concerning the flight, where we could make comments and observations to improve the service, and which was to be posted in the special box at the airport after landing. My colleague and I both made favourable comments, and when we landed duly posted the leaflets in the special box provided.

CHAPTER 3
Nairobi

After landing we moved into the customs hall to clear our luggage, and, although my gear turned up almost at once, my colleague was not so fortunate. After some frantic radio calls, an official came along to announce that there had been some mistake, and his gear was probably either in Cairo or Khartoum; or, more correctly, it *might* have been at one of these places, but had now vanished, and they were offering him very long odds against ever seeing it again. He was naturally very cross. Apart from losing his baggage, he had just posted a letter to say how well he had been looked after. Some more practised official then appeared, and produced a chitty which would allow goods to the value of £100 to be purchased, and a claim form which would cover any other replacement goods. He was somewhat mollified by this arrangement, and we continued through into the main reception hall of the airfield – another large tin shed full of people seeing people off, or waiting the arrival of their friends.

Along one wall stood a line of professional 'meeters' – a fairly cosmopolitan lot: some dressed in safari suits; well turned-out, uniformed government drivers; and a nondescript crew of individuals bearing aloft hand-printed notices declaring their company name, or the name of the passenger just arrived.

Our chap was well dressed in clean khaki drill, well-polished brass buttons, and with the company name on his hatband. We made ourselves known to him, and, after some difficult explanations about why there was only one set of luggage, he organised a porter, and we went out into the blinding sunlight of the car park. My luggage was loaded, and we climbed into the front of a Chevrolet pick-up truck

It was not until then that I noticed that his uniform was incomplete. He wore what used to be called in Highland regiments 'hose tops' – a pair of stockings without feet. This allowed the soldier to wear what he felt most comfortable in on his feet, without arousing ribald remarks from his comrades, or unkind comments from his sergeant major. The space between his boots and the hose top was covered by spats or puttees – a very sensible idea, as such garments helped to prevent small stones from getting down inside his boots.

I digress.

Our driver had the largest pair of feet I had ever seen, and after starting up the engine he wrapped his right foot around the accelerator pedal, and by alternately pushing it to the floor, and pulling it back again, produced a loud series of backfires. Satisfied by this noise, he gave a blast on the horn, and we set off into the traffic towards Nairobi.

This was my first experience of driving in Africa and, although it only lasted for about 40 minutes, it felt like a whole day at the Grand Prix. Our driver seemed to believe that if he made enough noise with the horn and the engine he generated a safety zone around himself, and that all other road-users would keep out of his way.

He seemed to be correct in this belief, though he was likely to encounter some trouble should he meet someone else with the same idea coming in the opposite direction.

When we arrived at the company headquarters, we were directed to the office of the chief engineer, who lived on the fifth floor of the building. Persons newly arrived at a high altitude, such as Nairobi, tend to suffer a marked shortage of breath for a few days, and by the time we arrived at the top of the stairs I was blowing like a grampus and pretty short of introductory conversation. I learnt later that this ploy was practised on all newly arrived visitors, as it worked wonders for the egos of regular high-altitude dwellers when new young men arrived breathless, and sometimes speechless.

We were warmly welcomed by the chief engineer, who advised that my colleague would go straight off to the Owen Falls that afternoon, and that it had been decided that I might be of more use on the construction of the new steam power station in Mombasa. I was given a train ticket for the overnight train the following day, and then taken round to the Torrs Hotel, where I had been booked in for the night. After another short trip with the demon driver, I registered in the hotel, and went outside for a quick look around.

Torrs Hotel was an old building in the centre of town, with frontage on Delamere Avenue – at that time the only stretch of dual carriageway between the Cape and Cairo. The avenue was lined with beautiful acacia and jacaranda trees, then in full bloom, and there were people sitting making and selling curios within the comparative safety of the central reservation. The traffic was speeding up and down in a manner that could be described as somewhat reckless.

Sometime later I worked out that they all suffered from overconfidence as to the likelihood of another vehicle approaching in the opposite direction.

The gardens and the entrance way to the hotel were full of people in for a sundowner; the immediate impression was of a lot of police and army uniforms, and a considerable thirst being slaked. Everybody looked incredibly healthy.

After some refreshment I returned to my room, which was on the first floor, and at the front. From the balcony I had a splendid view of the avenue and the activity below. I sat down with a cool beer to recover my breath, and when I woke up again it was almost dark, so I had a quick bath and went down for dinner.

After an excellent meal, served by white-gowned and red-hatted waiters, to the accompaniment of a three-piece orchestra, I went into the bar, where I was immediately recognised as a new arrival. I was advised that Mombasa was a hot and sticky place, and the demon drink was to be avoided there at all costs. This information was in exchange for the latest gossip from London, and as I had only spent one night there recently, this was a subject about which I knew nothing. I also learnt to my relief that the activities of the Mau Mau were pretty limited on the coast, probably because it was so damned hot and sticky.

I returned to my room after a nightcap, and was relieved to see that some local hero had unravelled the mosquito net, and that the light switch could be operated from within the net. Just as I was turning in there was the noise of a car driving very fast down the avenue, followed by a fusillade of shots. I rushed out onto the balcony to see if I could discover what was going on, just in time to observe a car disappear round a corner. Whether the shooting was at, or from, this vehicle I never discovered, but the thought did occur that I would be a good target, exposed as I was, for the resident population of insects. This sent me diving for the security of the net. Once ensconced, I conducted a brief but intense study of Swahili, with particular regard to insects. The generic word for all small beasties was 'doodoo', and, as far as I was concerned, they were all full of malice towards my person, and probably deadly poisonous.

Next morning, after a delicious breakfast, which included a number of tropical fruits previously unknown to me, I set out in a taxi to confirm my booking and sleeping car berth on the evening train. Once more I found I had put my life in the hands of another barefoot maniac of a driver, who also had adopted the practice of wrapping his toes around the control pedals and revving the engine unmercifully. I resolved that, when I had the choice of a personal vehicle, I would opt for a small armoured car.

After my tasks at the railway station were completed, I spent the remainder of the day in sightseeing, and making up my wardrobe with suitable colonial items. Had I but known, I could have made a much larger allocation towards riotous living before my departure, and still have had enough to start me off in Kenya. I also engaged in a bit of market trading for a leopard tail hatband for my bush hat and knocked about 50% off the asking price, but I chickened out at the end.

In the early evening I returned to the station and claimed my berth. The attendant explained the workings of the bed, which was folded into the wall of the compartment, and the mechanics of the fan and windows and the mesh screens. I walked along the platform to inspect the locomotive, an enormous Garrett engine, looking decidedly top-heavy on its narrow track, hissing and emitting small clouds of steam. As it had been running up and down this railway for some years, I supposed that it probably was quite stable, but I felt that I would have been happier if the wheels had been a bit further apart.

When I returned to my cabin my bed had been let down and prepared, the mesh screens had been set up, and the fan was running. The attendant came along and suggested that I might be more comfortable in the dining car, so I followed him along the train and into the bar. The dining car was set out with damask tablecloths, and the silverware and glassware were shining brightly. All very civilised, I thought, as I sat down in a corner with a cool beer in my hand, waiting for the train to start.

For the first few miles the tracks ran through the dreadful shanty towns of the suburbs, where the houses, or rather hovels, were made from flattened kerosene tins, rusty panels taken from long-extinct motorcars and lorries, pieces of packing cases and crates. Despite this awful squalor, cheerful children ran alongside the train, waving happily, probably in the fond hope that the passing travellers would throw them some small change from the windows.

We were soon out of town and running across the plains,

travelling slightly faster. The road ran alongside the railway on one side, and the train was being overtaken by a number of crazy-looking lorries, all with bright, painted designs and hopeful slogans daubed across the tops of the drivers' cabins. On the other side of the train, herds of gazelle, wildebeest and zebra were grazing amongst the odd-looking trees with curious horizontal foliage.

After yet another large meal I returned to my compartment for some more study of the language. I also reflected that I had now been travelling for nearly four days, and, apart from running up the stairs in the Nairobi office, had spent most of the time sitting down, either eating or sleeping. This might account for the slight feeling of constraint around the waistband of my trousers.

Apart from that, I felt that I was coping rather well.

CHAPTER 4
Mombasa

In the morning I was met at the station by the site manager, who took me to the hotel where I would be living, and then down to the site, where I was introduced to the staff and had a quick tour of the works. The new power station was in an early phase of construction. A large slice had been excavated out of the hill, and the main foundations were being concreted. A stores area had been created, with streets and avenues set out, so that goods could be delivered, parked and subsequently located without too much trouble. A large wooden shed had been set up for the more delicate goods, and this also was set out in a series of squares for easy reference. The office was perched on the edge of the excavation, and had a panoramic view over the site, and most of the quays in Kilindini harbour.

For the remainder of the day I was closeted with the drawings and programmes, and was returned to my hotel in the evening. I had been allocated a room in one of the annexes to the hotel – a small bungalow in the gardens, surrounded by thick bushes, probably full of snakes. The hotel was yet another relic of the flying boat era, and had been used as a transit facility for both crew and passengers, and was being maintained to the previous high standard. It was set upon a bluff overlooking the inner harbour, and the outlines of the track down to the beach were still visible.

When I returned to my room after dinner I found that my bed had been prepared and the net deployed. I turned in and continued my studies of the language, and must have fallen asleep, because some time later I was disturbed by a crunching noise, and turned to see an enormous insect – later identified as a praying mantis – devouring what was left of a substantial moth, while several potential next courses were buzzing around the lampshade. Some other horrors were perched on the net – mercifully on the outside, as far as I could see – and a number of smaller creatures were wriggling on the table under the lamp, apparently partly stunned from having flown into the hot bulb. With some difficulty I managed to switch out the light, and went straight back to sleep, despite the festivities in progress.

In the morning, all signs of violence had gone, and remembering my time as a Boy Scout I took great care to see that there were no scorpions lurking in my shoes. I can just imagine the alarm caused by tipping a live and disturbed scorpion out onto my bare feet, and just think if one had taken up residence overnight in my underwear! As a sensible precaution, I resolved to try to remember to take the clothes needed for the morning into bed with me, hung up under the net.

The work on the site was very interesting, and complicated by a number of dock strikes back in England, and a severe shortage of unloading capacity in the harbour. All the equipment had been documented as loaded onto specific vessels, but, because of the strikes, some goods had been transported by different ships, with longer passage times, and a lower priority for unloading in Mombasa – so the original works programme was seriously disturbed.

Even when we had notice of ships arriving in port, there was no telling how soon they would come alongside, or if they might be unloaded into lighters, and I was asked to rearrange the construction work to suit the delivery dates of the incoming materials, the availability of drawings and information, the best deployment of local staff and labour, and the delivery of locally supplied materials – an exercise that demanded great care and attention, and a lot of luck.

This gave me a good opportunity to get to know the staff, and also to find my way about the town, visiting the customs house, harbour offices, shipping agents, railway offices and local suppliers. I also visited ships both alongside, and anchored out in the stream, where I had to reconcile the description of the items against our order, the ship's manifest, bills of lading, and bundles of machinery in packing cases lying somewhere below in the hold. This was all daytime work, as it depended on access to the normal office hours of the various departments, and in the evenings I had to soldier on in the office, preparing site working drawings and instructions for the following day's work on the site.

The office was a prefabricated timber building, reassembled from a previous site, and set up on small concrete pillars. At this early stage, it had not been connected to the electricity supply – nor, for that matter, to any other service except the telephone. The welfare facilities were pretty basic, and comprised a Primus stove, a couple of tilley lamps, a bucket of clean water, a kettle and a few mugs. The Primus stove was no more difficult to ignite than normal, and the lamps gave out a good light, and a satisfying hiss when burning correctly. They also attracted the attendant flights of moths and flying beetles, which in turn encouraged small lizards and spiders.

During the day, entertainment was provided by hunting spiders, whose practice it was to lurk in the corners of the windows until some unsuspecting insect became trapped behind the glass. The spider would then estimate where the fly would be by the time he had moved along, and dip under the window sill, only to appear at the appropriate place, and make a mighty leap at the insect. This technique was generally very successful. Sometimes the prey would fly off with the spider hanging on below, but they seldom flew very far. Most moths and beetles were fair game, but large wasps and hornets were left seriously alone.

The men working on the site were a complete cross-section of the population. We had some old colonials; some new ones like me; an Italian who had been a prisoner of war; a white hunter; a South African who had been a crocodile-hunter, and who was quite the best bodger-up of broken-down machinery I have ever known; a couple of Rhodesians; some Arab crane-drivers; and a squad of Indian carpenters and masons. The main body of the men were from various local districts, with a few from Rhodesia, Nyasaland, Tanganyika (as it was then), and Portuguese East Africa.

As can be imagined, the language difficulties were formidable, and I must confess to numerous misunderstandings. Frequently, when listening to conversations between the Rhodesian and South African foremen, I was certain they were speaking Swahili, or 'Kitchen Kaffir', only to learn that they had been speaking what they believed to be English.

The Indian foreman carpenter was a splendid chap, who was fluent in Urdu, Gujarati, Swahili, Arabic and English, and was always willing and able to translate for me, and to promote my knowledge of Swahili.

CHAPTER 5
Language Problems

One member of the expatriate staff had stated that he did not intend to learn Swahili, but would conduct his business either in English or through an interpreter, and came a bit unstuck on one occasion at least.

We were lifting the upper drums of the boilers (each weighing several tons) into position by two large wooden masts, fitted with blocks and tackles for the lift, and with sets of guys and stays, rigged to allow the masts to be tilted as required. These stays and guys were worked by sets of blocks and tackles, each manned by about a dozen men. The operation of these stays and guys was a fairly tricky business, as, when one stay was slackened, the others in the pattern had also to be adjusted, and this called for careful supervision.

The drill was that, when instructed, the crew on one stay would lean on their rope, and – grunting and coughing, and singing some form of chant – take in or let out as directed, the end man ensuring that the end of the rope was always secure, in case they all fell down.

This work was not progressing quickly enough for our friend, and he came down and began to shout: "Faster, faster." As this encouragement seemed to be falling on deaf ears, and was having no significant effect, he turned to one of the spectators, and asked for the Swahili word for 'faster'.

Without batting an eye, the man told him the word for 'slowly'.

Thus misinformed, he went down the line shouting: "Pole, pole."

The crew, and the assembled multitude of spectators, all realised what was afoot, and chanted in unison, "Yes, sir; pole, pole," and continued their lethargic pulling on the tackle. This caused much mirth all round, and appeared likely to cause a case of apoplexy in our friend, until he realised that he had been taken for a ride and retired to the office in a foul temper. He insisted on a sort of court of inquiry on the incident the following morning, and the whole affair was put down to a slight misunderstanding. The man said he thought the bwana wanted to tell the men they were pulling too slowly, rather than telling them to pull faster.

No one believed this, but it had been a good laugh, and no harm had been done. The site manager took our friend aside and had a few words, and the matter was allowed to die away.

CHAPTER 6
The Pipeline

My first real duties in the field were concerned with the commissioning of the main freshwater pipeline, which had been laid about three or four years earlier, and ran in an arc for a distance of about three miles around the head of the bay, connecting into the new main water supply from Voi, about 100 miles away. The bush, which had been cleared during construction, had grown back thicker than ever, and consisted of short spiky bushes, barbed grass and thorns.

My tasks were to see that the pipeline had not suffered any visible damage, to check that all the valves were operating properly and were in the 'open' position, and then to operate the valves during the filling of the line. I set off, in my innocence, in my normal attire of shirt, shorts and desert boots, with a notebook, a bundle of drawings, some waterproof chalk and a bunch of spanners. Within a very few yards I was forced to retire hurt – scratched, stung, bitten, and with my stockings full of spiky burrs.

Some form of armour was required.

After a foray into the stores, I returned dressed in long thigh boots, thick gloves, and a heavy oilskin coat. This cladding was reasonably successful in protecting my person, but inside it was infernally hot, and it restricted my mobility.

The first tour of inspection went off reasonably well. As the pipeline ran across the face of a hill, it crossed numerous gullies, and crested intermediate small hillocks. The pipe was six inches in diameter, of black-painted steel, and generally set up on small concrete columns of varying heights. At the top of each crest was an air valve, an automatic device that would allow air to escape out of the pipe as it was being filled, or let air back into the pipe (to prevent a vacuum) as it was being drained. These valves operated on the principle of a small buoyant ball held in a cage, which would float up and seal when the water arrived. It was normally possible to see if the valve seats were clear, and if the ball was still in position.

At the bottom of each gully a wheel-operated drain valve had been fitted, usually on the underside of the pipe, with a short extension pipe pointing straight down.

This was not a good idea.

It was when we came to fill the pipe that the real entertainment began. The air valves were not too much bother, and generally sealed themselves after a bit of hissing, although the odd one or two were held open by small pieces of stick and twigs, causing a local waterspout, making access difficult, and requiring a personal effort to clear. The drain valves had all been set 'open', and as the water came into the section it carried all manner of vegetation and small creatures through the valve onto the ground below, and as the flow continued the surface became very slippery.

In many cases, where the water discharged onto was the only place available within reach to stand on while operating the wheel, and the ground was rapidly being washed away from under the feet.

Apart from the difficulties of access, language, dress and insects, I was aware of a background fear of snakes, and possibly other dangerous forms of the local fauna. There were known to be leopards in the area, and I had seen a number of crocodiles along the shoreline of the creek.

After the first two drain valves had been completed I realised that I would have to undertake the operation myself. My mastery of the language did not run to 'turn open' and 'turn close'. Nor was I then aware of the necessity of following each such instruction by "No, no; not that way. Turn the handle the other way." (Also beyond my command of Swahili at that time.)

I subsequently learnt, by hard experience, that failing to know which way to turn mechanical devices, such as screws, bolts and nuts, was quite common throughout Africa, and was a shining example of 'Sod's Law' – that, whichever way they turned it, it would be wrong.

After a couple of alarming descents, rather after the fashion of the Cresta Run without the bobsleigh, I retired to pause and consider.

FACTS.

All bushes so far located have spikes; sometimes on the trunks and branches, sometimes on the leaves, mostly all over.

All grass has spiky barbs, not dissimilar to the heads of barley, which migrate around the clothing, no matter where they enter.

Vegetation which does not harbour stinging, biting, flying insects is in a state of temporary grace, probably because it has just been invaded by ants.

All rustling noises in the bush are made by snakes, or small animals escaping from snakes (louder noises are made by crocodiles).

Leopards are reported as making no noise at all!

Thigh boots and an oilskin coat are some protection against thorns, but seriously inhibit mobility, especially at the bottom of a gully full of roaring water.

Also, it does take some time to get out of them when they become infested by ants.

SOLUTION.

Send one of the local crew down first. The noise he makes will probably be enough to scare off most of the nasties.

Go down myself on the end of a rope. Keep in mind the Swahili for 'pull me out. quick!'

(If there was any form of excitement, other than my gymnastic balancing acts while trying to operate the hand wheel, I was no match for my companion, encumbered as I was, and had to rely on being pulled out in an undignified manner.)

Wear heavy leather gloves.

This arrangement worked quite well, and I trust that, when the good citizens of Mombasa turn on their lights of an evening, they will appreciate that every drop of water which is made into steam that turns the turbines which run the alternators which make the electricity comes diluted with a small part of my blood.

CHAPTER 7
Mombasa Activities

There was a great deal of construction and development work in progress around Mombasa at that time. The new main water supply system from Voi was being commissioned, new dock and harbour works were continuing, the road and rail services were being improved, work was starting on the new airfield, and new houses, stores, shops and offices were under construction, together with all their services.

There was great camaraderie within the engineering staffs of all these departments, and the contractors associated with this work, and there was always a bit of borrowing and lending of small tools and minor items of equipment in short supply. I had gone off one morning to try to locate a spare part for a small piece of our pile hammer, which had become detached and fallen into the harbour, and was directed to the office of an old foreman working in the docks. Recognising a new arrival by the third-degree sunburn and the large lumps caused by insect bites and infected scratches, he went into a long harangue about the danger of sunstroke, and the need to wear a hat at all times. "It can addle your brains, laddie," seemed to be the burden of his song.

On further questioning, it turned out that he had never worn a hat, so he had to be immune. Perhaps the alcoholic fumes he was emitting were forming a sort of sun-proof halo, which diverted the rays onto more susceptible mortals.

We went into his store, and from the depths of a large wooden box located a tin full of ball-bearings, springs, screws, nuts and washers, and from this assortment managed to make up a contraption which would keep our hammer working.

I heard of some problems associated with the new water supply pipeline from Voi to Mombasa. It appeared that they had laid a spur line of four-inch pipe to supply a village some distance off the main line. A herd of elephants had been observing this construction, and found that they could dig the pipe out of the ground, break it, and make a marvellous waterspout and mud wallow.

Various remedies were tried, without success, and then an elephant psychologist was summoned. He found some means of diverting the elephants' attention, and a new pipe was laid at a

26

greater depth on a different alignment. The problem seemed to have been solved, but whether the psychologist did any good, or the elephants were just passing through and had tired of the sport, we will never know.

Some people were not always so obliging.

We were expecting the delivery of a mobile crane from Nairobi, and had been advised that it had been loaded onto a railcar and was no longer in the Nairobi marshalling yards. I had been asking around the railway company offices, and the best advice available suggested that it was in transit – wherever that might be.

In the course of other visits about the town, I happened to see our crane in its distinctive company colours, still on a railcar, in one of the sidings in the railway yard. On my return to the site I informed the site manager, who then telephoned the railway goods office to try to expedite its delivery. After some shunting about on the telephone explaining our urgent need for our crane to each individual official, the last man announced, "It is now half past three; that is the end of my business for today. I suggest you call again tomorrow and I'll see what I can do," and then hung up.

For a few seconds there was a stunned silence at our end, then all hell broke loose. He tried to call everyone remotely associated with the railway, from the general manager to the second under-porter, but without any success. They had all knocked off for the day. I was dispatched to try a bit of bribery and corruption, or threats, but either I was speaking to the wrong people or my price was too small – or my threats insignificant. I failed dismally.

Something of our annoyance must have rubbed off somewhere, as our crane was delivered early next morning with a mild form of apology for a delay in transit.

I formed the impression that this 'transit' place must be like the undiscovered place where the elephants go to die, but full of rusting machinery and abandoned and rotting foodstuff, all infested with ants.

On my first free weekend in Mombasa I was still without transport, and after a leisurely breakfast I walked down a steep pathway to the beach below. The slope was of bright red laterite, with frequent deep gullies scoured out by small streams, and the whole area was covered in tufted grass. The path looked as if it had been the access for the flying boat service, and it was fairly wide and not yet overgrown, and where it came out at sea level there was a large flat area of sandy beach.

Being an inveterate beachcomber, I wandered along this shore and found some strangely shaped shells and a number of other interesting pieces of flotsam. While scuffing about in the sand I uncovered some pieces of flint, which I believe might have been Stone Age arrowheads. I had made a collection of several dozen when it occurred to me that they would be very heavy to carry back up the hill, and that I had probably broken half a dozen laws concerning unauthorised excavation, treasure trove, disturbance of relics, interference with tribal burial grounds etc. etc., so I scattered them back onto the beach, and went in for a swim before returning to the hotel. By the time I made it to the top of the hill I was feeling the heat, and after a long cool bath rested on my bed. When I awakened some time later, having missed my lunch, it was nearly dark. The room servant told me that there would be a dance in the hotel in the evening, but due to the limitations of the phrase book I was unable to ascertain whether it would be a grass skirt and feathers local jamboree, or a 'one, two, three – hop' sort of performance.

Ready for anything, however, I dressed in my new lightweight tropical dinner suit and went down to join the fray. Rather to my relief, the band was playing conventional-style dance music, and I was invited to join with another member of the site staff and his wife for dinner and the cabaret, which was scheduled to follow later.

After a few days I was provided with a Land-Rover, and spent most of my spare time at the yacht club, where I was introduced by some of the people from the local shipyard, whom I had met up with in the course of my travelling around the town. Most weekends, I sailed as one of the crew of a large dinghy.

On the evening of my first weekend in the club we were sitting out in the garden having a drink before dinner when there was a sudden loud metallic noise. Conversation froze, and in the silence a quiet voice asked, "What was that?"

A much louder voice replied, "Oh, it's only a coconut falling onto the roof of some poor bugger's car." There was a general surge of viewers into the car park to see whose car it was, and, by some mischance, it turned out to belong to the man who had made the comment.

There's poetic justice for you. The car had a substantial dent in its roof, and I noted in my mind that, in addition to all the previously listed hazards of living in Africa, I should add 'keep away from under coconut palms'.

The racing at the yacht club was governed by the tides, so every other weekend the races were inside the harbour, and the alternative weekend we raced outside.

On one of these outside races, as we were returning at the head of the fleet, we had the misfortune to capsize the boat in the main channel into the harbour. She rolled over onto her side, and began to fill. Just as she went under, one of the passengers turned to me and announced that he could not swim.

"You had better hold onto my arm, then," I said, "and we will float off directly."

The boat vanished beneath us, and we struck out for the shore, about 200 yards away. After about five minutes' swimming we were picked up by a launch, just off the beach.

George was very good; he just laid back in the water and hung onto my arm. He remarked afterwards that there was no way he would have let go, but, as we seemed to be managing fine, he was not all that worried. Next morning I discovered that my arm was all black and blue, and the two handprints were quite easy to distinguish.

George enrolled in a swimming class after that and became quite proficient.

It was while sailing on this boat that I had first met Jean, the young lady who was later to become my wife, and our first evening out together was shortly after the sinking. I had arranged dinner at the hotel, and all seemed to be progressing smoothly until we entered the restaurant.

It was the time of the sausage flies.

These strange insects were about an inch and a half long, about as thick as a pencil, and came in droves. They swarmed around the lights, and when they struck the hot bulbs fell down, buzzing frantically. Some landed on the lampshades, and hung there for a short time before they too fell down. The whole place was covered in these obscene insects; on the floor, on the tables, in the glasses, on the plates – everywhere.

Dinner was quite impossible, and just imagine what it would be like in the kitchens. We retired to the yacht club, where we could eat out in the garden, well away from coconut trees, but in clear air, and without any flies in our soup.

CHAPTER 8
The new Boat

I learnt of a 12-square-metre Sharpie-class yacht that had been looted by the navy from Kiel, and was now lying at the Admiralty depot on the other side of the harbour, the previous owner having been demobilised. Another young engineer from the shipyard and I decided to see the boat, and we set about obtaining the necessary permission to visit the depot.

We first obtained permits to enter the naval yard in Mombasa, and from there obtained a pass to go across the harbour in an East African Navy launch. Then we needed another document to allow us to land and enter the depot, and a note to the commandant to allow us to inspect the boat. Armed with all this paperwork, we set out one Saturday morning to go and carry out an inspection.

The commandant was very pleased to see us, and after a few drinks in his quarters we went down to look over the boat. It appeared to be in pretty good order, and, as far as we could see, most of the equipment was there, and serviceable. After some further drink we agreed a price, and were ferried back to the Mombasa shore, having arranged a further visit for the next weekend to complete the formalities and collect the boat.

During the week we made contingency plans, and collected some additional items – i.e. a bucket, some old inner tubes, some spanners, an oil can, a screwdriver, and some short lengths of cord – and went through the documentation procedure for our next trip. On the Saturday morning we set out with all our gear, and on arrival at the depot, and after the mandatory drink with the commandant, persuaded some of the staff to assist us in getting the boat down and into the water. We then scrounged all round the yard to see if there were any useful items of gear the navy did not seem to be using, and collected all the bits and pieces for the boat.

When we returned to the dockside the boat was about half full of sea, having been stored under cover for some extended period of time, and dried out in consequence. We were not too bothered by this evidence of the rising damp, and bailed her out in about ten minutes. This was a comfort, as we reckoned that the passage back to the club would take 40 minutes, and as we could bail out an hour's worth of leak in ten minutes we would be all right. In any

31

case, we had brought along sufficient extra buoyancy in the form of inner tubes to ensure that we would not sink.

After making our number with the commandant again, and a few drinks, we returned to try to fit all the bits of stick and string, and after a final bailout set off back across the harbour. We had a good sail back to the club, which by then was quite busy, what with late lunchers, people preparing for the afternoon race and family parties swimming from the beach, and we planned to make a spectacular landing. The plan was that we would drop the jib, turn into the wind, and Arthur would step out into the shallows and catch the bow as I lowered the mainsail. The first part worked fine, but when Arthur stepped off we seemed to have misjudged the water depth, as he disappeared completely, and bobbed up just as the stern was going past.

This manoeuvre was much appreciated by the audience, who delivered a hearty round of applause. Shortly after we landed we were approached by an elderly African, who told us that he had been the boat boy for the previous owner, and knew the boat well. He produced numerous testimonials, and we agreed to take him on again at the going rate for boat boys. His duties were to make sure that the boat was rigged and launched ready for racing, or for other expeditions as required, and to have her washed out with fresh water, and pulled up under a shed when we returned. He was a cheerful old man, and gave us very good service.

Having our own boat opened up a whole new area for entertainment. We raced most weekends, and still found time to go off on fishing expeditions and picnics.

At this time, scuba activities were practically unknown in Kenya, the highest levels of technology being a snorkel and a pair of fins. A few people had elastic-powered harpoon guns; others just carried a trident on a long pole. As with most sports, it all looks very simple, and providing you keep calm you can come to little harm. It is alarming, however, to say the least, when you are halfway through a deep breath and the air stops coming in. This is probably because you have swum through a small wave, or have turned your head slightly – but it could just be the first indications of being swallowed by a shark.

On one of our first expeditions we had anchored in about ten feet of water over the reef, and were admiring the variety of colours in the coral, and the brightly coloured small fish.

Someone swimming past gave a friendly tug to another's flipper, and immediately confusion reigned. What had been a relaxed and gentle drift was transformed into a frantic dash back to the boat.

There was a general theory that sharks do not come in over the reef, but I, for one, do not believe in theoretical sharks, and always kept a good lookout.

Before we acquired a harpoon gun our favourite form of fishing was by a baited hook, which could be dangled in front of any particular fish. This was fairly successful, but generally disappointing, as, due to the magnification effect of the water, what appeared to be a reasonably sized fish, about the size of a mackerel, turned out to be like a sardine when pulled out.

My first attempts with a harpoon gun gave me a bit of a fright.

I had been swimming round and round a coral head, pursuing a good-sized fish, which kept just out of my range. I tried a short cut by swimming over the top of the coral head, but my fish obviously knew about such tricks, and stayed well away. I did see another fish, which went in under a ledge, but some of its back was still exposed, and I fired a shot into it. When I reeled it in the wretched fish came up to the surface, and began to inflate itself, to the size of a beach ball. The harpoon was right through it, and it was making a puffing noise as the air escaped. The skin was covered in small spines, which stood out in a menacing manner. I was later advised that this inflation was part of a defensive stratagem, designed to frighten off potential swallowers, and it certainly scared me to death.

I had heard tell of people who had harpooned poisonous fish, and when they had held the harpoon up to display their catch the fish had slid down the arrow and stuck its spines into their hand – very painful indeed.

I started to swim gently for the shore, keeping the thing well away from me, at the full extent of the line to the harpoon, and it was following about six feet astern. I settled down into a sort of a sidestroke, so that I could keep my eye on the fish, just in case it started up towards me, and then I noticed that there was some blood coming out of it. Folklore took over again. Blood in the water equals an attraction for sharks, and I was off like a one-handed torpedo, with one eye on the fish and the other on the lookout for sharks.

When I landed in shallow water the balloon fish was still

attached, and swimming in a short arc, still firmly skewered by the harpoon, and still puffing away. Much to my relief a couple of local fishermen came over, and, without any fuss, cut the line, pulled the harpoon out and returned it to me. They had been watching my efforts with some curiosity, and a lot of amusement, and were very happy to accept the fish, which they assured me was not of the stinging spine variety.

Another fish that often caused a bit of a fright to unsuspecting swimmers was the garfish. This creature grew up to about three feet long, with an extended nose. It was very thin, and normally swam on the surface. When snorkelling, the main interest is downwards; this also reduces the chances of a stiff neck, but it can be very alarming to come on a yard of long green fish six inches from your face when you happen to glance ahead.

There were numerous stonefish and lion fish swimming close to the rocks. These were the poisonous ones, and the stonefish was often quite difficult to see until it moved, as it was well camouflaged. The danger from them was that you might tread on them inadvertently, and their barbs were very venomous and also difficult to remove. The lion fish were quite small, but with an array of beautiful feather-like fins. They generally hung motionless in the water, but if brushed against gave a nasty sting. On the sea floor were many spiky sea urchins, which had long nasty spines – well able to penetrate the rubber of a flipper. These spines caused painful wounds and were difficult to extract. If left unattended, they would always turn septic.

We found that the juice of a papaya was soothing when applied to a stinging bit of foot or hand, and it also made the removal of the spines a little less difficult.

As an interesting piece of additional information, the juice of a papaya can also be used for tenderising otherwise tough mutton or goat. It makes little difference to the taste, but does reduce the amount of chewing required.

CHAPTER 9
The Ferry Boat

One weekend we decided to go on a picnic by car, along the coast to the north of Mombasa. Our party was made up of me, Jean and her future bridesmaid, plus her boyfriend. We set off on the metalled roads through the town, crossed the old bridge past Fort Jesus (which gave a fine view of the old harbour, full of large sailing dhows just arrived from the Gulf), and on for a couple of miles on tarmac until we reached the city limits. From there on the road was a laterite track, worn down by traffic to a depth of about three feet below the surrounding country. Apart from restricting the view to either side, which was pretty monotonous anyway, the dust from other traffic tended to hang in the air over the roadway.

The alignment of the road appeared to follow the original path worn by donkey traffic, and apart from frequent meanders had never been graded. This meant that, when you came over a small hill, on the ascent all you could see was sky ahead, and then suddenly you were over the crest and pointing downwards into the next valley. After the first flying take-off on a crest, such hazards were approached much more carefully. We soon developed a technique for negotiating these bumps, and, after a few errors of judgement, settled down to a routine approach.

This was too good to be true.

On the next crest we sailed over we found that, instead of a valley ahead, we were on the short, steep ferry ramp – and the ferry was on the other side of the creek! We managed to stop before we reached the water, but it was a close-run thing. Realising that following traffic might have the same problem, we backed into the side of the ramp, well out of harm's way, and sat down to wait for the ferry to return.

This ferry was a pair of barges fastened together, with a deck between them, and a pair of ramps at each end. It was secured to a wire stretched across the creek, and was moved back and forth by the crew pulling on a heaving rope. By adjusting the angle of the barges to the securing line, the ferry could be driven across under the weight of the current, and only needed pulling to get it into or out of the loading areas, or at periods of slack tides, when the ferrymen arranged their lunch break – or contrived some other form of breakdown.

Shortly after our arrival we were joined by an American couple, dressed as tourists, hung all over with cameras, spare lenses, light meters etc., and also in possession of a tape recorder. They said that they had heard that the ferrymen sang as they pulled the boat across, and they wanted to record them, as one of the quaint customs of the colony.

Having experienced other ferrymen around Mombasa we could assure them that the crew would probably sing, but we did not tell them what the subject of their songs might be. They had an uncanny knack of knowing which of their passengers were fluent Swahili speakers, and their songs were arranged to make fairly ribald comments about the dress and habits of the non-Swahili-speaking visitors.

When the ferry arrived at our side of the creek, the captain came ashore to inspect his next cargo. We were allowed to drive on, and were secured for the passage. There was then a discussion as to how much the Americans would pay to take the recordings, and, anyway, there would be a delay, as the tide was running too fast – so the discussion could be prolonged. We suggested that they had better improve their offer, or we would be there all day, and – somewhat grudgingly – they doubled their bid, and honour was satisfied all round.

The men cast off and began hauling us across, singing that this was an easy prey, and they should have started their bidding at twice that amount, but, anyway, they had had a rest, and a bit of fun, and they would probably get them again on their way back.

We continued up the coast on a road equally terrible as before, and after a few more dusty miles decided that this bit of shore was as good as any other we were likely to find. It had been much the same all the way from the ferry, so we turned down a narrow track, which came out on a long expanse of shining white sand, with the ocean breaking on the reef about half a mile offshore and a beautiful green lagoon of still water just waiting to be swum in.

We parked the car in the shade of some casuarina pines, spread our towels on the sand, and, after a quick dip to wash off the dust from the journey, settled down to some serious sunbathing.

On our way home in the evening we proceeded with great caution, trying to raise as little dust as possible, but this was a futile attempt, and we arrived back covered in red dust.

CHAPTER 10
The Power Station

Despite all the social activity, the work on the power station continued. Part of my duties consisted of supervising the construction of the cooling water system. This comprised a pump house, set in a closed box of sheet piles, with a pair of large-diameter pipes running back to the station. The first task was to complete the driving of the sheet-piled box, to allow work to continue in the dry. This was formed from interlocking panels of steel plate, with special corner pieces, which were hammered into the seabed by means of a compressed-air-driven pile hammer. We had just made the final connection at the last corner, by a lot of good luck and a bit of clever engineering, when we dropped a cylinder of acetylene gas over the side and into the harbour. At the end of the day's work we tried to fish for this cylinder with a grab fitted to the crane – more in hope than in anticipation of success, I must confess, having had some experience of operating model cranes in amusement arcades without much success. At least, there, you could see what you were doing.

On our first scoop we brought up the expected tangle of bits of fishnet, mud, broken pieces of wood and a few shells. On the second attempt we caught up on something pretty substantial, and, after a bit of heaving and slewing, up came a large piece of heavy machinery.

This caused us to pause and consider, and, after some discussion with the oldest inhabitant, he remembered that an old tugboat had been grounded there a long time ago, and had eventually rusted away. We must have been driving our piled wall right alongside this wreck, and could have been in a lot of trouble if our pump house had been designed a few feet wider, or on a slightly different alignment. We marked the wall where the cylinder had gone over, just in case there was to be an insurance case, and abandoned all hope of recovering the cylinder.

Having completed the box, we then had to excavate all the silt and mud from within the structure to prepare the foundations for the pump house, and, after an inspection by a diver, we had to place a large dollop of concrete underwater right across the floor, to form a rough base and to hold the bottom of the piles in position. The placing of this concrete was by means of a Tremie pipe – a long steel pipe with a hopper on top – the object being to keep the pipe full of concrete, and to lift it gently to allow the concrete to come out at the bottom into the pile of recently placed concrete, without losing all the cement into the adjacent water. This procedure took a lot of careful operations, coordinating the filling of the hopper and the lifting and moving of the pipe, and was necessarily a very slow process.

We arranged a shift system to run a round-the-clock operation for this concreting, and estimated that it would take about three days' and nights' continuous work. Because of the critical nature of the work, and the problems that might arise if the work was stopped for any reason, I felt obliged to be present most of the time, and – apart from meal breaks, and the odd nap – saw the whole performance through.

We had been invited to a production by the local dramatic society of *Man and Superman* on the evening after our concreting was completed, and I dashed off to my hotel, had a quick wash and brush-up, collected my partner, and arrived at the theatre just before the curtain went up. Despite frequent proddings and nudges by a very sharp elbow, I was unable to stay awake, and was ushered out at the first opportunity and told to go to bed.

I never was much of a fan of Bernard Shaw, anyway.

Despite my dormouse act, I must have been forgiven, as on our next night out we became engaged. It probably appeared that I was in need of looking after.

Our next task on the construction was to pump out the water from inside the box, and, to prevent the water pressure from collapsing the piles, we set in a number of heavy timber frames, floating on top of one another. As the water was pumped out and the level reduced, the top frame was wedged into position, and so on downwards.

We found that the leakage through the gaps in the sheet piles was more than our pumping capacity could overcome, so we organised a gang to go around hammering caulking material into the gaps as they were exposed by the tide. This worked quite well, but – once again – we reached stalemate.

Then I remembered an old trick of dumping boiler ash round the outside, and letting the pressure of the water draw this into the gaps. We set up a walkway all round the top and arranged a procession of men with head pans to dump the ash over the side into the sea. This worked like a charm, although there were numerous altercations when some dozy man dumped his head pan full of ash onto the crew who were going round the outside at water level, hammering in the caulking from a small raft made up out of empty oil drums. "They were not there a minute ago, sir."

I suppose they were lucky that they had not had a head pan dropped on top of them as well as the ash.

On one occasion the raft became detached from the structure, and began to float away on the tide. The men aboard were all from up-country, and had never seen the sea before, never mind being cast adrift on a raft. They gathered to one side, causing the raft to list, and by all paddling in the same direction induced a spinning motion without making any direct progress to return to the site. They went further and further out on the tide, to cries of encouragement and derision from the ash men, until I had to go off and borrow a launch from another contractor and tow them home.

When they returned they were full of tales of their adventures on the ocean, about Sindbad, and Jonah and his whale, and the wonders of the deep.

CHAPTER 11
A Picnic gone wrong

One weekend we arranged to take Jean's bridesmaid and her young man for a picnic to one of the beaches across the harbour. We set out in a fair breeze, with the boat loaded up with food and drink, towels, sun oil cream, the survey umbrella for portable shade, and our glass-bottomed box for underwater observations.

About halfway across, David asked if he could have a turn at sailing the boat, and as he seemed to be reasonably intelligent, and had said he had been sailing before, I said, "All right; but, if the breeze gets up and the boat heels over too much, just let everything go." We sailed on for some time, apparently under complete control, when a slight squall arrived, and the boat began to lean over. Instead of letting go, the clot was trying to pull the boat upright by heaving on the tiller and the mainsheet, and although by then I was running along the side of the boat, trying to get back and knock him unconscious or something, the water began to come in over the deck, and we gently capsized, tipping us all into the water.

There was no danger of the boat sinking, as we had plenty of buoyancy aboard, and as I began to bail out the water, while the others remained in the sea, someone came alongside in a dinghy with an outboard motor. There was only room for three others in the dinghy, so I elected to remain with the boat, and they all climbed in.

I was hanging onto the transom of our boat, assisting with the steering as best I could, as we made for the shore a couple of hundred yards away. This was fine until I thought that there might be sharks about, so I collected one of the sails and wrapped it around my legs, allowing a lot of it to trail out behind. If I had felt the slightest tug I was sure I could manage to walk on water at least until I was inside the boat again.

We arrived on the beach without further incident, and had a quick check to see that the boat was undamaged, and that we had not lost any vital pieces of equipment, which might prevent us from sailing back. We thanked our rescuer, and he departed while we sorted out the picnic supplies. Everything was very wet; of all the food, only the cold chicken was possibly edible. All the remainder – bread, biscuits, sponge cake, salad – had to be abandoned. The only items undamaged were a small tin of boiled sweets and some wine.

We bailed out the remainder of the water in the boat and hung out all the towels etc. to dry, and to counter our possible dehydration we decided to drink the wine, eat the sweets and – as there was little point in a picnic without food – set off back to the club, where at least we would not starve.

CHAPTER 12
An Encounter with Elephants

The next weekend we decided to keep off the water, and arranged a trip to the game park at Voi. This time, David declined to accompany us; perhaps I detected a slight cooling-off between us. We set off in the long-wheelbase Land-Rover fairly early in the morning and drove up the bumpy and dusty road to the park, about a hundred miles up-country to the west. When we arrived it was nearly noon, and very hot, so we parked under some trees and ate our packed lunch.

There was not much movement of animals as far as we could see, although there was evidence of elephant in the form of large cylindrical droppings all along the track. After lunch we drove slowly around, and saw a lot of small game, some baboons, a couple of giraffe and a number of small groups of zebra, gazelles and wildebeest. Later we came upon a herd of elephants, grazing on some small bushes about a hundred yards away.

Susan climbed up onto the roof to take some photographs, and the herd moved slowly nearer. Suddenly, one large elephant wheeled out of the line and came directly towards us. I shouted to Susan to hang on (probably rather redundant advice), and turned the vehicle away. The elephant seemed to lose interest, and returned to his browsing. I stopped and got out and asked Susan if she had taken any good snaps, but she just collapsed on top of me like a rag doll, and insisted that we put some further distance between us and the elephants. After that, the remainder of the expedition in the park was pretty tame, and we were soon on our way back to Mombasa.

About half way down the road in the gathering dark, we came across what appeared to be a pair of eyes reflecting our headlights. As we got nearer, I became a bit concerned, as they seemed to be about ten feet above the road level, and pretty far apart, with a large shadowy shape behind them.

I stopped to pause and consider, and then decided to approach with caution. Much to my relief, what I had imagined was a very large animal turned out to be the back of a pantechnicon-type lorry, stopped on the roadside, with reflectors high up on the back doors. I looked in to see if the driver was in attendance, but there was no response to my shouting and banging on the doors, and it appeared that he had just abandoned his vehicle, without lights, at the side of the road.

Night driving was not recommended, as there were frequent reports of accidents caused by people running into the back of unlit abandoned vehicles, and there was more than a sporting chance that you might run into a large animal going about its legitimate business. Besides, the thought of changing a wheel, being bitten to death by insects, and with your back exposed to the bush, where God alone knows what fierce creature is might be sitting, licking its chops in anticipation of its next meal, is enough to discourage most people.

—

CHAPTER 13
The Governor's Ball

We had received an invitation to attend the governor's ball in one of the larger hotels in Mombasa. It was a very formal affair, with everyone in their best boots and spurs, and to begin with, anyway, on their best behaviour. The music was provided by the band of one of the garrison regiments, and everything was in very strict tempo.

During an interval, when the band had gone off for some refreshment, the stage was overrun by some of the company of one of the regular Union Castle mail steamers, which happened to be in port at the time. They struck up with some very trendy jazz and swing music, such as 'Blue Suede Shoes' and 'The Green Door', followed by 'See You Later, Alligator' – music experienced only by recent arrivals in the colony, or bright young things with access to a good wireless set.

The whole place went demented, and what had been a very staid sort of affair turned into a frantic mêlée of gymnastic dancing. Some others of the crew of the liner were expert at jitterbugging, and this art was now being practised by all the senior citizens of the town, with varying degrees of success. Before long, many of them had retired hurt, or exhausted, and had adjourned to the bar for rest and repair.

When the regimental band returned, they could see that all was lost, and joined in the fun with great gusto, and a most energetic and memorable evening was had by one and all.

I don't know if the governor and his party enjoyed the evening, but it set the tone for all future dancing nights at that hotel, at least.

CHAPTER 14
The Tintagel Castle

Sometimes, when I had been working late, I went into town and stopped over for a drink in the club, or in one of the many bars, before going back to the hotel.

One evening I met up with some of the officers from the *Tintagel Castle*, and after some entertainment and general conversation I offered to drive them back to the ship. There was a bit of a party going on aboard, and I was invited to join in, so I parked the Land-Rover out of the way at the end of the quay and climbed up the gangway. The party was in aid of the junior third engineer, whose birthday it was (or something), but most of the ship's officers were present in a tiny cabin.

When we arrived the overcrowding was recognised, and we moved to a more spacious room, belonging to the senior second engineer. The party was going very well until it was interrupted by an anxious young officer who asked, "Is that your Land-Rover at the end of the quay? Because, if it is, they are backing a train into it now." This rather broke up my participation in the party – probably just as well, or I might have been shanghaied, and woken up in Aden, or Dar es Salaam, or somewhere. No harm was done; I recovered the car before they had shunted the train into it, and made my careful way back to my hotel.

Many years later I was in the West Country, and visited the real Tintagel Castle, but it was a poor comparison with the ship of that name, and I went away somewhat disappointed.

CHAPTER 15
Misunderstandings

By this time I was becoming reasonably fluent in Swahili, and could order drinks with supreme confidence, and could also direct men to: "Hammer this" and: "Lift that" without direct reference to my phrase book. Some of the worst confusion, however, often arose due to direct misunderstanding, and general bloody-mindedness among the English, most frequently in government departments.

Part of our construction called for the building of two houses, for the station manager and his assistant, and work was well in hand towards the completion of these buildings. There was also a railway spur to be laid into the site to permit the delivery of fuel oil, but the construction for this small piece of line was delayed until the completion of a new railway bridge across the end of the harbour.

One day I was visited by a very excited young railway engineer, who announced without further ado that our houses would interfere with his railway, and we would need to knock them down and rebuild them elsewhere. I was pretty sure that they were in the correct position, according to our drawings, and, if they were not, it would be much easier for him to deviate his track by a few inches to obtain the required clearance; anyway, he had practically the whole of Africa to move his track into.

Positions were adopted, and some serious surveying was undertaken, checking of survey stations, measurements, angles etc.

During the checking operations, I discovered that our houses were in the correct position as indicated on our drawings, but that there was a slight discrepancy between two official survey stations, and as a result of using one station rather than the other our buildings would just protrude onto the railway land.

I also discovered that the prospective station manager had instructed the architect to add a small balcony along the front of the buildings, and it was only a small part of this balcony that actually breached our site boundary.

By this time, the problem had escalated well above my humble level, and while the exalted members of the railway company and the Colonial Survey were deliberating, I organised the chopping off of a few inches from the end of the offending balcony, and face was saved all round.

I suppose that there would only be a train passing about once a fortnight, and this would not cause a lot of trouble to the residents. Anyway, the station manager would probably allocate that house to his assistant.

CHAPTER 16
The Efficiency Experts

My partner in the boat was an engineer in the local shipyard, and they were undergoing the attention of some so-called 'efficiency experts'. Arthur called me over to see what was going on, and on the day of my visit the experts were taking the measurements of the individual fitters and mechanics, with a view to having them located at benches that would match their height.

Some time later I was back in the workshops, and was delighted to note that all the fitters were perched upon their variously levelled benches, working away as they always had previously. I suppose that the efficiency was gained by them not having to climb up as high onto the benches as they had before.

So much for the experts.

In the yard were a number of elderly Chinese shipwrights, and on one occasion I saw them measuring up for some repairs on the bow of a small coaster which appeared to have run into something pretty substantial. The vessel had been pulled up on the slipway and the shipwrights had set up a bit of a platform around the damaged area. They took measurements by means of a giant set of compasses, and while the damaged pieces were being removed they set to work to make a replacement section.

I happened to be visiting again when the replacement section was being offered up, and it fitted like the proverbial glove. What a pleasure it is to see such fine and skilled workmanship.

CHAPTER 17
Changing Times

Things were catching up on me. The work on my section of the power station was drawing to a close, the wedding day was approaching, and I had been asked to go up to Nairobi concerning future work.

Jean lived in a large flat that was situated over the bank, which had previously been the manager's house. It was a large, airy apartment, which allowed free access to a family of bats living in the roof space. The house was set on one side of a tree-lined open square, and the gardens around the buildings were well planted with fragrant shrubs. These were the haunts of some anvil birds, which made a regular dinging noise from dusk until very late. What with the birds and the bats, visiting this establishment was often quite an adventure, although I never yet had to untangle a bat from anyone's hair. There were a number of ceiling fans installed (controlled by an adjustable wall-mounted switch), one of which had become the residence of a small lizard, and when the fan was started a strong smell of toasting cheese was evident. Later investigations revealed the remains of this poor creature. Perhaps the ants knew about electricity, as they ate almost everything else about the place.

The bank manager had moved out to a new house, fitted with air-conditioning – a novel concept at that time – but due to delays in the shipping arrangements the machinery had yet to be installed, and it was a toss-up as to whether the promise of air-conditioning was a more comfortable arrangement than the old flat, complete with transiting bats.

He had kindly agreed to give the bride away, and had offered his house as the location of the wedding reception.

We also attended at the local government office, purchased the licence, published the intimations, and attended the cathedral for some words of advice, and made the formal arrangements for the service.

In the midst of all these preparations I had to go up to Nairobi to be briefed about my next assignment, and went up on the evening train. Next morning in the office I took my time in mounting the stairs, and arrived in the engineering office only slightly out of breath, and after the introductions the meeting commenced.

There was to be a comprehensive study of the water-power potential throughout Kenya and Tanganyika, and a survey party would be formed of me and a geologist, with others attending from London from time to time. For the next few days we worked out lists of equipment, a proposed route and a programme of events, and then prepared more detailed arrangements, making out letters of introductions to landowners, government departments and managers of local offices of the power company, and setting up arrangements to draw cash, recruit local labour etc.

When all these preparations were in a sensible state of readiness, I returned to Mombasa, and assisted with the final arrangements for the wedding.

The start of the survey was still about two months away, and it was agreed that, after a short holiday, I would return to the power station in Mombasa and assist in writing up the various manuals of operation, complete the 'as built' drawings, and try to tidy up the accounts into their various compartments. This exercise gave me an insight into what is now known as 'creative accounting', but as it was performed after the deed was done it became, rather, a process of diluting the blame for minor disasters, and spreading the successful bits fairly thinly over the remainder.

Our wedding was a great success; it was a beautiful day, with a cool breeze blowing in off the sea, and the ceremony went without a hitch. I can still remember the minister giving us a small piece of advice about married life, to the effect that we should not become too bothered about either of us squeezing the toothpaste out of the middle of the tube; and then, to a great ringing of bells, we were out in the sunshine again, and on our way to the reception.

This also went off very well, and we departed early in the afternoon, full of high spirits, and with a reserve supply in the back of the car, to the yard where I had secreted our Land-Rover (to try to circumvent any monkey business, like tying old fish on the exhaust system, etc.).

We changed into our travelling clothes and collected our luggage on the way round, but when we arrived at the yard some silly person had taken the keys of the vehicle from the gatehouse, and locked them in the office, where we could see them hanging on a hook. After some consideration about breaking the window, and fishing the keys out on a long pole with a hook, it was deemed better to work the short-circuit trick on the electrics of the vehicle, and after a few mild shocks I found the appropriate wires and off we went again, heading up the road to Voi, where we arrived just before dusk.

The hotel at Voi was a sort of overnight stopping place for people driving between Nairobi and Mombasa, and was also developing into a base for visitors to the game park, but had not yet quite made the transformation. It was comfortable and the food was good, and we retired early with some of the champagne we had liberated from the wedding feast.

In the morning we were off quite early and arrived at the national park near Nairobi with plenty of time in hand before dark, so we made an excursion into the park. There was a fairly impressive entrance off the main road, but the park itself was relatively undeveloped. The roads were all dirt tracks, and we were given a pretty basic sort of map, with some suggestions as to where some of the larger animals might be found. The other visitors were mostly in Land-Rovers, or a few private cars, and we discovered that where there was a collection of cars it was likely that there were also some lions sunning themselves, or sitting in the shade.

Nearby there were herds of wildebeest and zebra, and also many gazelle, and from time to time these animals would suddenly rush about in a bit of a panic, with the springbok leaping up and down like jacks-in-the-box. We sat about, hoping to see some other activity, and then got bored, and drove on into the bush, where we encountered some baboon, ostriches, warthogs, and a pair of giraffes. We did not see any men on foot in the park, and surmised that, as the official rangers would generally be near their vehicles, all others were either poachers or wandering members of the Mau Mau, so we resolved to keep well away if any loose men were observed. Towards evening, we drove back out of the park, and on into Nairobi, where we had booked into a hotel for the night, and on the following day we went on to the Rift Valley, where we stopped in a very pleasant hotel in the hills, with individual small cottages, complete with fireplaces and well-tended gardens, and even a thin skin of ice on the puddles in the mornings.

We made daily excursions into the nearby country, before going on again into the foothills of Mount Kenya, and to a splendid hotel at Embu. Once again we did the circular tour, driving all round the mountain, and visiting Thompson's Falls before returning to the fleshpots of Nairobi.

A few days later we were off again, this time heading south for Mount Kilimanjaro, where we spent a cold night in an old German hotel, with genuine bearskin bedcovers, halfway up the mountain. We went sightseeing again, and came across a number of large mission stations in the foothills, where they were growing coffee, amongst other cash crops.

Later we went on to Moshi and Arusha, where we stopped for a couple of nights in a small hotel that had been recommended to us. We found this place about three or four miles out of town on a wide, sweeping plain, with the now familiar layout of a central block with public rooms, and a number of small bungalows set in the gardens.

We were warmly welcomed and shown into our bungalow, which was the last one in the complex, and had an open vista of a lot of Tanganyika. It was the normal design of two bedrooms separated by a central living room, with a large veranda to the front and the bathroom to the rear, of timber construction set up on small concrete-block pillars, and with a corrugated iron roof.

After settling in and unpacking, we went along to the bar, where we were advised that the dining room would close at ten o'clock, and the electricity generator would be shut down at eleven, but that there were kerosene lamps and candles provided in the rooms. We had a pleasant meal, went back to our bungalow and turned in before the lights went out.

Some time during the night we were awakened by a tapping noise at the window. I got up to investigate, and found that the main double doors onto the veranda had blown open and that a strong wind was agitating some rose bushes, planted close to the house. Some of the branches were scratching against the windows and along the roof, and this must have been what had woken us up. There was not much to be done about that in the dark, but, by then, Jean had got up and was dismayed to find that the main doors were wide open, and there was no sign of a key.

I closed the doors, and set a couple of heavy chairs against them, and we returned to bed. What I had failed to mention was that our barricade was on the inside, and the doors opened outwards, but this was only realised later, when a smiling houseboy came along with the early-morning tea. I had a word with the management and they trimmed the bushes round the house, and produced a massive key, which, even if had fitted, would have been better as a doorstop to keep the doors closed.

The manager was a bit surprised about our concern for security, as they had never had any trouble from thieves, and most animals seemed to keep clear. I explained that my wife had lived in Nairobi for some time, and things were not nearly so civilised there.

After a second more restful night we were on our way again, through Moshi, crossing the border into Kenya once more. The road followed the railway line to Voi, and then we were back on more familiar territory on the road down to Mombasa.

We had arranged that we would stay in a small hotel on the outskirts of Mombasa until the survey was ready to start, and I returned to the power station to complete the documentation and accounting, and our social life blossomed again. We reverted to the weekend sailing and picnic routine, and this happy state of affairs continued for about a month.

One weekend we had agreed to babysit for some friends, and set out in the morning to a small hotel about 20 miles south down the coast. The hotel was right on the beach, and had shady changing rooms, a few thatched cottages, and an excellent bar and kitchen.

We had brought along a glass-bottomed box from the boat, and also had an inflatable air-bed, and spent a long time with the children, paddling about over the reef, observing the myriad small fish and the brightly coloured coral from a position of some comfort and little danger. We had a wonderful day until it was time to pack up and return to town – when I discovered that, despite my normal caution, I had become badly scorched on the backs of my legs, and was forced to drive all the way home in a standing position. This is quite difficult in a Land-Rover, as the roof is low, and the seats are fixed, so eventually we arrived back with me in a bit of a state. The moral of this tale is that the sun can burn you even through a few inches of water.

The time drew near for our departure, and we farmed out all our heavy gear with friends, and prepared for the first stages of the survey. We had some canvas covers made up for our suitcases, and performed numerous trial packing and discarding activities, until we could fit all we needed into the Land-Rover, with space for the surveying equipment and other safari stores.

I had been asked to drive a different Land-Rover back to Nairobi for overhaul, as a new vehicle was waiting for me in the office, and we had only gone about 20 miles when the silencer fell off. There was not a lot we could do about it, so we carried on for the remainder of the journey, making a deafening noise, and scaring all the animals for miles around. I abandoned this wreck in the yard of the hotel, and next morning summoned a driver to collect it, as I had no desire to appear before the local magistrates, charged with disturbing the peace.

My new vehicle was a much more elaborate machine, with some pretence of insulation, and even had lining on the doors and roof. What luxury!

In the morning we assembled for the final briefing, and met with our travelling geologist, and after loading up the surveying equipment set off towards Kericho – our first location.

SECTION 2 – THE GREAT WATERFALL SAFARI

CHAPTER 18
Kericho

Our route took us through the Rift Valley, where the company had a detachment drilling for geothermal steam, and the geologist thought it would be a good idea to call in on them as we were passing. We drove up a small track off the main road, and eventually found the drill rig set up on the side of the hill.

It seemed that they had had some trouble a few days previously, and the drill rods had broken about half a mile down the hole. They had tried to connect with the broken section, but without success, and a visiting American drilling expert had been called in. He had visited the site a few days earlier, taken an impression and some measurements of the broken section, and had just returned from Nairobi that morning with a special fishing tool that he had made.

When we arrived he was just about to try to connect, and we watched, fascinated, as he instructed the driller. "Down a little bit, a little bit more; OK, try a couple of slow turns; OK, down a bit more, and keep turning gently. Stop. Let me feel it myself."

He went over and tried the drill, and, apparently satisfied, instructed the driller to heave up, came over to speak to us, and, saying: "That's got it," threw his hat into his car and drove off.

What a display of confidence.

Mind you, he had probably done it all before, and there was no way of inspecting the break until the half-mile-long section of drill stem had been brought up to the surface, which would take the rest of the day at best.

Much impressed by this performance, we continued on our way to Kericho, which was to be our base for a few weeks, as we conducted the first part of our survey.

The little town of Kericho was the centre of the tea-growing industry in Kenya. The elevation of the general area was about 5,000 to 6,000 feet, and the climate was excellent, with cool nights and pleasantly warm, sunny days. While we were there, it rained for about half an hour every afternoon – just enough to lay the dust, and keep the golf course fresh and green.

The various tea estates were scattered in the surrounding hills, each one with a small community of expatriates, and their local labour force. The Brooke Bond company owned the majority of these estates, and they had opened up the Kericho Hotel as a centre for visitors to the area and a main base for their operations. The hotel was set in well-tended gardens, with a number of small bungalows in the grounds, and we moved into one right at the end of the garden, surrounded by rose bushes, with lush lawns and overlooking tea gardens where they were experimenting with different varieties of bush and shade trees.

We settled in and began our work. First we marked upon the map the various estates, and arranged to visit each one in turn. Generally, we attended in the office in the early morning, and then went round the estate, making notes on the terrain, power needs, and existing power supply systems. Next we examined the existing system of power generation, and found that there were a number of small hydropower stations in the area, backed up by a few diesel-powered installations. From this information, and projected future loadings, we marked out the catchment areas for rainfall, and derived estimates for the volume of water likely to be available for power generation.

We then had to select suitable sites for dams and power stations to deliver the supply system that was most economic overall, and from all this desk work we were then able to go out into the field to measure the flows in the existing rivers, and decide on the location and height of each dam, the volume of water it could be expected to hold, and the locations of power stations that would make the best use of the available storage. These calculations were based on the simple proposition that a small amount of water coming from a great height might generate the same amount of electricity as a vast quantity of water falling through a smaller distance.

In effect, we had to consider a number of propositions for each river, and an overall consideration for the diversion of some rivers by means of aqueducts or tunnels into adjacent catchments.

We had an instrument for measuring the flow in rivers, which was like a small torpedo, and was hung on a wire. It had an electrical system that counted the number of turns of its propellor, and this could be related to the speed of the current. What we had to do was to select a suitable site, stretch a marked cable across the river, and then run the machine at various depths, and at measured distances from the bank, so that we had a picture of the area of the section and the average speed of the stream, and could then calculate the flow.

In the Kericho area, most of the streams were fairly shallow, and we were able to select our measuring sites near areas of cultivation, so we were not too troubled by thorny bushes, and we were assured that – apart from the odd leopard – there were no large animals, such as rhinos or elephants, in the district. We soon developed a routine for measuring these rivers, and apart from the occasional ducking, where deep holes had been scoured out in the river bed, the work progressed well.

Kericho boasted an active social and sporting life, with tennis, golf and fishing as the main outdoor pursuits. In addition to the regular bridge at the club, there was a thriving amateur dramatic society. We managed a regular game of golf every afternoon, some tennis in the early evenings, and after dinner returned to our calculations and report writing, with the usual emphasis on the expenditure of the petty cash.

The dramatic society were playing *Blithe Spirit* while we were there, and, to my untrained eyes, it was an excellent production, with a very few minor calamities, which only served to make the performance more interesting. Would he manage to open the door the correct way next time, or would the oil lamp in the corner set fire to the curtains before the end of the act?

A special programme of events had been arranged over the Christmas period, with dances at the club and the hotel, a car rally, a treasure hunt and various unusual competitions all being organised.

Jean had decided that she would add a bit of colour to her hair, and had bought some dye in one of the local shops. It came in a bottle with a picture of a young lady on the cover, complete with flowing, shiny, chestnut-coloured hair, and with a comprehensive set of instructions. While the application was in progress I was sitting on the veranda sipping a sundowner, admiring the scenery, listening to the birds, and reading a book. This peaceful scene was disturbed by a piercing scream from within. I hurried in, expecting at least to be confronted by a small spider in the bath. What met my eyes was much more serious.

Apparently, the last part of the hair dye instructions concerned wrapping the damp hair in a towel for five minutes. This towel had been the first intimation of disaster. It looked as if it had been soaked in permanganate of potash – and Jean's hair was much the same hue. The scream had been uttered when she had looked in the mirror. Not only was her hair bright red, but the dye seemed to have run down onto her forehead and over her ears, staining the skin to a deep tan.

And all this before dinner, too.

Other than trying not to laugh, there was not much I could do to reduce the intensity of the colour; all I could do was to try to restore a little sanity to the situation.

It was dark by then, and the dining room was not well lit, being full of discreet little corners with red-shaded table lamps, but no amount of persuasion would induce her to go down to dinner, and we had to resort to eating in our little bungalow, courtesy of the waiter service, with herself doing a vanishing act at the sound of approaching footfalls.

The night wore on with almost continuous washing of hair and the application of cream to the skin, but, whatever the dye was, it was pretty permanent, and had set in well. After the first couple of rinses no more colour was coming out in the water, and the remaining towels were hardly coloured at all. All the towels used in the earlier stages were by now soaking in soapy water in the bath, being stirred at regular intervals, but without much success. Further action that night was abandoned, and we retired.

In the morning, in the bright light of day, Jean's hair was still a strong shade of purple, and I was dispatched to the local shop to purchase some bleach, and told to arrange for our meals to be delivered to our bungalow, until further notice.

In the evening the barricades were up and the blinds drawn when I arrived home to change. In the dim light inside I could see that the colour had faded slightly, and the tan stains had almost vanished. I was unable to persuade her to come out onto the golf course, however, even with a scarf over her head, and once more we had dinner in the bungalow, and the washing and bleaching continued.

This situation was getting a bit out of hand. I had a quiet word with the manager's wife, suggesting that she might telephone with some words of comfort and advice, and perhaps suggest an antidote. After a few days of intensive washing and bleaching or toning, or whatever, the purple colour faded, however, and we were once more able to eat in the dining room, as long as we stayed out of the bright lights.

WHAT YOU SHOULD KNOW ABOUT TEA

Tea bushes, which are a variety of camellia, are grown in rows across large, rolling fields, with substantial shade trees at regular intervals.

Different groups of employees are engaged for cultivation, planting, pruning and harvesting, together with services personnel who maintain the roads, the irrigation ditches, and the machinery in the sorting and packing sheds. The harvesting is normally done by women, who pick the ripe shoots and throw them into a large sack that they carry on their backs, and these are later collected and weighed.

Our surveys in the area revealed a number of interesting facts concerning the selection and preparation of tea. The leaves were weighed and roughly sorted in the collecting shed, and set out in heaps by size and quality, and credited to the picker.

In the sorting house the leaves were graded and stacked in separate piles around the sorting house floor, and these were then packed into boxes for further examination and drying. All the remaining leaves – small, discarded and unselected – and any which had escaped from their earlier heaps were swept up, together with insects, to be used in tea bags.

The final blending and sorting was completed at a later stage, and the leaves were then repacked in conventional tea chests, and these sent off to a central collecting depot prior to transport out of the country.

Meanwhile, we had been collecting information resulting from our surveys, and were then preparing a number of proposals for further consideration, and, resulting from these deliberations, gradually reducing our ideas. The scheme we recommended was based on two small reservoirs, with a series of aqueducts, and a tunnel leading to a single power station. Having reached this conclusion, we then had to make a more detailed study of the various sites, take material samples, dig trial holes, and walk round the proposed top water levels of the reservoirs to see if we were likely to flood out any of the local inhabitants or valuable tea-growing land.

At the end of this study we had to produce our arguments, justify our selection, make a cost estimate and probable programme for the development, and then produce reasonable drawings for presentation. As our allocated time on this area was drawing to a close we had to work long into the nights, and the social life began to suffer. The food was excellent, the rooms were comfortable, and the weather continued fine, however, so we just had to put up with it and carry on.

Before leaving Kericho we decided to make a trip to see Lake Victoria, and set off early one morning, bound for Kisumu. This journey took about three hours, along very bumpy and dusty roads, but it was downhill most of the way. The pleasant wooded country round Kericho gave way to dry and dusty plains, with scrub bush right down to the outskirts to the town. The railway line ran to Kisumu, which served as one of the major ports on the lake, and there was the usual industrial sprawl around the port area (with the odd workshop), which we were able to include in our report.

Kisumu was also the centre of the gold-mining area in eastern Kenya, but most of the mines had been closed down due to the Mau Mau operations, and only a few individual prospectors were still busy in the vicinity. We did see some transactions over the bar, where small sacks were exchanged for drink, but whether this was genuine or just an act put on for the visitors we never found out.

Our geologist took a particular interest in these affairs, and we had a bit of trouble dragging him away to continue on our journey back to our hotel.

We were leaving Kericho the next morning, so had no time to do a bit of prospecting ourselves, and set off early, back to Nairobi, to deliver our reports, and prepare for the next stage of our expedition.

CHAPTER 19
Moshi

Our next area was to be around Moshi, in Tanganyika. We loaded up and were off in the early morning, and, once clear of the town, were travelling across the Kapitti Plains. This was a vast landscape of rolling hills covered in small scrub bush, which later opened out into the Nyiri Desert – a pretty awful place of dusty, sandy scrub. The only inhabitants we came across were small groups of Masai gathered around their thorny hedged enclosures, and numerous herds of cattle guarded by small boys. In the distance we could see the massif of Kilimanjaro, with its snowcap gleaming in the sun. At last we came to the border, marked by a small thatched hut by the roadside – and no one to check our papers, or examine our baggage for contraband. Not that we had any.

The scenery continued unchanged until we reached the small town of Longido, basically a row of shops in the middle of the desert, with the ground rising to the top of a little mountain about 2,000 feet above the surrounding plain. As we went on, the ground rose gently towards Meru, another significant mountain about 15,000 feet high, with tree-clad slopes. The desert gave way to more fertile country, thicker bush and more frequent smallholdings and settlements in the river valleys.

We skirted the mountain to the west, and were shortly into Arusha – quite a civilised little town at the end of the railway. To the north the background was dominated by the bulk of the wooded slopes and snow-covered dome of Kilimanjaro. All to the south stretched the barren plains of the Masai Steppe, a vast desert of small scrub and thorn bushes.

We settled into a tiny hotel, and next morning made our number with the local manager of the power company, who took us round his parish, and explained the workings of his various power stations. We had a look at a number of small waterfalls, took some measurements in the forests surrounding the town, and after filing our report moved on to Moshi, about 100 miles to the west.

The road followed the line of the railway, with the high ground of the foothills of Kilimanjaro rising gently to the north.

The town was quite large, and was the centre for the agriculture of the area, with products including timber, grain, coffee and livestock. We had booked into the government rest house, as we had reports of the poor quality of the local hotels, and were soon comfortably settled in. The buildings were a number of small timber bungalows centred around a main house containing the mess hall, reading room, lounge, kitchens etc., and the whole complex was situated in a pleasant garden with shade trees and well-kept lawns. All the buildings were set up on small brick pillars fitted with anti-termite ledges – a feature we were to find frequently throughout east Africa. Each house had a corrugated iron roof, a large veranda to the front, and was equipped simply with comfortable locally made furniture. The cooking, carried out in the main building, was by means of wood-burning stoves, and was generally very good

We met the local manager of the power company and obtained his blessing to call on any of his staff throughout his territory for assistance and advice, and then set about formulating a plan of campaign.

The new Moshi hotel was under construction nearby, and nearly ready for occupation. We thought it was a good place to start our report, as part of the general facilities in the area, and arranged for a visit. I think the manager must have been misinformed, as we were treated as visiting royalty, shown round the rooms and grounds, and introduced to all the senior staff. We inspected the kitchens, and were invited to partake of the first meal prepared in the new kitchen, which was served in the new dining room, with the new cutlery and crockery, by the new waiters – all done especially for our benefit. For a first attempt, it really went off very well, and we noted in our report that the new hotel had considerable potential!

The building itself and the furnishings and equipment were more than adequate, but the grounds outside were a shambles of builder's rubbish. No doubt, in time, that would all be transformed into a pleasant garden. The view out of the front of the hotel was quite magnificent, across the gently rising ground towards the snow-capped mountain.

We started our investigations at the small local hydropower station on a tributary of the Pangani river, about 20 miles out of town, and recruited a number of men locally to assist us in our operations. Our main target was the Pangani river gorge, where the river had scoured a deep valley through the extensive plains to the south of the town. We drove out across the plain, sandy and rocky ground covered by small spiky trees and bushes, until we came suddenly to the edge of the gorge. It fell away steeply down to the watercourse some 200 or so feet below, and about 100 yards across. The sides of the gorge were very steep, and at the top there was little vegetation, but nearer the river level there were fairly large trees, whose top branches were almost up to the level of the adjoining plain.

We drove carefully along the top of the gorge, looking for a suitable place where we might measure the flow in the river, and also where we might be able to climb down with all our equipment, and after some discussion on alternatives settled on a possible route down. We parked the vehicles on the edge, leaving the drivers in attendance, and organised our porters and survey crews for the descent.

The order of march was quite important, as we had a lot of heavy equipment by way of a small boat, anchors, ropes, the usual surveying gear of level, theodolite, ranging poles and staffs, chains and tapes, arrows, bundles of wooden pegs, pots of paint, and – of course – our current-metering machine with its wires and cables.

Whoever went first had to consider the route for the others more burdened behind, but had also to be on the lookout for snakes etc., anthills and other livestock, such as bees, and also had to be aware of the possibility of loose material, and occasional porters, falling down from above. After considerable difficulty we reached the floor of the valley, and then had to make our way along to the site selected from above as most suitable for our measuring operations.

This proved to be quite difficult, due to the dense bush and large trees, some of which had fallen and now snarled up the access. The river was about 40 feet wide, with clear, sparkling water running swiftly between rocky banks. On each side there was a sort of tidemark of dead bushes, all tangled up with one another, and lodged in the growing vegetation – the habitat of all manner of small winged insects, most of which bit when disturbed.

We managed to find a suitable place to launch our boat, and by means of ropes and anchors moved it to a place where we could measure the flow, and see a bit of the banks, so that we could estimate the shape of the valley which would be immersed in time of flood; from this information we could make an estimate of the maximum storm water flow.

After the actual flow measurements were completed, we had to select a suitable site for a possible dam, and this involved further scrambling about in the undergrowth, taking samples of the rock in the river bed, and trying to envisage possible means of access and construction. After about an hour of this tiresome struggle, we decided that we had done enough for the day, and began to make our way back to the path we had blazed earlier on our descent.

By now we were additionally burdened by a few hundredweight of rock samples as well as all our original survey equipment and the boat; and a short break was indicated before we began. By this time the sun was shining directly down into the gorge and it was becoming quite warm, so we all had a quick swim in the river, without a thought for crocodiles and piranha fish – or their African equivalents.

Refreshed by our rest, we started upwards. This time there was no doubt that the safest position was in the lead; no one could dislodge stones and branches on you, and if you slipped and fell your fall would be cushioned by the remainder of the party struggling below. We toiled upwards, making frequent stops to rest, and reallocate the loads, and eventually reached the top of the gorge.

The Land-Rovers were still there, and, just as we were about to come over the top, I noticed the two drivers, perched on top of a small thorn bush about ten feet up, making frantic signs for us to keep out of sight and be quiet. I called up one of our local guides and asked for an explanation. He cautiously put his head above the edge, and by a process of intuition and sign language with the drivers came back to inform us that there were two rhinos in the area, and that it would be wise to wait out of sight for a little, and hope they went away.

Some time later there was another exchange with the drivers, and it was then decided safe to move out. The drivers came down out of their tree, all scratched and cramped, and said the rhinos had come along very shortly after we had left, so they must have been stuck up there for about five hours. When asked if they thought that their elevation of ten feet was sufficient, they replied that that was the highest tree for miles, and they were still alive, weren't they?

They implied that rhinos have horizontally limited vision, probably because of the great weight of the horn on the end of their noses, and charge only at persons or vehicles below eye level.

This I found hard to believe.

Armed with this knowledge, we loaded our gear with an eye to the nearest tree (thorn bush), or a quick dash over the edge of the gorge. Having suffered from thorn bushes in Mombasa, I knew which way I was about to run.

It also occurred to me that there was another dilemma in progress. Should we work away as quietly as possible, and try not to draw attention to our presence, or should we bang things about, and shout and make a lot of noise, in the hope of scaring the animals away? This problem was soon solved, at it is almost impossible to have a crew of weary porters and surveyors do anything quietly, as this might delay their return to their homes.

For the next few days we ranged up and down the top of the gorge trying to select alternative suitable sites for a dam, and the best location for a power station for each site, and on completion of this recce we selected what we considered to be the most promising and carried out a more detailed study of that area. This involved regular descents into the depths, but as we were interested only in the topography and geology there was no need for the boat, although it did entail swimming or fording the river.

The river at our selected location ran almost due south, so the gorge was shaded and quite pleasant until about eleven o'clock, but soon fired up when the sun came round, and we were obliged to stop for a swim and a picnic lunch during the heat of the day.

The flow in the river was quite erratic; the levels could rise and fall by a couple of feet during the day, so we were always conscious of the possibility of being marooned on the opposite bank. This would have made for an ascent over unknown territory, and a very long drive round to the nearest bridge for someone to come round and pick us up, if they could find us.

We visited the local small hydrostation, measured the manager's river for him, and ran a rough survey to see if there was a possibility of increasing his catchment area by canals or aqueducts, and having exhausted the potential of the district and written up our report we prepared to move on. By now we considered ourselves to be seasoned travellers, and could pack up and load the Land-Rover with all our worldly goods and all the survey equipment in a matter of minutes.

We had also learnt some of the tricks of the road. The roads were generally unmetalled, except for short stretches through town, and sometimes for about a mile either side. The traffic had produced a series of bumps on top of the usual corrugations, which tended to increase in violence as you progressed, and then suddenly vanished. This could be explained by the periodic reaction of the springs, when the last violent bump caused the driver to slow down.

At about 40 miles per hour, the corrugations and the bumps were at their worst, but we found that by driving on the wrong side of the road at about 60 the ride was not too bad, and the effect of the ascending bumps was much reduced.

The journey time was also reduced.

Driving on the wrong side of the road was no great problem, except that the Public Works Department, in their wisdom, carried out so-called maintenance. What they did was to run a grader down each side of the road, dumping the loose material into a ridge about a foot high down the centre. This could cause some few seconds' worry when approaching vehicles were sighted.

Another hazard was the occurrence of the so-called level crossings. Frequently the railway ran alongside the road over long distances, and crossing places had been provided where it took the railway engineer's fancy. The roads had been worn down by traffic, weather and the efforts of the PWD until they ran in what amounted to a continuous depression. It was usually possible to see or hear a train travelling in the same direction, but oncoming trains were practically invisible until the last moment.

The majority of level crossings were marked by poles, and these served to reduce the speed of prudent drivers some distance ahead, as what happened was that the road, with its attendant banks, suddenly turned off at 90 degrees, presenting you with a bank of earth, often surmounted by the remains of wrecked vehicles. Going over the rails was also very disturbing, as they sat up about six inches from the surrounding surface, and could wreck the suspension and stall the engines of inattentive drivers.

Most of our travelling had been in the dry season, so our main complaint was of the dust – but just wait until the rains.

And so, we went on to Iringa.

CHAPTER 20
Iringa

The distance between Moshi and Iringa is about 300 miles, and this was taken in two stages. We travelled back on the now familiar road towards Arusha, and then south towards Kondoa and Dodoma. The route took us through some high plains, and there were a number of potentially interesting waterfalls we passed and noted as deserving further investigation at a later stage. On then to Dodoma, also in the pleasantly high country, and boasting a railway station on the line between Dar es Salaam and Kigoma, on Lake Tanganyika – or that was its name in those days. (I believe it still rejoices in that name today. How could you call such a fine body of water Lake Tanzania?) We stopped overnight in Dodoma, in a very reasonable rest house, somewhat disturbed by the noise of frogs in the hotel garden, and a plague of flying ants as we went down for dinner.

During our journey to Iringa we were run into by a large wild pig. We were travelling through fairly open bush, at our normal cruising speed of about 60 mph, through what seemed to be quite an empty landscape. Some short distance ahead I could see some large animal off to our right, travelling on what seemed like a collision course with the front of the Land-Rover. I slowed down to try to get a better look at it, and to allow it to pass clear ahead on the road, but it also altered course, and was still coming at us at great speed.

I suppose we were doing about 50 when the collision occurred. There was a heavy bump on the right-side front wheel, which almost wrenched the steering wheel out of my hands. I stopped as quickly as I could, about 100 yards further down the road.

As the only weapons in our possession were a couple of pangas, and some steel-tipped surveying poles, I was a bit windy about getting out of the vehicle to inspect the damage, just in case an enraged and damaged pig was still intent on attacking us, or some of its relatives might be looking for vengeance, when I noticed two young men, who had magically appeared out of nowhere, prodding at something at the side of the road.

I reversed and spoke to them, and they said I had just done a marvellous job, and thanks very much for the supper. I climbed down and saw the remains of a very large wild pig lying at the roadside, with most of the side of its head seriously damaged, and thought, my God, if it has done that to its head, what can it have done to the vehicle?

A close inspection of the front of the Land-Rover disclosed a slight kink in the front bumper beam and a lot of blood and fur on the front wheel nuts. It must have hit the centre of the wheel spot on, and been bounced clear.

How lucky can you get?

During our travels we were often surprised by the sudden appearance of people from what we took to be uninhabited areas. When we stopped for lunch, or to change drivers, even only for a short time, it was astonishing how children would appear within a few minutes, and stand nearby, smiling and saying: "Jambo" – the traditional Swahili greeting.

It was in the evening that we arrived at Iringa, where we had arranged to meet up with two others of our survey party, fresh out from London. The hotel in the town was very rough, and the rest house fully booked, so we had to make the best of it.

Jean had a colleague from the bank working as manager in Iringa, however, and he invited us to stay with them in some splendour in their house just out of town.

The town itself was pleasantly laid out at the foot of some high mountains, and was well positioned for our surveying work, as there were numerous waterfalls in the hills all round, and passable roads in all directions. We met with the local power company manager and arranged for guides and survey assistance as required.

The area round Iringa was good mission country, with a number of extensive townships and their supporting lands, run entirely by the fathers. One particular station had a small hydropower plant, originally installed before 1914, when Tanganyika was a part of the German Empire. The mission ran on very egalitarian lines, and the blessing of God, with the clergy taking their turn at being responsible for all the various trades and disciplines on the basis of a three-year tenure of office.

We had arrived shortly after such a rotation, and the Father Fundi, the previous incumbent of the office of mechanic and electrician, had been banished to some region of outer darkness in the parish, while the newly appointed Father Fundi, the man currently in charge of all matters electrical and mechanical, including water supply and drainage, was a young man who had no experience in these fields. He had been a sort of apprentice to the Father Farmer, and was quite knowledgeable in haymaking, milking cows and other matters agricultural.

The Father Superior asked us if we could offer some assistance and advice, and took us down to the power station.

The operating instructions were a translation from the original German into Flemish, with a few Swahili words thrown in just to confuse, but by dint of searching about in the offices we found the original German notes, and drawings, and with great difficulty produced more up-to-date instructions in a mixture of English and Swahili.

The station was very clean, with all the copper and brass fittings highly polished, and one of the sets was running quietly. The wiring of the station was in a bit of a state, and appeared to have been modified at random. It looked as if every time they found a fault, or were unable to switch in one of the other sets, they had run new wires, and made suitable connections within reach.

Hydroelectric power is basically a very simple concept.

You make a pond to store water, connect a pipe from this pond to a water turbine some distance downhill, and, by opening a valve, the water flows down the pipe and turns the wheel of the turbine. This wheel in turn spins an alternator, and electricity comes out of the wires. Once the system is set up and running, it is fairly self-regulating, as long as the water level remains reasonably constant.

Technology has produced equipment to regulate the process, and to adjust the flow of water in the pipe as the electrical load varies, and these mechanisms have not changed much since the station was originally commissioned. All the equipment installed was original, very robust, elegantly simple, and readily accessible.

The previous incumbent had left one set running, and the new man was unwilling to switch in one of the other sets, having been warned of the consequences of mistiming the synchronisation. Instant darkness, with flashes coming out of the switchgear, and some penance to pay for annoying the Father Superior.

We studied the manuals and were then able to demonstrate how to run the other machines up to speed, and how to switch them in and out without blowing all the fuses every time. We came in every day for a short visit and the new Father Fundi was soon quite expert.

The river systems around the town were very complicated, and we had to carry our boat and equipment several miles to reach some of our selected sites. This was easy enough over open fields, or in light bush and scrub, but there were some places where the only path led through elephant grass, about ten feet high. The paths twisted and turned, forward visibility was practically nil, and sideways the view was the wall of grass right by your elbow. These conditions called for some alterations in the order of march (a military term indicating the position of each man in the column).

We had a local guide, who marched in front, usually armed with a long spear. He was followed by the man carrying the recording machine in its large, heavy box, the theory being that, when the guide had been run over by a buffalo, it would trip over the box and give the remainder of the party time to dive off into the side grass. Then came the men carrying the boat over their heads, followed by the geologist and me, then the men with the oars, ropes and anchors, and the rearguard being the chain men with the surveying equipment (also large, heavy boxes and tripods to trip any predators approaching from behind).

In open country the men would chant in unison on various topics – probably about our persons, as they sang in some local dialect – and there was a lot of laughing from time to time, but in close thickets the men carrying our poor little aluminium boat beat hell out of it, and the chanting reached a frenzy. I suppose all this noise was intended to scare off any large animals, such as buffalo or rhino, and as we were never troubled during these excursions it must have worked. I often thought that, if I had been a large animal peacefully grazing in the bush, I would have been disturbed by all this racket, and come down fast to try to have them stop.

We were advised that there were crocodiles in the rivers, but we never saw any while we were working. We did come across many snakes of various colours and sizes, however. The discovery of a snake was always good for about ten minutes' rest and entertainment, not to mention a certain degree of alarm. In some cases the snake was hunted and killed, ready for someone's supper, and at other times the excitement seemed to be designed to frighten the snake away, with no serious attempts being made to catch it. Without appearing to seem too bloodthirsty, I must confess that I prefer a dead snake than the thought of one which has just been scared off the path, and is now lurking in the bush alongside, ready to bite the next passer-by.

While we were in Iringa we met up with some others conducting a study of the Rufiji river basin, mainly with a view to controlling the flooding, and providing irrigation water. They had established a number of river measuring stations, and we exchanged information quite freely. Their operation was being run by a Polish engineer, who was away on leave during our visit to the area, but who was the subject of many tales of dedication in the face of floods and wild animals.

On one occasion he was returning to town from a hill survey during the rainy season and came to what was known as a 'Tanganyika bridge'. This was no bridge at all, but merely a local name for a ford, where the watercourse crossed the line of the road.

Sometimes the bridge would be paved in concrete, and marked with posts driven into the river banks, which might also give some indication of the depth of water that could be expected. These were designated as 'improved Tanganyika bridges'. All seasoned travellers sent a man ahead, sometimes attached to a rope, with a stick to check out the depth of the water and the presence of large potholes before risking their vehicles.

This time the river was in flood, and, in the pouring rain, one of the Land-Rovers had been swept off the roadway, and was being trundled downstream by the current. While the main party were attempting to recover the vehicle, the boss had paced out a length of river bank, and stuck small pegs in to mark the flood water level. He was then observed to throw sticks into the flow, and time their passage over his marked-out distance.

No doubt, somewhere in the records, there is a report of a traffic accident to a Land-Rover, damaged in the course of duty in a flood, and somewhere else in the records there is a series of calculations which estimate the flow in the river on that day, and from which it could be deduced how many cubic feet per second it took to dislodge a fully laden Land-Rover.

This might turn out to be a *vital statistic*.

During our stay in Iringa we managed to have a few games of golf on the local course. On one particular hole, the tee was perched on a small rocky outcrop on the side of a steep, bush-covered hill, with a narrow path leading down to the fairway some 100 yards ahead, extending for another 100 yards to the back of the green, which was also surrounded on three sides by pretty dense bush.

I had driven off, a bit wayward perhaps, and was surprised to see the caddy going straight up to the edge of the green. When I arrived I mentioned that I had thought that my ball had gone off to the left a bit, and would still be in the dense bush along the side of the hill. He insisted that it must have hit a rock and bounced out, for there it was, lying in a small depression at the edge of the green.

What a lucky bounce, I thought.

Later, in conversation with some of the members in the clubhouse, I was informed that this was a very common experience, as none of the caddies liked to venture into that particular part of the bush. To this end, they all carried a selection of balls in their pockets – and anyway, they were out to please.

When we had completed our reports and stayed as long as we could, enjoying the comforts of living in a pleasant house, we set out for our next location, the small town of Mbeya, about 180 miles to the south-west.

The road led through the cultivated plain outside Iringa, onto a stretch of desert land, with bushes and small scrub trees; it was pretty desolate, practically uninhabited, and a very bumpy road, until we began climbing into the mountains around Mbeya, where the trees grew tall on the steep slopes. This was tea country again, with a number of large estates, each self-sufficient, with its own accommodation and services.

CHAPTER 21
Mbeya

The small town of Mbeya was set in a long valley in the hills of the Southern Highlands of Iringa province. To the south the mountains ranged up to 10,000 feet. The slopes were tree-covered, and there were many good rivers flowing down from the heights.

The hotel was an old timber building, set into the side of a small wooded hill in well-tended gardens, whose water supply was augmented by the discharge of the clients' bathwater, which fed into a system of small canals and regulating weirs. This certainly improved the productivity of the garden, but the presence of all this water attracted all the frogs for miles, and at nights the noise they made had to be heard to be believed.

In our investigations we visited the nearby site of the gold rush of 1905 or thereabouts. The valley was full of derelict buildings, and curious items of mechanical equipment, long since abandoned, and robbed of all their moving parts. Our geologist had a careful look about, but we were unable to find any gold left lying on the surface. Every time we measured a river, however, we had a few minutes with a head pan, looking for alluvial gold that might just have been missed by the earlier prospectors – but without much luck, or colour, as the experts call it.

I had a cousin who was working in the copper mines in Northern Rhodesia, as it was then known, and had written to him earlier, suggesting that we might meet up at the border at Tunduma one Sunday. We drove over, a distance of about 80 miles, and were confronted by some very officious persons, who would not consider us crossing their line without the production of passports, visas, health certificates, and formal documentation for the car.

We waited about for an hour, in the fond hope that my cousin might appear, and we could arrange to meet up half a mile further west, where there were no such restrictions, but no one arrived, so we set off and had our picnic in the shade of some tall trees on the side of a hill, where we could monitor the traffic on the road into Tanganyika. After a pleasant lunch we made our way back to town. It transpired that my cousin had been away on leave at that time, and only received our letter when he returned, about a month later, by which time we had moved on.

Our investigations took us through the mountains and down to the shores of Lake Nyasa, as it was called then, where we stopped for a brief swim on a very hot and dusty day. Most refreshing it was, too. We then inspected the port at Mwaya, had a spot of lunch at a very basic hotel, and continued on our way.

During our stay in Mbeya we were graced by a visit from Princess Margaret. The whole town was dressed overall, with flags and flowers at every conceivable location. All the residents of the nearby tea estates had come into town, and a public holiday was declared. We went down to the airfield to see the visitors arrive, and as each aircraft landed the assembled schoolchildren waved their little flags, cheering wildly. When the official car passed with HRH aboard, they stood in silence, as a mark of respect, and as soon as she had passed they followed on behind, cheering and dancing all over the roadway.

We met some of the people from the tea estates who had come in for the entertainment and formal reception, and who had stopped overnight in the hotel. One couple told us of their experience recently, when they were having their daughter christened on their estate. The logistics surrounding this ceremony were pretty formidable, what with arranging for grandparents coming out from the UK, relatives and friends from elsewhere in east Africa, the celebrant himself, food and lodgings for the multitude – it had, we were informed, cost a fortune. The straw that had nearly broken the camel's back was a bill from the priest for holy water. "The bloody man just filled a jug from the tap! 'Holy water' indeed."

The town was served by two power stations: a small diesel station in the town itself, and a hydrostation about ten miles out. A new manager had been appointed about a month previous to our arrival, and we had met with him in his office in town and made arrangements to hire local assistance from his usual contractor.

He had told us of a forthcoming visit from the general manager of the Power and Darkness from Dar es Salaam, and suggested that we might like to join them for lunch.

After a pleasant meal we all returned to his office, and then inspected the diesel station in the town. The station had been scrubbed and polished spotless, and when we came out the GM jumped into the local man's car and said, "Let's now go out to the hydro."

There followed a dreadful pause. The wretched man had never been out to the other station, and knew not which way to go. He hadn't even enough sense to follow the overhead wires out of town. (As there was only one through road, he had at least a sporting chance whichever way he turned.) When he admitted that he did not know which way to go, and that he had not yet visited his other station, we were asked to "Carry on for a bit," and the pair of them retired into the office.

Some time later the GM came out and asked us to take him out to the other site, and when we returned the other chap had cleared off, never to be seen again.

The full title of the company was the Tanganyika Power and Light Company, but it was generally better known as the Power and Darkness, due to the not infrequent interruptions to the power supply. The reasons for these failures were often from causes not generally encountered in normal domestic and urban supply, some of which are enumerated as follows: bush fires; animals knocking down power poles while in use as scratching posts; termites eating the heart out of power poles; poles being chopped down for use as roof timbers for houses; wires and cable being stolen for making bangles and other ornaments; etc. etc.

In general, the customers took these interruptions in good part, as they were all beset by similar problems in the course of their business, and, in any case, they were well provided for by way of oil lamps and candles for lighting, and most cooking was undertaken by wood-burning stoves.

After a very busy but enjoyable stay in Mbeya, we prepared for our departure to the north again, to Kilosa, nearly 400 miles away, where we would meet up with other members of the survey party, out from London.

We had been advised that the standards of hotels in Kilosa were rather basic – this translates into 'bloody awful' on the Egon Ronay scale – but we made arrangements for at least one night's stay in what was suggested to be not quite as bad as the other hotel, and set off early in the morning.

There had been some rain overnight, and the roads were very slippery, with large areas covered in puddles. The main problem now was to guess how deep the puddles were, and after a number of severe jolts we found that we had to reduce speed to about 20 mph to have any sort of comfortable ride.

A further problem was evident. The wash of thick red water raised by vehicles coming in the opposite direction completely obscured any forward vision, and. if perchance the side windows were open, it came into the interior in large volumes, leaving a residue of thick red mud on everything and anyone aboard.

CHAPTER 22
Kilosa and Morogoro

We retraced our route to Iringa along the edge of the Southern Highlands, and stopped overnight in the local hotel, which had not improved since our earlier visit. Next day we set off towards the north-west, through some pleasant wooded hills with cultivated valleys and open plains, towards Kilosa – another railway town on the line between Dar es Salaam and Uiji, on Lake Tanganyika; but no doubt it has another name these days.

Around Kilosa there were many sisal estates. Sisal is a hemp-like plant which is allowed to grow for about seven years. Each year, the thick spiky leaves around the base of the plant are cut off and processed in what were called gin mills, the resultant products being a foul-smelling yellow liquid, which was discharged into the nearest stream, and long strands of yellow fibres, which were hung out to dry in the sun on extensive fences. Many of the estates had small rope works, where the sisal was stranded and twisted to make ropes of varying diameters, but the bulk of the crop was made up into bales, and sent off on the railway.

Most of the estates were little townships in their own right, with a power station, water and sewage systems, and their own private small-gauge railway system, with diesel locomotives, workshops, offices and shops. They normally had an English manager, Greek or Italian mechanics, a few Asian tradesmen, and about 100 local Africans, all with wives and families in residence.

We rushed through our work from Kilosa, as our London visitors were a bit put off by the somewhat primitive living conditions in the hotel. We decided we could do as much from Morogoro, which was not very far away, and boasted two hotels – the Savoy and the Ritz.

It was a bit of Hobson's choice, as they were both terrible; owned and run by Greeks, not at all clean, and with awful food. The countryside round about was marvellous, however, with thick wooded hills, steep-sided valleys and excellent waterfalls, and we were soon busy scouting around for sites.

One day we came on a village about three miles out into the plains, with a large waterfall visible in the hills, falling about 200 feet straight down, and an obvious site for more extended surveying.

We spoke to the headman of the village and arranged for a guide to be available to take us there the following morning, and returned to base to make our preparations.

In the morning we set out, and were welcomed by the headman, who introduced our guide – a small, wizened man renowned as a skilled hunter, who in turn introduced his gun-bearer; this was another old man, carrying an ancient .22 rifle, polished and burnished to perfection – obviously his pride and joy, and probably very effective for shooting pigeons.

Jean had agreed to assist with our survey, and remained by the village, attended by our drivers (and most of the population) to make regular observations on our aneroid barometer. We carried a similar instrument, and by comparing the readings we could come to a pretty fair estimate of the difference in elevation between us.

The introduction of armed men into our party gave us some thought concerning the order of march. These two would obviously lead the way, but, if some wild animal came down the path towards us, by the time he had been handed his rifle, cocked it and fired, whatever it had been would probably have been halfway down the column by then, and we would be in the line of fire. Mind you, there would not have been much of a column by then.

We had decided that, for the first visit, we would not need our boat, so the party formed up with the guide and his bearer, the survey chain men with their burdens, the four Europeans, and the men carrying the water-measuring box in the rear. At first our path led across the plain, mainly scrub bush, but as we got nearer to the base of the waterfall the ground grew steeper, and there were many tall trees in the denser vegetation.

After we had been going for about an hour, the guide held up his hand, and gestured for silence. Then, without further warning, both he and his bearer shinned up a large tree.

We all stood transfixed.

Were they trying to see something?

Had they been frightened by something?

What the hell were we doing, standing about, waiting to be run down by some monster?

After what seemed like a very long time, our guide and his colleague climbed down again, and announced that there were elephants about, but we could proceed quietly now, as they had probably gone away (?). Sure enough, when we came around the next corner in the track there were great heaps of steaming elephant dung all over the place. Apart from the chirping of the cicadas we could hear no other noises, and continued very quietly, moving individually from tree to tree, just in case we might have to do a bit of tree-climbing ourselves.

Soon we were near the foot of the waterfall, and the noise it made obscured all others, and while keeping a good lookout all round, and upwards as well in case of leopards, we reached the spot for our surveying at the foot of the fall. The ground was very wet and slippery, and visibility restricted by the many trees and bushes, but we contrived to take some measurements, and made a few sketches before climbing out to one side and continuing our ascent to the top of the falls.

Away from the actual waterfall and the area affected by the spray, the ground was fairly dry, and we made good progress through the spiky bushes and thorn trees, disturbing clouds of insects and several flocks of birds, and hearing the furtive rustling of small creatures in the grass around our feet.

We came out quite suddenly over the rim of the hanging valley, where there were a few trees, with many open spaces of grassland, and the river flowing quietly before it plunged over the edge of the falls. We were about 500 feet above the general level of the plains, and had a panoramic view of the entire area, enclosed by further mountains in the far distance. Below us, the dark green foliage around the foot of the falls thinned out until it became a narrow ribbon going out across the plain, the colour of the grass changing from green to brown, and then red in the distance. The valley behind us had gently sloping sides, and twisted away into the hinterland, with not much sign of cultivation.

We had a short rest and some lunch, and then set about measuring the flow, selecting sites for a dam, taking samples of the rock and completing the first stages of our survey, before diving into the river for a swim. After we had dried off, we re-formed, and began our descent. Once more we encountered signs of the recent passage of elephants, and adopted our tree-to-climbable-tree routine, until we were well out into the plain, and there were no more suitable trees, only scrub bush.

During this passage, about halfway back to the village, our guide stopped, snatched his rifle from his bearer, and fired into a small bush some ten yards ahead. There was some commotion within the bush, and a short hairy animal came out, travelling like an express train, and went crashing past the column to vanish into the scrub behind us. After a few moments of stunned silence, we all fell about laughing, but tempered with some caution, keeping a sharp lookout just in case some of its relations were still around.

Our guide reloaded his firearm, and with his bearer went ferreting about, looking for bloodstains, or something; more in hope than anything else, as the animal – whatever it was – had come out at such a speed that, even if it had been hit, the first drops of blood would have had to have been about 50 yards away before they reached the ground. After some further discussion, but no more evidence, we decided that it was probably a warthog, and we were assured that it would have made good eating in the village.

No one mentioned the damage it might have done to the survey party...

We re-formed again and continued our march, still keeping a good lookout until we returned to the village, where Jean was waiting, reading her barometer and still surrounded by the village children.

After a short conversation with the headman, we boarded our transport, with our guide aboard, as he had requested a lift to the next village on our road. We persuaded him to unload his weapon, and sat him in the back of the Land-Rover, and when we came round a corner in the track and saw a herd of buffalo on the side of a small clearing he became quite agitated. Even he realised that his rifle was as much use as a pea-shooter on these huge animals, but he indicated they were generally unfriendly, and we were not to tarry. We signalled to the other car behind us, and made off at some speed, bumping down the road, comforted by the thought that the buffalo would need to pass the other vehicle before they caught up with us, but slightly worried that we might meet the remainder of the herd coming towards us in the opposite direction. Anyway, we arrived safely at the guide's village without further incident, and were then on the main road back to town.

We had met up with some mining engineers in the hotel, and they had invited us out to their location for the weekend for some party they had arranged. Their camp was about 30 miles out in the bush from town, and we set out in two Land-Rovers, over some terrible roads, to their mine. Coming round the shoulder of a hill we could see the large mound of discarded material in the distance, and as we approached the extraction plant and main buildings of the mine came into view.

The camp was in another small valley nearby, out of the way of the noise, and the permanent cloud of dust that was coming off the top of the spoil tip. We were met by the whole community, and allocated billets for the night, and after a quick wash and brush-up we were escorted round the workings and plant, then returned to the bar in the club.

During our short absence, an entertainment programme had been prepared and posted up on the noticeboard. There was to be a dinner, followed by a treasure hunt round the precincts, a film show, singing and dancing, and supper, and the bar would stay open all night.

I suppose that, with us in attendance, there were about 50 people present in camp, but it was the most cosmopolitan gathering I have ever attended. There were English, Scots, Welsh, Irish, Swedes, Danes, Poles, Dutch, Italians, Swiss, Germans, French, Americans, Canadians, Rhodesians, South Africans, and some others of indeterminate nationality. The language problems in operating the mine must have been dreadful, but after a few drinks, some food, and a smattering of Swahili most conversations continued loudly, with apparent understanding all round.

During dinner, the teams for the treasure hunt were posted, and by great skill the organisers had managed to have only one of each nationality in each team. The clues had been written in Dutch, with an 'English' translation supplied to further confuse the situation. As the clues were all related to locations on the site, local knowledge was at a premium, and we spent a happy hour driving round, wakening all the watchmen, and disturbing some chickens near the kitchen compound. This last event was in pursuit of a clue indicating the Swahili word for 'kitchen' – 'jukoni' – which, when spoken in Dutch, and with some drink taken, might easily be mistaken for 'chicken'.

Before the hunt ended we had sent out two search parties for other teams, and one other search party to find one of the search parties, but eventually all returned safely, and the evening continued, more or less as planned. In the morning we foregathered for a somewhat frugal breakfast, and after our farewells set out on a very steady journey back to our hotel.

After Morogoro our party split up: the London contingent took the train to Dar es Salaam, to fly back to London from there; the geologist drove off towards Nairobi, to spend a week on holiday in Kericho; and we set off back towards Mombasa, with a planned short spell on leave in Lushoto, a small hill station in the Usambara Mountains about 50 miles inland from Tanga.

We arrived in the early evening to a warm welcome from the hotel. Our accommodation was in a bungalow in the grounds; there were tall trees all round, a golf course, tennis courts, a swimming pool, wood fires in the bedrooms and public rooms, an excellent club, and a fresh scent of pine trees. After all our travelling it was great to be at rest for a spell in such pleasant surroundings.

Apart from the rest and relaxation, some time was spent on completing our reports, parcelling up the numerous rock and soil samples, and trying to work out a reasonable story to explain away our expenses and balance the petty cash.

At the end of our short holiday, we set out once more on the road to Mombasa.

CHAPTER 23
More Mombasa

We returned to Mombasa after our short holiday in Lushoto, and on the completion of my final report on the surveys and propositions I returned to work in the power station. We moved into a small hotel on the south side of the harbour, and I went to work across the ferry each day, completing the documentation of the power station, stores returns, disposal of plant and equipment, and the inevitable accounts.

Our social life around the town reverted to our previous routine, with sailing at the weekends, swimming in the hotel pool every evening, visiting our friends, and attending the odd hop.

During our travels, Arthur had become very keen on racing the boat, and to save weight he gave away one of our paddles, and as we had saved a bit of money on our safari we decided to buy a new sail – just a jib. We visited an Indian tailor in the bazaar, who had a reputation for making sails. He took the old sail and together we selected the material he would use, and he set to work at once, promising delivery in a couple of days. Right enough, the sail was delivered two days later, together with his bill for 100 shillings. At that rate we could nearly afford a new mainsail as well.

In a rash moment we decided to dye the sails red, and procured the materials in the town. We followed the instructions implicitly, dunking the sails into a dark red mixture in an old earthenware sink. After the required period of immersion we took them out and hung them out to dry in the sun. We had expected a deep red, and judging from the colour of our arms, and the stains on the floor, this seemed a reasonable assumption – but when everything had dried out we found that we had a suit of pale pink sails, with darker edges where the material had been doubled over.

There was not much we could do about it, as we were racing the next day, and duly turned out in our pink sails. The boat had kept her original German name of *Eine Piek* (the *Ace of Spades*?), and was thereafter known as *Eine Pink*.

With the work on the power station drawing to a close, I was summoned to Nairobi to be advised of my next posting. I was informed that I was to go to the site of a proposed new hydrostation on the Pangani river, some distance out of Tanga, and carry out the preliminary work and surveys for a new dam and tunnel that would replace the existing power station at Pangani Falls. There was a house available on site, and I should make a recce and see what was required to make it up to standard by way of furniture, decoration etc. – and I should start next week.

When this news was imparted to Jean she was a bit apprehensive, as she had never really lived in the bush before, except on our travels, but looked on the bright side, noting that our own house would be a lot better than some of the hotels we had stayed in.

SECTION 3 – PANGANI FALLS

CHAPTER 24
Moving in

We arranged to go off over a weekend, and set out back into Tanganyika, this time in a lightly loaded Land-Rover. We stopped in Tanga overnight, in a pleasant hotel a short way out of town. Its main purpose was as a rest house for passengers embarking on the mailboat home – who liked to arrive in the port a couple of days early, just in case there were some problems on the roads – and it had a very relaxed atmosphere. The building was in a sort of Edwardian style of architecture, with large airy rooms, wide verandas, and a charming garden.

Next morning we travelled up to the site and met with the station manager of the existing plant. He lived in a fairly new, large bungalow, and took us round to our house, about 200 yards down the track. It had been built about 20 years previously, for some of the construction staff on an earlier extension to the station, and had been occupied until quite recently. We made out a list of our requirements and asked that the place be repainted before our return, and set out on our journey back to Mombasa in a pensive mood.

All our possessions were in care of various friends about the town, and we had to collect them and pack them for the journey. Also, we had to buy some essential items, such as everyday crockery, and clothes pegs and a host of other domestic goods, then pack them in order of need, and make a first trip with some of the heavier items.

During our previous trips into Tanganyika we had crossed the Kenya/Tanganyika border without encountering any signs of officialdom, apart from the tsetse fly control point. This consisted of an askari (a local police constable) and two operatives, who lived in a small thatched shelter at the side of the road. The askari would stop the vehicle, and the two operatives would then ask the occupants to step outside while they sprayed some insecticide all round the inside of the vehicle, and then politely sprayed the occupants before they were allowed to re-embark.

It seemed a bit of a pointless gesture, as there we were, all fumigated in the middle of the bush, with countless animals, birds and off-the-road pedestrians passing in all directions all the time!

I later discovered that the same rigorous fumigation was conducted on all air traffic into Tanganyika. The first man onto the plane as soon as it landed was the 'flit gun' operator, and no one was allowed to leave their seats until they had been treated.

Once clear of the control point we continued to Pangani Falls, where we dumped all the heavy equipment in the house and conducted a quick survey into the renovation work in progress before returning to Tanga for a night in the hotel – and then back to Mombasa for the next load.

CHAPTER 25
Living in the Bush

The house had been painted, and really looked quite smart, even though the garden was a bit overgrown.

At the time of our arrival at Pangani Falls the expatriate staff consisted of the station manager and his wife, who lived in a fine new house, and one other European couple, who lived nearby in one of the old construction cottages, similar to ours. These people moved back to Dar es Salaam shortly after our arrival, and we never really got to know them.

The Indian engineers and their families lived in a small compound of eight houses on the other side of the road, and the main labour camp for the station employees was about half a mile away in the other direction.

Our cottage was of a fairly basic design, of three main rooms, a sitting/dining room and two bedrooms, with the kitchen and bathroom to the rear. At some time a large veranda had been added all along the front of the house, and this had a shiny concrete floor and low walls all round; the space between the roof and the walls was screened by mosquito netting. There were roller blinds on the outside, which, if you were smart enough, could be lowered to keep most of the rain out.

All the so-called windows were merely openings fitted with mosquito screens, which also had outside roller blinds, and – for some strange reason – the doors were solid timber. The roof was made of corrugated iron sheets, and when it rained the noise could be quite impressive. There was a short covered walkway to the outside kitchen, where there was a sink, a storeroom, and both an electric and a wood-fired cooker.

Our garden was bounded at the rear by the road to the manager's house some 30 feet away, and marked by a hedge of hibiscus bushes. There were several pineapples and a couple of banana trees growing quietly, courtesy of the previous occupants. In the front, the garden sloped away as far as we wanted, overlooking the bush and hills below us for about 40 miles.

We had moved the main items of our furniture and worldly goods into the house over a series of trips from Mombasa, staying overnight in the hotel in Tanga, but on our last journey we dumped the final consignment in the house and drove on to the nearest town of Korogwe, about 20 miles away.

This was a big mistake, as the Korogwe hotel turned out to be quite the worst place we had ever found. The hotel was built round a square, with the bedrooms (cells, more like) round three sides, and all the facilities – showers, dining room etc. – on the fourth side. The square was covered in mud, it was pouring with rain, the food was awful, the beds were lumpy, and the place teemed with mosquitoes. Egon Ronay might have something better to say about it nowadays, but at that time it was 'not recommended'.

In the morning we hurried back to Pangani after a quick trip round the local market, and on our arrival found a queue of about a dozen potential cooks and houseboys waiting for us. The news gets around pretty quickly by bush telegraph. Each man carried a bundle of testimonials, carefully wrapped up, extolling his virtues and capabilities, and after a quick scan through these, with reference to a list of unsuitable characters supplied by the station manager's wife, we selected one man as the houseboy and another as the gardener, and set them to work inside the house shifting the furniture about and putting the place in order.

Our status required that our houseboy should be properly dressed, so after some rudimentary measurement we sent off his dimensions for a 'khansu' (a long, white nightshirt kind of garment), a red sash, and a small red hat. We thought that the white gloves might come along later. I measured out a distance of about 50 yards from the front of the house and stuck in a few sticks to mark our boundary, and the gardener was soon at work cutting down the long grass and planting a hedge.

There was a lot of electric detonator wire lying about as the result of some earlier quarrying operations, and we set up some aerials for our wireless with it, fixing the end of the wire through the mosquito netting on the veranda. We had the gardener walk about with the wire on a long pole, and where we found the best reception from each station we stuck the pole in the ground. Inside the veranda we had three or four wire ends, which we could connect into the set as necessary, and listen to the world.

The word had gone round that I would be taking on some men for the preliminary survey and clearing work, and we were besieged with applicants outside the house the following morning. Once more with advice from Ernest, the station manager, a small selection was engaged, and we started work, first on the labour lines for the temporary camp.

I had located a suitable site for the permanent camp by then (no trouble from the local planning committee), and selected another place nearby, convenient to water and the roads, and set my men to work clearing the bush, while I made out orders for materials and equipment for the temporary camp from the local stores in Korogwe and Tanga.

The establishment of our temporary labour lines was not a subject to be found in any of the company manuals in my possession, and I had to make reference to an old copy of the Royal Engineers' recce pocketbook – which I just happened to have to hand – to sort out what was required by way of water supplies, latrines etc.

The book laid down essential quantities of water per man per day, and, as my camp was probably to be occupied by married men and their families, I had to make allowances for at least three times the specified quantity.

The selection of the site for the latrines was also determined, and during the considerations of this problem I was reminded of a lecture given by a senior medical officer during my military training. The subject had been field hygiene, and it had been illustrated by some pretty gruesome photographs showing the possible effects of failure to provide adequate facilities. The lecturer had gone on at some length, speaking in language more appropriate to medical students, referring to excrement etc. by all the recognised scientific terms, and then he brought the house down by concluding: "After all, gentlemen, you just can't have your soldiers shitting indiscriminately all over the place."

Within a couple of weeks we had some of the camp occupied, and had established the necessary services, one of which was an office, as pay day was almost upon us.

The pay parade was conducted along very military lines. The foreman mustered the men in tally number order (on engagement, each man was allocated a tally disc, which he had to hand in before starting work in the morning, and collect again in the evening). The pay clerk called out the man's name and number, and he was presented by the foreman to the pay table. I handed over his pay in an envelope while the pay clerk had him sign for it, or mark the payroll with his thumbprint. The man then stepped down with his money, into a crowd of creditors and wives, and much haggling and animated discussion followed. At the end of this performance I had to reconcile any money left over with the amounts that had not been collected and signed for.

I later realised that this was a means of not being seen to have any money in hand – a stratagem which was not lost on the more persistent wives, who would follow their husband back to my house, and stand there, demanding payment from the unfortunate workman.

The routine was that a full pay parade was held on the last Thursday of the month, allowing the travelling men a long weekend away at home, and a supplementary pay parade was held in the middle of the month to maintain the men's cash flow.

When the camp had been completed the work settled down into a routine.

I had marked out lines through the bush at about 100-foot intervals, and some of the men were set to work to clear swathes about ten feet wide, for surveying. Others were digging trial pits, and making access roads for future use. We also conducted a major triangulation survey for the layout of the new dam site, the tunnel, and the proposed new power station, and more detailed surveys over each of these particular areas. In addition to this routine work, I was asked to carry out some river measurements some distance upstream every fortnight, and, all in all, I was kept pretty busy.

During our travelling days I had taken on Christopher, a Nyasalander, as my driver, and he was happy to stay on in Pangani. I was constantly practising my Swahili with him, and got to be fairly fluent. His dialect was slightly different from the local one, however, and my instructions were passed out in my Swahili, and in cases of doubt reiterated by Christopher in his Swahili. This also gave me time to think while he was rephrasing any questions to me, and I could then answer directly. There was a little book on Swahili that I always carried around, and I made frequent reference to the vocabulary section. It contained what were alleged to be common phrases (?), obviously listed to illustrate some fine point of grammar, but hardly likely to be encountered in normal conversation. It did not advise that the postilion had been stricken by lightning, but one phrase, which I never heard in anger, went like this: "Please, sir, will you advance me 30 shillings, as my house has just been destroyed by a rhinoceros?"

Possibly the inexorable rise in house prices had raised the going rate to such an extent that there would be no chance of the present price being advanced, even by the most sympathetic employer.

CHAPTER 26
The Case of the Vanishing Ducklings

Our neighbours, the station manager and his wife, had a large garden, which was tended by a very old man who had been in the German colonial army in 1914. He was a cheerful and industrious gardener, and was always very proud of his work. Our friends had recently imported a dozen Aylesbury ducklings, and the old man had built them a small compound with a pond and a little thatched hut. He came past our house each morning on his way to work, and reported on the progress of his ducklings, and one day he was in some distress. One of his ducklings had vanished.

We all trooped down to visit the scene of this calamity. There was no evident damage, no signs of a struggle, not even a spare feather. The old man was perplexed, but came over later to report that a very large bird had been seen, and this must have been what had taken the duckling. He reported that he had set up some strings and strips of canvas over the run, and was confident that this would not happen again.

A few days later another duckling was reported missing. A second investigation was conducted, and the blame this time was laid on a large python, which, with great cunning and skill, must have reached over the fence and snatched the little bird. It must have been a snake, as there were no incriminating footprints.

The old man neglected his other duties and, together with his assistant, set up a 24-hour watch over the compound. Despite this surveillance, another ducking vanished.

Another inquiry was convened, and this time there was evidence. The fence had been disturbed, and there were footprints of some animal in the mud around the pond. We were then informed that a large baboon had been seen in the area. Such a large baboon, indeed, that it seemed to have three legs. The fence was repaired and the watch re-established.

We were enjoying a quiet drink on Ernest's terrace one Sunday morning, and it was casually mentioned that Hamish, their small Scottie dog, had been off his food for the last few days. This minor malaise was put down to the heat, or the coming rains, but he seemed to be fit enough, and not in need of special attention.

On our way back to our own house, who did we see but Hamish, digging away in the garden, and exposing the remains of a badly decomposed duckling. No wonder he had been uninterested in his usual dog-food hash, feeding as he was on best Aylesbury duckling. We managed to keep this from the old gardener, and complimented him on his vigilance over the remaining birds. We asked him about the three-legged baboon, and he assured us that it was still about in the area, but that he had scared it off the ducks.

We saw baboons quite often about the gardens, and they were well organised in their stealing operations. They posted a lookout in a tree, which would signal if there was any danger about. Ernest had an elephant rifle as part of his household equipment, and when the baboons were active in the gardens he would come out with it. As soon as the lookout saw the gun he would start to scream, and the other baboons stopped whatever they were doing, and grabbing their young dashed back into the bush.

These baboons were very destructive, and would tear out carefully nurtured cabbages and lettuce plants, and just throw them away. They were also bad news for banana trees, and Indian corn, which they tore off and sat devouring in full view of any spectators. They were able to distinguish between a gun and a rolled-up umbrella, and took no notice if the umbrella was directed at them, but as soon as a gun was produced they were off back into the bush.

CHAPTER 27
The Lion's Tale

Ernest went out one day to scare off some baboons from a small hollow at the end of his garden, and worked himself into a good position for a shot. As he was preparing to fire, one of these baboons rolled over, and Ernest discovered, much to his horror, that these baboons were not baboons at all, but a pride of lionesses, taking their ease in the sun.

Very sensibly he withdrew quietly, believing quite correctly that the lionesses would do more good scaring the baboons away than ever he could. Anyway, they were not disturbing the vegetation. We warned the old gardener and his assistant but they did not seem to be much bothered.

Shortly after this episode I had a minor encounter with lions myself.

Each evening the watchman came around and reported his presence, usually collecting a cup of tea from the houseboy. He then went off on his rounds and met up with the other watchmen, discussing the state of trade, the latest gossip and the coming rains. About 9:30 he would return to the garden, and walk about, shining his flashlight. Most evenings he spotted a leopard raking in our compost heap, and he came to call me out to see it. The compost heap was about 20 yards from the end of our drive, and I usually walked over with him to confirm that he had indeed spotted the leopard, and praise him for his watchfulness.

That night we walked over as usual, but when he shone his light two large yellow eyes were reflected back at us – not the greenish ones normally displayed by the leopard, and much further apart.

"Simba," he shouted, and as the lion roared we turned and fled back to the house.

"You were not out very long tonight," Jean remarked. "No, dear," I replied, "I think it might be going to rain," and, thinking of the thin mosquito netting all round the veranda, suggested we might go to bed early.

At least the window openings were much smaller than the expanse of the veranda netting.

CHAPTER 28
The Watchman

A week or so after this event we had gone to bed, and were settled down for the night, when there was a lot of shouting and banging on the back door. I rose and went into the kitchen, and through the window I could see the watchman on the doorstep, bleeding profusely and making a lot of noise. Certain that he had been attacked by either a lion or a leopard, I grabbed the bread knife (the only available weapon), whipped open the back door, pulled him inside, and slammed the door shut again.

On closer inspection, it was obvious that, although he was seriously wounded, the damage had not been inflicted by an animal. By this time Jean had come through, and together we washed down his head and shoulders. He told us, rather incoherently, that he had been fighting with another watchman, and they had been going at one another with their short spears. Our chap was still bleeding profusely from a number of deep gashes on his head, and was so bad that I thought it would be best to take him into the local hospital in Korogwe. We washed him down, gave him a clean shirt, bandaged him up with the contents of my first-aid box, and loaded him into the car.

As we drove past the main compound I spoke to one of the other watchmen there, to tell him to collect his comrade and take him into the first-aid post in the camp, then drove on to our own camp, where I knew that our watchman's father lived, and knocked on his door. The old man came out and I explained that I was taking his son to hospital, and asked if he would like to come along.

The patient appeared to have made a miraculous recovery under my ministrations, for he jumped down out of the vehicle, and ran off into the camp, which was now aroused. There was no trouble locating him – bandaged as he was, his head stood out like a lighthouse – and he was soon rounded up and returned to the Land-Rover. He refused point-blank to go to hospital, and having seen him dash off into the camp I reckoned he would probably last the night in the tender loving care of his parents, but told his father that he should bring him to the first-aid post in the morning. We turned the car round and drove back, stopping to see the other contestant suffering from the attentions of the first-aid man, whose main remedy seemed to be a liberal application of iodine.

In the morning I called at the first-aid post and saw the two watchmen together, comparing their wounds and the treatment they had received. Both were badly cut about the head and shoulders; the other chap had been stitched up overnight with something like 20 stitches, and our man was just about to undergo a similar treatment. I learnt later that he needed only a dozen stitches, and by this score he thought he had won the argument.

Later in the day a court of inquiry was called, and our two wounded watchmen were marched in. They were both very sorry for themselves, recovering as they were from the attentions of the first-aid post attendant, and called on the mercy of the court, pointing out that the argument was on account of some woman, of no great account in the light of day, and that they felt that they had suffered enough.

The attendant also gave evidence to the effect that he had treated one man in the night, and had seen both of them in the morning, and, apart from a few scars, was of the opinion that they would both recover in about a week.

Both Ernest and I then delivered a stiff lecture about the duties of the night watch, and they were both ordered to be suspended from their duties, without pay, for ten days.

For the remainder of the period they were to be seen together morning and night, attending for treatment at the first-aid post, and bragging about their fortitude under the care of the attendant, whose motto appeared to be 'If it ain't painful, the cure can't be working'.

CHAPTER 29
A Crocodile

Part of the preliminary work was to survey and make up access roads into the sites for the permanent work, and for this we needed to find a source of gravel and sand. I had located a suitable gravel bank in one of the smaller rivers nearby, and arranged a meeting with the local headman to negotiate rates and royalties for this material. My foreman and I went along to his village, properly dressed with collars and ties for such formal ceremonies, and met with the old chief. He also had turned out in all his finery, and wore an old army greatcoat, with all the buttons shining, a tall hat with a long feather stuck in the band, and a pair of thigh-length sea boots, turned down below his knees.

We sat down in his courtyard under a thatched awning, and had a long, rambling discussion concerning the weather, the state of the crops, the number of scholars in the school, and the prospects for employment when the main work started on the site. We then came to talk about the gravel in the river, and I explained that we would be wanting him to arrange for his men to collect the material in boxes (which we would supply) and carry it up to the side of the road, where we would load it into our lorry, and keep a tally of the number of satisfactory loads.

This proposition seemed to go down well, and we then discussed the price per box. At the mention of money, the attending crowd, who had sat silent until then, broke into wild speculation, with bids being thrown from side to side amongst the prospective diggers. Having listened to these arguments for some time, I suggested to the headman that we should come to a private agreement between ourselves, and that he should sort out the breakdown of the payments, as I had no desire to interfere with his authority. He seemed reasonably happy about this, but just as we were about to clinch the deal he said that there might be a small problem.

He explained that the gravel bank was in a part of the river occupied by a sacred crocodile. (The possibility of there being crocodiles in these small streams had never crossed my mind during my earlier prospecting.) I assured him that we did not want to offend his crocodile, but asked if there was a way we could somehow appease the creature. Following some discussion with the elders of the village the chief announced that we would need to consult with the local wizard, and he could arrange for an interview with this dignitary for the following day.

Next day, again suitably attired, we attended for a session with the chief and his wizard. Once more this was to be conducted in full view of the assembled villagers, and, as before, the old man was in his best boots and funny hat. After a repeat of the previous day's preamble about the weather, etc., he introduced his wizard. An old man came in, wearing a brightly coloured blanket, a tall, conical felt hat, and very little else, followed by his apprentice who carried a box, and had a bundle of rags slung over his shoulder.

I explained that we wanted to dig out gravel from the lair of his sacred crocodile, and felt sure that he could see his way to appeasing the crocodile for our mutual benefit. There followed a long harangue between the chief and the wizard, in some dialect that neither my foreman nor I could understand.

When they had subsided, I asked the chief what had emerged.

"He says, 'Yes.' It can be arranged."

"Will you ask him what he needs to perform this ceremony?"

Another long discussion; this time the apprentice was called in, and some items produced from the box. These were set out on the ground, and stirred around a bit, with much muttering, and then another long speech to the chief.

"What was all that about?" I asked.

"He says he will need two days to prepare, and that you should bring a bottle of gin and some Andrews Liver Salts when you return."

This did not seem unreasonable, so I agreed, and after the usual salutations and farewells we made our escape.

We arrived – as appointed – two days later, with the ingredients, and met in the chief's compound in the village. Once more the whole population was in attendance, but the wizard and his mate were not present. We exchanged the normal pleasantries, and were served warm lemonade out of very strange glasses. Declining any more of this beverage, we set off down to the river bank – followed by the whole tribe – and met with the wizard and his mate, now properly dressed for the occasion, with rows of coloured beads, and carrying a long stick rather like a shepherd's crook.

They had built a small grass hut by the edge of the water, and we were invited to sit down outside while the ceremony proceeded.

I handed the gin and Andrew's over to the chief, who passed them to the apprentice, who gave them to the wizard, and they both went into the small grass hut. There was a bit of chanting and rustling from within, and a short time later they emerged, somewhat unsteadily, and hopped about on the long grass by the river's edge. After a few moments the wizard raised his stick and made a proclamation to the crocodile, and then he emptied the contents of the bottle into the river, followed by the contents of the tin of salts, which fizzed in a desultory fashion. He then gave the river a bit of a stir with his crook, and – with some support from his apprentice – approached the visiting party, and declared that the crocodile was now happy and we could get on with the digging of the gravel in perfect safety.

To show our appreciation, my foreman then produced a goat, which I presented to the chief for his assistance, and we all went happily on our way. Before we got back to the Land-Rover, there were already half a dozen boxes of gravel sitting on the roadside waiting for collection.

This should make interesting reading for the accountant when the petty cash entries note that gin, Andrews and a goat should be allocated towards the cost of gravel from the river.

CHAPTER 30
Shipping the Boat

We had made arrangements with Arthur to buy out our share of the boat, and to have it shipped down to Tanga courtesy of a fellow member of the Mombasa Yacht Club who had a tanker business, and the boat was duly loaded as deck cargo. The word came through that the ship was in port.

We set off early one morning and as we came along the front we could see the tanker lying off, with *Eine Piek* tied up astern. I scrounged a lift out in one of the pilot launches and soon was aboard, having coffee with the captain; then I transferred all the boat's gear, set up the mast and rigging, and set sail for the yacht club, where Jean was waiting for me on the beach.

While we were pulling the boat up the beach into its allotted parking space we were approached by two red-faced and irritated officials, one from Port Health and the other from Customs, waving sheaves of formal documents and making threats of official condemnation concerning unauthorised imports, port hygiene, decontamination, tsetse fly clearance, etc. etc.

Somewhat taken aback by this confrontation, which I suppose was perhaps warranted, I argued that this had all been done in complete innocence, and if I had brought the boat in on a trailer behind my Land-Rover, other than the odd squirt by the fly control (if they had been awake at the time), no one would have known. Furthermore, if they were all that concerned, I would take the boat back to Mombasa by tanker, and do just that. This would have led to even further documentation (jobs for the boys), as it would compound the felony.

Before the situation got completely out of hand, however, a member of the club, who just happened to be the senior officer from the Port Authority, came along, and on his suggestion we all trooped into the club bar, where these two clowns were told to simmer down.

Couldn't they see that this was yet another boat for the club? And have another beer?

CHAPTER 31
The Poll Tax

As a result of this brush with the powers that be, I realised that we had not formally registered as being resident in Tanganyika, and our passports still gave our address in Kenya. On our next visit to town we called at the Immigration Office and had our address changed to Pangani Falls. Apart from the nominal fee for re-registering, we were advised that there was an additional tax of ten shillings per head on all persons resident in the territory, and falling due in about three weeks' time.

At the same time I had a notice from our head office in Dar es Salaam that I would probably be approached by the local headman to collect this tax directly from my men, and that this practice was not approved, and some other method should be developed to keep the company out of local tax affairs.

Sure enough, a few days later, the district chief came to my office and asked me to collect his tax from my men on his behalf. I explained that this was not possible, but that I could arrange to have an even ten shillings made up in a separate envelope, and pay this over at the next pay parade, and that he could have his collector present at a separate table to extract the money and issue receipts.

When the next pay day came round the taxman arrived, escorted by two local askaris, and they were set up at a table nearby. Word of their presence must have got around, for by the time I arrived with all the cash and documents the two askaris had been well and truly nobbled, and could barely stand up to salute my presence.

The procedure commenced. The first man was called, came up, received his two pay packets, made his mark on the payroll, and suddenly dodged round the back of the office, with the two askaris in lurching pursuit. Meanwhile, about ten more men had been paid and escaped, before one poor individual was collared and obliged to pay up; and while this transaction was being undertaken by the clerk another half a dozen or so had been paid and escaped, much to the delight of the assembled crowd of wives and onlookers.

At the end of the parade I suppose they had collected tax from about ten of my men out of my strength of about 60, and probably thought they had been quite successful. I advised them of the date of the next pay parade and arranged for them to be taken back to their base in my Land-Rover.

At the next parade, two different askaris had been mustered, and the same performance was repeated, but this time with the complication of them collecting a man who had paid the previous time, and who had a receipt to prove it. While this argument was in progress, the majority escaped, untaxed as before. This happy state continued for about another couple of months, with different askaris attending and getting plastered, and a few unfortunates caught and taxed.

If Mrs Thatcher had known about this experiment, she might have had second thoughts about introducing such a tax in the United Kingdom.

Whatever else it did, it caused me a lot of bother making up the pay envelopes for my men, some of whom had to have two packets, with an even ten shillings in one of them.

CHAPTER 32
The Union Man

A young man came to my office one evening and announced himself as the local union representative. He asked if I would call a meeting of the men, so that he could address them with a view to having them join his union, and I replied that he could speak to the men before they started work at 7:30 the next morning, or after they had finished for the day.

He was not pleased with this response, and set off muttering about lack of cooperation and typical capitalist opinions. I called him back and explained that this was not a totalitarian state, and I had no authority over the men except during working hours, or in disciplinary matters in the camp, and as I was paying them to work I was not prepared to have any interruptions during working hours.

After some further conversation, I learnt that he had been a schoolmaster, but had given up teaching to go off to Moscow on a course for union officials. He had just returned, and been allocated my district, and as we were one of the major employers in the area he had come along himself. I suggested that he could better serve the country by resuming his schoolteaching, helping with the education of the people rather than stirring up trouble with my workforce.

It turned out that he had been promised a car by his union if he was able to recruit sufficient members – a prize far beyond the aspirations of the average village schoolmaster.

Anyway, the following evening he collected about 30 of my men after duties, and held his meeting. My foreman informed me that he had said that if each man joined up, and paid in his subscription of six pence, he would get a car.

He phrased this very carefully, and there were a number of critical questions asked.

Some of the men joined up on the spot, however. I told the foreman that he could put me down for membership and my sixpence if he could guarantee me a car as well.

The young man returned about a month later, still on his bicycle, and attended a very unsympathetic meeting in the evening. He came along to my office afterwards to complain about the state of apathy in the working class, and was a bit put out when I suggested that it was probably due to a lack of basic education, caused by trained teachers going off to Moscow and neglecting their scholars at home.

CHAPTER 33
Snakes in the Roof

We came back from a visit to Tanga one evening and our car lights must have awakened the watchman. He came out to greet us as we climbed out of the car, and as he was escorting us across the short covered walkway between the house and the outside kitchen a small snake crossed the concrete pathway and vanished into the grass.

I shouted to the watchman to kill it, and he went round the back of the kitchen and returned with one of the props for the washing line – a pole about eight feet long, which he held at its extreme end, making it completely unwieldy, and prodded around in the grass and shadows. On my suggestion that he might be a bit more active with his stick if he changed his grip, he replied that he had a great fear of snakes (I felt the same way), and that they could run quickly up the stick, and bite him.

As I tended to agree with him on this matter, we went into the house, and told the watchman to stay in the outside kitchen with the door closed, and the light on, and we would have a good look around in the morning. He would not like me to think that he normally spent his watching hours in the kitchen with the light on after we retired.

During breakfast another small snake appeared, seeming to come down off the roof. This created something of a diversion during the meal, and I sent a runner down to my office to summon the foreman, not wanting to leave Jean alone with the house under siege. The foreman arrived with a few men and they set about clearing all the grass from the foundations of the building, and generally making a lot of noise.

Here was a fine chance to demonstrate what fearless snake-hunters they were, but, even so, they all took great care to keep their weapons at arm's length. Two more small snakes were found and chased away, to much shouting and banging of sticks, and the place was then declared clear. The foreman suggested that they might have come from a nest in the roof space – about the only place we had not searched.

There were no volunteers to investigate there.

As chance had it, we had bought a couple of smoke bombs to fumigate the roof space that weekend, and I supposed that, if we killed off all the small insects, the snakes – not having any food –

would probably depart. From a quick reference to the label on the bomb package, it was advisable to light the blue paper and retire quickly, so it might even poison the snakes as well.

The theory was: stand the bomb on a saucer; light the fuse paper; stand on the table and lift an access panel in the ceiling (this might just be the nest of a family of vipers); place the fizzing bomb on its saucer onto one of the rafters, where it might be less likely to set the place on fire; replace the ceiling panel; and get out of the house. All without taking a breath. The lurid instructions on the label suggested that a period of six hours should elapse before returning.

Having carried out the operations described above without coming to any direct harm, we stood about at a safe distance to see what effect this had.

There was no violent evacuation, but faint wisps of smoke could be seen coming out at the eaves, and we set off on a picnic to allow the prescribed time to elapse.

When we returned, the house had not burned down, and there was no doubt that the overhead was much quieter, and there were no more small snakes to be seen descending from the roof.

Instructions were issued to the gardener to make sure that the grass around the house was kept trimmed right down, so that there would be no place for lurking snakes to find cover.

Having had the whole area cut down already, he was quite happy to be seen to swing his cutting instrument with apparent vigour, meeting – as it did – no resistance worth mentioning. The grass-cutting instrument was of local manufacture, and comprised a walking-stick-type handle, with what looked very like someone's discarded table knife bound on at an angle. This could sometimes cut an area of about three inches wide by 12 inches long at each stroke, if he was really trying, and knew he was under observation.

CHAPTER 34
Ants in the House

Some time after our encounter with the snakes around the residence, we were disturbed one night by the presence of one or two small ants in our bed. When we turned on the light we discovered that there were some more ants visible on the outside of the mosquito net, and on further investigation we could see that there was a column of ants – of various sizes – marching across the floor; they were coming from the window, passing under the bed, and continuing out of the bedroom door towards the kitchen. The ants in the bed and prospecting across the net appeared to be outriders or scouts, presumably looking out for food, or the presence of potential foes that might cause the main column to divert.

After a short pause for consideration, I jumped out of bed, taking care not to disturb the column, and dashed into the kitchen, where I found my wellington boots and the Flit gun. This was well before the times of aerosol insecticides, and our defences comprised a small drum for holding the liquid and a hand pump, which, when operated, vaporised the Flit liquid, which then appeared as a fine spray from a small nozzle on the drum.

Mindful of our food supplies, I examined the food cupboard. This was a large wire mesh box, standing on long legs that were immersed in small cans full of water, with a skim of Flit on the surface. It seemed to be untouched, and I observed that the ants were still coming on through the kitchen, and out towards the back door.

In response to some questions from the bed, I returned to the bedroom and directed a concentrated blast of Flit all round the foot of the net and the legs of the bed, and then attacked the point of entry at the window. They were coming in through a small gap under the window sill, marching about six abreast, and coming straight down onto the floor. Where they met up with the milling crowd resulting from my efforts around the net, a certain degree of confusion was evident. In pretty quick time the column reorganised, and now it was marching by another route towards the kitchen. A further blast of Flit into this grouping caused more panic, and a more sustained blast by the window reduced the incoming numbers very considerably. During these operations, I was relieved to discover that ants do not appear keen to climb up wellington boots, especially if they are liberally splashed with Flit.

Having stemmed the tide, I collected a bucket, and swept up all the dead ants from the bedroom and kitchen, and then recharged the Flit gun to be ready for the next phase of the invasion. As there were no more ants coming in at the window, I thought it prudent to investigate where they were going instead, and, armed with a torch and the Flit gun, and with hardly a thought for our friendly neighbourhood leopard, I went out round to the outside of the window. The situation there was not really surprising. The Flit had contaminated a small area of the wall under the window, and the column had diverted locally. It was now going up the side of the window frame, and vanishing into the roof space.

In response to enquiries from within, I made some reassuring noises about the matter now being sorted out, and went back into the kitchen. At least there were no new ants evident there, but on further investigation outside, on the other side of the house, the column was coming out of the roof and descending in a vertical line onto the house foundations, and then going off into the garden, still about six abreast, and appearing undaunted.

Back inside, I made some anodyne comments and completed the sweeping up, eventually collecting about half a bucketful of Flitted ants; and so to bed.

Apart from poisoning the ants, and other insects, the Flit had an unpleasant and penetrating smell, but other than bringing tears to the eyes in strong concentrations it did not seem to have any serious effects on the person.

In the morning a further inspection revealed that the ants were still marching up the wall, through the roof, and down the other wall. After some discussion with the garden boy we splashed a lot of Flit all round the foundations, but allowing the end of the column still in the roof to escape. After some further milling around the remainder of the column re-formed, set out on a detour around the Flitted area, and continued round the building until it rejoined its original line of march on the other side of the house. This procession continued for another two days.

As a small precaution, we set off another smoke bomb in the roof space, and retired to Tanga for the remainder of the day.

During my surveying operations in the bush nearby, our work was often interrupted by what might be called 'environmental events'. It was not uncommon to see the chain-men (the technical term for survey labourers) hopping about (as active as they ever were), usually indicating the presence of ants, or bees. Sometimes they vanished from view, the excuse being a snake or a crocodile; more likely no more than some rustling in the undergrowth. It all added a bit of excitement to the day's work.

Sometimes it was slightly more serious, when the noise was close by, or our setting up the instrument had disturbed a termite mound, or an ants' nest, and the excitement was closer at hand.

As an aside, the material taken from termite mounds was used as foundry sand, and there were experts who could select the correct grade of material for different castings by reference to the shape and size of the particular mound, or by the particular type of termite in residence.

Is there no limit to man's ingenuity and skill?

CHAPTER 35
Jasper

On one of our visits to Mombasa we acquired an Alsatian pup, which we named Jasper. He came of a very good family, and was highly intelligent, very active and good-natured. He enjoyed our place in the bush, and proved to be an excellent watchdog. Sometimes I left him at home with Jean during my working hours, and sometimes he came along in the back of the Land-Rover, where he could observe all that was going on from a position of some security.

One day I had stopped in the village to arrange a meeting with the headman, and Jasper was left sitting up on my seat. There was a commotion among some chickens in progress, and they came towards the vehicle. This was the first time Jasper had seen chickens in action, and he took a great interest in the proceedings. When the senior chicken vanished under the car, Jasper decided to investigate further. He jumped down, and began to follow the bird – at a safe distance, it must be said – and, together, they went all round the village square.

Suddenly the chicken turned and flapped its wings. This caused the dog great surprise, as the change from following a fairly small, running chicken to being confronted by a large and angry creature with its wings extended, more than doubling its apparent size, was more than he had bargained for, and with commendable prudence he turned and leapt back into the Land-Rover.

A few days later we were visiting our friends in Tanga, and their garden boy appeared, claiming that Jasper had chased all his chickens, and made off with one of them.

"Never," I replied. "This daft dog is afraid of chickens. We will go round to your garden and demonstrate."

I felt reasonably confident that Jasper had not chased his flock, as he had been with us most of the morning, and I also remembered his affair in the village, so we all trooped round to the garden, where the chickens were pecking about in the dust.

Without a moment's hesitation Jasper was off like a rocket, and scattered the chickens all round the garden, where they came to rest out, of his reach, up various trees and bushes. There was little else I could do but pay up some reparations for the earlier bird, which turned out to be a prize leghorn – whatever that might be – and very expensive too.

Later in the afternoon the garden boy came back, and said that another big dog had come into the garden and taken another of his flock, and it had not been Jasper at all in the morning.

I commended the man on his honesty and told him to keep the money.

When we were having dinner later that day I became aware of the houseboy standing at my back, about to serve up the soup. He seemed to be transfixed, and stood there for some time. Then I noticed that Jasper, who had been lying under my chair, was licking the steward's feet, much to his alarm.

I gave the dog a prod with my shoe, and said, "Don't worry, Jumo, he quite likes Africans." This was obviously the wrong thing to have said, and the poor man lurched away, in a state of shock.

CHAPTER 36
The Leopard Skin

We had been visiting our neighbours, the station superintendent and his wife, for dinner one evening, and admired a large leopard skin that he had collected during his time in Uganda, where leopards were considered to be a pest, and there were no game regulations to protect them.

Next day there was a great barking and snarling from the dog, and a strange African with a bicycle was standing at the end of the garden. I locked the dog up in the house and spoke with this man. He unwrapped a large parcel from the back step of his bicycle, and produced a leopard skin, indicating that he had heard that I might be interested. Realising that the skin was of no further use to the leopard I began the normal bargaining routine, and eventually we agreed on a price. The skin was repacked and placed out of reach of termites in the outside kitchen, the man was paid, and the dog released. He came rushing out, with the hair on his back standing straight up, and pranced all round the outside kitchen, generally in a very agitated condition.

We looked in the Settler's Manual, and various other books of reference, for instructions on how to cure skins. As the taking of game was illegal in Tanganyika at that time, we could not go to a recognised stuffer and curer without encountering all manner of officialdom, probably compounding a felony, and generally getting into a lot of bother.

On our next visit to town we obtained the necessary ingredients, and, on our return, set about the job of curing the skin, which by then was nailed onto a large board and making more than a little smell. Every time any operations were conducted on this process the dog had to be kept well away, as he was still extremely sensitive, and would go about on tiptoe, with all his hair on his back on end.

After quite a long time the skin was cured, and reasonably dry and supple. Most of the smell of leopard had been overtaken by the smell of the curing ingredients, and we folded it up in greaseproof paper, sealed it all round with sticky tape, and packed it away ready for our departure home.

When the time came to depart I went down to Tanga to sort out the documentation for our luggage on the voyage home. I was given a sheaf of papers to fill up, listing the contents and value of each box, and there was also a note setting out prohibited items, and possible fines and jail sentences for those found to be failing to comply.

By this time the skin was at the bottom of a large box, underneath various items of household equipment, and I felt reasonably confident that no one would bother to search for it, but Jean was adamant that we had to leave it out and not risk the displeasure of the law. Reluctantly, I excavated the box and recovered the parcel, which we left with our neighbours, on the understanding that they might be able to bring it home under the 'household goods acquired in Uganda' umbrella. (Or just plain smuggling.)

Unfortunately, we never met up with them again, so we don't know to this day whether we have a skin in England, or if the curing was not quite successful, and it rotted away before they came home.

During our subsequent travels we lost our neighbours' address in England, but, if they ever do happen to read this, we would be delighted to meet up with them again.

CHAPTER 37
The Protest March

On one of our fortnightly trips to Tanga to collect the money for the men's wages, and our household shopping, we came across a procession of workmen from one of the sisal estates marching down the road towards the town, still about 20 miles distant. We slowed down to negotiate our way through the marchers, who were all very polite, and learnt that they had some palaver with their estate manager concerning their pay or allowances, and were marching to town to complain to 'someone in authority'.

As we were passing through the marchers a Land-Rover appeared from the other direction carrying the district commissioner from one of the up-country districts.

As he passed they all stood aside, and many doffed their caps and shouted, "Jambo, bwana." He stopped and spoke briefly to one or two of the men, who probably came from his district, and then went on his way.

I supposed he might have called back to town from the first available telephone, but, anyway, we continued on our journey and passed the word to the local DC.

He said, "Thanks," and told us not to worry as he would soon sort this situation out.

We learnt later that he had dispatched a lorry to the scene, with directions to get stuck in a ditch; after pulling it out the marchers decided that they had done enough for that day, and, as it was hot and sticky, persuaded the driver to transport them back to their station.

Someone must have got the message, and sorted the situation at its source, as there were no signs of any disaffection to be seen on our homeward journey.

CHAPTER 38
A loose Crocodile

One evening we were sitting in our screened-in veranda, listening to the wireless and having a quiet drink, when a messenger arrived to say that there was a loose crocodile in the labour camp, and asked if I would like to come and assist.

Ernest, the station super, arrived shortly thereafter, armed with his gun, and we set off to the camp. The camp comprised two double rows of houses, with a large space between them, often used as a football field, and we parked the car and walked out into the middle of this field. Most of the inhabitants had foregathered in this space, and the remainder were going about making a dreadful noise, beating cans and shouting, in an attempt to keep the crocodile well away from them. (A very sensible procedure, I must agree.)

As they seemed a bit disorganised, and appeared not to realise that, if it went away from any one of them, it would probably go directly towards someone else, who would doubtless chase it back again, we managed to establish some sort of order on the proceedings. The majority were shepherded to some higher ground on one side of the site, and a detachment of beaters was sent in to try to chase the beast into the open field, where Ernest and I were standing, he with his gun, and me with a powerful torch. It did occur to me that there was not much I could do with the torch if the crocodile came close; maybe I could jam it in his mouth when he came to take a bite at me. The best idea, other than ignominious flight, was to keep Ernest between me and the target.

After a lot of shouting and banging, the crocodile was chased out onto the field, obviously very annoyed, and probably quite confused. He came towards us in a series of short runs. The difficulty was that he was between us and the beaters, who were well in the line of fire from any stray richochets, but had to keep fairly close to prevent him from turning back into the camp.

When Ernest judged that it was near enough for a shot, and the beaters were reasonably clear, he fired and then fired again. The creature fell down and lay on its side, obviously badly damaged.

All was silence.

We approached cautiously and Ernest put another shot into it, without any apparent reaction. A great cheer went up from the assembled gallery, and several brave spirits came up and poked at it with long sticks. As these assaults seemed to provoke no reaction, it was pronounced dead, and after a further short wait it was lifted up onto the shoulders of about a dozen men and carried in procession around the camp, accompanied by much singing and chanting about how brave they all had been, and what a good shot Ernest was. (No mention of me and my light.)

They brought it back and, just as they were about to put it down on the ground directly in front of us, it gave a bit of a wriggle – probably just the nerves relaxing or something, but it had a miraculous effect. Ernest and I were left alone in an instant, with the crocodile at our feet, well within biting distance. We backed off carefully, and Ernest fired off another two rounds just to make sure. This appeared to have done the trick, as the crowd re-formed, and began the preparations for a large meal of crocodile stew.

Ernest let it be known that he might like some of the skin for a pair of shoes for his wife, but I think this must have fallen on deaf ears as, by then, most of the animal was already in the soup.

CHAPTER 39
Prospecting for Gold and Diamonds

Part of my duties on the preparatory work for the proposed new power station were the preparation of surveys and the examination of ground and foundation conditions at various proposed locations for the works. In some places the rock was on the surface, and readily accessible, but elsewhere there was a depth of overburden, sometimes up to 15 feet deep, and these areas had to be investigated by trial holes. I organised a squad of men with shovels, picks and ladders, and a series of buckets, and set them to work on these excavations.

In general the ground was of laterite, and steep-sided excavations were satisfactory down to a depth of about five feet. Beyond that I had to arrange for timber sheeting and internal supports to allow the work to continue in safety, and I made a practice of visiting and inspecting these supports at frequent intervals throughout the day.

In one of these holes, about 15 feet down, I observed a band of shiny pebbles, showing all round the excavation. I noted this in the log of the hole, and went off for lunch, thinking that it might just be possible that these stones could be diamonds; so I collected a number of small tobacco tins, and when the men had gone home for the day climbed down into the pit again and filled about six boxes with selected stones.

In the evening these stones were carefully washed and examined, and various experiments conducted. As we had no glass in the windows, we tried scratching the stones on the mirror, with some success, and further runs were made on various pint mugs, all of which indicated that, whatever they were, these stones were very hard indeed.

In the course of a roundabout conversation with Ernest, who had done most things mineralogical in his time, he agreed to have a look and give his expert opinion. The stones were duly examined again, and pronounced to be not diamonds – yet – but, if I liked to wait for a few million years, under the right conditions they might develop into commercial diamonds.

With great disappointment, the stones were duly scattered around the house, and made a pretty pattern on the other gravel. The log of the trial pit was adjusted to note that the band of pebbles was not diamondiferous. Ernest pointed out that there might be diamonds in the area, and the most likely place to find them would be at the bottom of the waterfall, so for the next few Sundays I sweated it out digging handfuls of gravel out of holes in the rocks at the foot of the waterfall – but, alas, without any success. Any likely specimens were examined by Ernest, and none were found to be of any value, but soon the paths around the house were all full of shiny stones.

During one of our examination sessions, Ernest suggested that it might be more fruitful to try panning for gold in the river, as we were not having much success with the diamonds. This seemed to be a good idea, so for the next few weekends we went panning for gold in the gravel of the river banks. This was much harder work than looking for diamonds, and as we did not seem to be having much luck, and the sailing season had returned, we gave up on the mining operations and spent most of our free time on the boat in Tanga.

This decision was justified by the thought that, if there were all these diamonds and gold lying about, what were the people doing, growing sisal and raising cattle, etc.?

CHAPTER 40
A Visit to Moshi

Our routine at Pangani was disturbed as the result of a very heavy rainstorm on Mount Kilimanjaro. This storm had caused a lot of damage on the flood plain of the river, and had washed out a small dam that was the source of water for the hydroelectric station supplying the town of Moshi. I was asked to go and report on the damage, and make some proposals for immediate and long-term repairs to the dam and surrounding works. This was at the start of the rainy season, and I thought it better to go off by East African Airways rather than drive the 200-odd miles over pretty terrible roads, and we went off to Tanga to make the arrangements.

Jean would remain at home in Pangani, and I would travel by air to Mombasa, Nairobi and on to Moshi, and, when my work was completed, return via Dar es Salaam and other places west and south. When all this was confirmed, and I had made arrangements for a car to be available in Moshi, and some other equipment on the site, I left a couple of days later.

Like most of the East African Airways fleet, the aircraft was a Douglas DC-3, and served as general bus and household transport for the up-country regions. Apart from the passengers, there were day-old chicks among the cargo, travelling in a large box behind the pilot, and making cheeping noises throughout the trip.

After leaving Nairobi we flew on to Moshi, where I was greeted at the airfield, taken to the local office, and advised of my booking into the rest house. The car that had been arranged for me was an open Land-Rover – no roof, no windscreen, and a bit bashed about, but the engine sounded fairly healthy. I drove round to the rest house for dinner, and made ready for an early start in the morning. The weather had reverted to the clear and calm conditions usual just prior to the start of the rains, and I had a quiet night.

When I arrived on the site in the morning I met with the station manager, and we went along to the dam, which had been pretty well destroyed by the flooding.

I spoke to the gatekeeper, whose job it was to operate the sluice gate into the canal, and who had been on duty during the flood. He told me that there had been continuous heavy rain and strong winds all that day. The river had risen, and he had been kept busy clearing bushes and small trees that had been uprooted by the flood and carried down on the stream, so that they would not interfere with the working of the sluice.

At this time, the water was ponding upstream of the dam and flowing over the spillway in an orderly manner, but suddenly a large tree, complete with its root system, went over the spillway and then lodged in a narrow part of the river channel slightly below the dam. Within a few minutes the tree had collected all manner of other debris and the river was backing up; soon it overtopped the dam.

The gateman, by then very alarmed, climbed up the structure of the sluice gate, and watched as the water level continued to rise until it was just below his feet.

Then, much to his relief, the tree was carried away, and the water level began to fall rapidly. The force and speed of the flow undermined the centre of the dam, however, and the whole centre section, including the spillway, was washed out. He climbed down from his perch on the structure, as the water level fell, and reported back to the station that there was no water in the dam, and the canal was empty.

From this graphic account, and the evidence of flotsam all along the banks of the river at a level of about 12 feet over the top of the dam, there was not much anyone could have done to avoid this disaster, and I set about to take some photographs and a series of measurements, which would form the basis for the design and construction of a replacement dam, with a longer spillway and a wider passage downstream to ensure that this particular event could not recur.

After a couple of days completing my survey I arranged my trip back to Tanga, and on my last evening in Moshi I had an excellent dinner with the station manager and his family on the site.

Driving back to town later that evening, when I was passing through a wooded area where the track was shaded by large trees I became aware of an eye reflecting the car lights, some way up a tree, and as I bumped along I wondered why it did not turn towards me and present two eyes reflecting. As I got a bit nearer, still thinking of a one-eyed bird, or perhaps a bushbaby, I suddenly realised that this was no such creature, but a very large elephant, standing right at the side of the road, and by then about ten feet ahead of me.

I ducked down under the dashboard and jammed my foot flat to the floor on the throttle, still holding the steering wheel, and in fearful anticipation of being lifted out of the back by a long, sinuous trunk. Fortunately, the road was reasonably straight, and fairly well rutted, so I managed to maintain my course, but remained crouched down for a considerable way until I was out of the wood. For the remainder of the journey, across the open plain, I was very careful to keep away from reflecting eyes, and very happy to get back in one piece to the rest house.

Sitting in the bar afterwards, someone related the story of a drive in the dark across the plains between Nairobi and Nakuru. This was an unmarked track, with no visible edges, and scrub bush on either side from time to time. One of the passengers woke up and shouted out, "Look out! You are into the bush!" or words to that effect. What had happened was that a herd of giraffe had crossed the road in front of them, and all the passenger could see in the headlights was their legs, which he mistook for the trunks of small trees.

Another score for the demon drink.

CHAPTER 41
Mpanda

In the morning I bade my farewells to the local manager and set out, care of East African Airways, once more. We landed at a couple of places, and on the last take-off seemed to bump a bit on the tailwheel. Our next stop was at Mpanda, a small mining development that had been operating for only a very short time, and was still in the course of setting up camp, stores, workshops, etc.

When we landed there was a heavy bump from the rear, followed by a loud grinding noise, and the aircraft skidded about a bit on the runway before finally coming to a stop halfway along. The airport staff arrived in a Land-Rover, and we all disembarked. The tailwheel had broken, and the aircraft was unable to move, so we walked back to the airport buildings. These comprised two small, round huts – or, rather, shelters – with low walls and grass roofs carried on sticks.

One was labelled 'Passengers', the other 'Staff Only'.

The captain soon joined us to advise that a spare wheel was now on its way from the stores in Nairobi, but would not make it to us until noon the following day. He took a note of all our destinations, and went back to radio on to the various offices that we were all safe, but would be delayed until late the following day.

While he was gone, another vehicle arrived at the airport buildings and took some of us into the mine area, where we were offered drinks and lunch, etc. The remaining passengers and crew returned shortly with our baggage and we were allocated a place to sleep overnight. All the ladies were fitted in with families who had spare accommodation, while the rest of us were shown into the extensions of the club, which was then still in a very early state of completion.

When we had dumped our baggage we returned to the club, where arrangements had been made to have a special visitors' night. Additional rations were located and the fridges stocked up with cold beer, all at East African Airways' expense. We had a splendid evening's entertainment, with singing and dancing in most European languages, and eventually retired to our various beds.

My accommodation was in a small hut, about 12 feet in diameter, with a tin roof, and neither doors nor windows, as these had not yet been delivered. I packed all my gear into my bag and set it on the chair provided, pulled down the net and was soon fast asleep.

I was woken some little time later by the grandfather of all thunderstorms. The lightning flashed, the thunder rolled and banged, the wind blew straight through the hut, making the net billow out like a flag – and then the rain came. Within a few seconds the whole place was awash, and the electricity had been blown out.

We all stumbled across to the club, where someone had managed to light up some kerosene lanterns, and the fire was still burning, and by common consent the bar was reopened and a good time had by one and all. During the night, someone managed to restore power to the generator, and by the morning all signs of the storm had passed and everyone had survived, albeit with damp clothes.

After breakfast we had word that our spare wheel was on its way, and soon enough the small de Havilland Rapide had landed. Volunteers were requested to go out and assist in lifting up the tail while the experts fitted the new equipment, and by noon we were ready to continue on our way. Our flight continued uneventfully to Mbeya, where we flew over the area I had been surveying shortly before, and we could still see the evidence of our work where we had cut survey lines through the scrub.

Our next stop was at Mtwapa, which had been developed as the southern port for the great groundnuts fiasco, and, from the air, not much was to be seen, apart from the rows of broken-down tractors and other heavy equipment. Then it was on to Dar es Salaam, where we had another overnight stop, this time in the splendour of a large and modern hotel, with a genuine palm court inside the main hall, and an orchestra playing.

All very civilised.

In the morning we were subjected to another bout of the tsetse fly spraying before boarding, and soon we were off, bound for Tanga and home again. We flew along the coast, and could make out the island of Zanzibar to our right before we started our descent over Pangani, where I could see our house, and all the area of my surveys – the river, two dams, and the power station – and we followed the road right down to Tanga.

Jean was waiting for me on the tarmac; we had lunch, filled the car up with food and drink, and set off on our journey back to our little house in the bush.

There is nothing like a trip across Africa to make you appreciate the comforts of home.

Next morning it was back to work in my own parish again, checking on all the work which had been undertaken in my absence, and I was very pleased to see that it had been well done.

I spent the next few days completing my report on the dam at Moshi, and posted it off by the following weekend.

CHAPTER 42
River Fishing by Dam

During the normal operation of the hydroelectric station, the silt and mud carried down by the river collected in the head pond. Special arrangements had been made to allow for this material to be removed, as it could cause damage to the turbines, and also reduce the capacity of the head pond.

As there was no suitable channel to divert the river past the dam, a second low dam had been built about a mile upstream. This dam had a system of removable hinged washboards, and when these were closed the water ponded behind them, thus cutting off the flow in the river below. In normal dry weather conditions, this dam would hold about three days' river flow, which normally gave sufficient time for the water below the dam to drain away through a sluice in the main dam, and for the silt to be dug out of the head pond.

News of this operation was always welcomed in the locality, as it provided an excellent opportunity to catch all the fish in the dried-up watercourse, and people came from far and near to participate. The operation was normally arranged over a weekend, and for a few days before the scheduled closure camps were established, and parties gathered with their equipment in readiness for the fishing.

Due to some unforeseen difficulties elsewhere, the first weekend closure was cancelled, and the fishers departed, not without a certain amount of muttering. The date was set for the following weekend, and once more the clans gathered. Once again, some difficulties elsewhere in the power supply system made it imperative that the station had to continue running, and once again the campers moved out. There was even more muttering; it looked as if the natives were becoming restless.

Great care was taken to ensure that all would be in order for the following weekend, as the thought of a third-time failure was not to be contemplated. The clans gathered as before, and late on the Friday evening the small dam was closed, and the main sluice opened. In the morning, the river had been reduced to a series of large pools, and the fishing was in full swing.

Family groups had adopted a pool each, and the men were floundering about with loosely woven baskets, trying to scoop up any fish still swimming in the water and throwing them to their wives, who stood around on the banks, fielding the catch. The children then took them to their camp area, where the older women gutted them, and hung them out to dry on racks.

At the main dam, excavation was in progress, with men and women loading head-pans and depositing the silt into the main watercourse of the sluice, which was still running fairly strongly. Some sort of fish trap had been set up in front of the sluice, and as fish were washed into it they were scooped out and flung ashore.

This fish trap also acted as a safety measure, as it would catch any overenthusiastic fisher who had slipped and fallen into the current. As the diggers in the head pond were all station employees, some form of common accounting was established, and the catch divided up fairly. At least, that was the general intention.

Meanwhile the pools were draining rapidly, and what fish remained were burying themselves in the mud, making the basket catching much more difficult. This had been foreseen, however, and a revised technique was introduced, using tridents. In each pool there were perhaps half a dozen men, prodding around in the mud with their tridents. Accidents were not uncommon: some operators were more careless than others, and speared their neighbour's foot; others were cut on sharp rocks; some of the fish were armed with sharp spines on their backs, and treading on them could cause a lot of trouble, and as the water drained away the pools reduced in area, crowding men and targets into a much smaller area. Due to tiredness, the accuracy of the throwers diminished, and fights developed ashore as to whose husband had thrown which fish to which woman.

By the Sunday evening all the silt had been excavated from the head pond, most of the catchable fish had been taken, and the water was almost up to the top of the washboards on the small upstream dam. Various warnings were issued, and an inspection undertaken to make sure that there was no one sleeping at the edges of the pools.

The washbboards at the upper dam were removed, and water allowed to come back into the river bed, and by the morning the river was flowing as normal, the station operational once more, and the work of clearing up the area of the campsites was well in hand.

CHAPTER 43
More River Fishing

Another form of fishing occurred every seven or eight years, at the lower end of the waterfall and turbine discharge, where the river was open to the sea some 40 miles away. The first we knew of it was the arrival of a large number of people, who came and set up their camps along the river below the stilling pools of the turbines.

A few days later the river was alive with small fish, about the size of whitebait. They got everywhere, even into the flushing system of the toilets in the power station.

The main method of catching them was by dragging a basket through the water; this could be expected to come up almost half full of these small fish. Apart from those eaten at once, they were set out in the sun to dry, and the smell was pretty overpowering.

A few days after this first arrival, some larger fish began to appear, about the size of a herring, apparently feeding on the smaller fish, the numbers of which did not seem to have diminished by any great extent. The basket trick was still fairly successful, and the smell increased ashore. Over the next few days, the size of the fish in the river increased still further, with fish of about five pounds quite common.

During periods of low demand on the electricity supply in the town, the discharge from the turbines was greatly reduced, and the water flow in the river practically dried up, leaving a number of ponds still full of fish; this is when the trident technique came into its own, the fish being speared and then thrown to catchers on the bank. Some few days later, much larger fish came along – and then suddenly they were all gone. The fishers packed up and departed, leaving only the smell, and a few days later even that had gone, probably eaten by termites.

After all, they eat everything else.

I made some enquiries about this strange fishy behaviour, and it seems that, due to some peculiarity of the tides and river flows, and the migration of small fish, every now and again a channel is opened at the mouth of the river, and all these small fish ascend, followed in turn by their predators.

The remarkable thing to me was that the fishers knew in advance, and had all their preparations established well before the fish arrived.

For all I knew, they had also fixed the market. I wouldn't have put it past them.

CHAPTER 44
The giant Crayfish

We had fallen into a routine of visiting our friends in Tanga most weekends.

They had a fine house right on the shore, with a large garden, and it was only a short walk from the yacht club. John took a half-share in the boat, and we often raced on Saturdays, while on Sundays we usually went out for a sail and tried our hand at fishing. John was a very fine swimmer, and expert with his elastic-powered harpoon gun. Between us we could usually manage to catch enough fresh fish for our dinner, and seldom came ashore empty-handed.

There was a reef quite close to the shore, where there were always good-sized fish, and slightly further out were a series of isolated coral mushroom heads, some of which were over 20 feet in diameter, with the top about ten feet down, and the bottom nearer 30.

We had gone out one day and swam over these heads, looking for our supper. I was following a good-sized snapper when I heard John's harpoon being fired.

I turned round and could see him coming up to the surface towards me, and noticed he had left his gun some way down in the water. I swam slowly over to see what was happening, and saw him diving down again towards the bottom of the coral, following his line, which appeared to have become stuck under the rock. He reached his gun and gave it a strong pull, and out came an enormous crayfish, speared through with the harpoon, but still full of life.

Due to my position over the rock, and its attachment to John's harpoon gun, it was coming straight towards me – much to my alarm. From my languid observation posture to going full-bore backwards out of its way took about a split second, and I just managed to get clear before it reached me. When the panic had abated, and I returned to a more normal breathing rate, I could see John making back towards the boat, with the giant crayfish in tow, swimming in short arcs at the end of the harpoon line, and by a bit of circuitous navigation I managed to get on board before he and the fish came within range. Then I was able to assist him aboard, and to land the creature, which we unhooked, and set sail for the shore immediately. It was still flapping around in the bottom of the boat, and looked as if it could do one some serious injury if it got the chance.

We beached the boat and could then take a proper stock of the situation. The crayfish had hidden itself somewhere up in the bows, and had curled up with its back towards us. By standing alongside the boat we managed to reach in with a long stick, and pulled it out into the open, there to catch it by its antennae. It was still pretty game, and flapped itself up and down quite violently, escaping back onto the beach in the process, where a smart blow to its head with the paddle from the boat put it out of its misery.

It was only then that we could see the size of it; when holding it at chest height by its antennae close to its head, the tail was still on the ground, and it was as thick around as my thigh, and covered with barnacles.

When you realise that its size was magnified by about one and a half times underwater, you can understand my anxiety to be well clear when it was coming up to the surface.

We unloaded the other parts of our catch and returned to John's house, where we had to find some cooking pot of sufficient size for the crayfish. As there were no missionary pots within sight, we commandeered a half-barrel, currently in use as a piece of garden furniture and containing some vegetation, and, after scrubbing it out, set it up over a bonfire in the garden.

The remainder of our catch – two or three small crayfish of about a pound and a half each, and a couple of snappers – we cooked in the kitchen on conventional equipment, and sent round an invitation to the yacht club for anyone looking for a feast to come along. There were no instructions available concerning the boiling time for large crayfish, but we guessed that about 30 minutes would be sensible, and this was convenient for the arrival of our guests.

Much to our regret, the meat was all stringy and mushy, and had very little taste, even when liberally sprinkled with curry powder, and was deemed a complete failure. I resolved that, when fishing for the pot, only reasonably-sized targets were sensible, as it was not often possible to commandeer the odd barrel as a cook pot, and – in any case – the taste was much better.

Later that night we were disturbed by the most appalling screeching coming from outside the house. Believing at least that some murder was being committed, I dressed and rushed outside.

The noise ceased as I shut the door behind me. With extreme caution, I crept all round the building, expecting to find a body round every corner, but all was serene. I shone my torch into all the dark corners, and still there was no further noise or movement, nor evidence of murder having been committed.

Back at the front door, I thought I saw something on the tin roof, and swung the beam of my torch upwards. There sat two bushbabies, huddled close together, their enormous eyes looking about the size of pennies, reflecting back the light from my torch. I made a sudden movement, and they separated and vanished across the roof like shadows.

In the morning I remarked to our hosts that we had been disturbed by the noises in the night, and they commented that this was the bushbaby serenade season, and sometimes it went on all night. Thinking about it later, I suppose I was lucky not to have trodden on a snake, or woken up the nightwatchman, who might have mistaken me for a ghost and vanished back into the forest whence he came, never to return.

Africa is full of strange noises in the night; what with anvil birds dinging, bushbabies screeching, frogs croaking, the odd lion grunting, and the cicadas going at it hammer and tongs, a quiet night is to be relished.

This was all long before the time of air-conditioners, the noise of which now drowns out all these natural sounds, and the effects of which demand the closure of windows, so that not even a whiff of the smell of Africa is permitted inside the building.

Perhaps that last statement is a bit incorrect, as one of the common smells in Africa is that of burnt toast, and even modern technology can still be outwitted by a cook/houseboy.

When we were packing up to return home at the end of our tour, one of the earliest gadgets to be taken out of service and stowed away was an electric toaster, a wedding present, which had an excellent 'jump up when finished' mechanism. I went into our kitchen the morning after the toaster was packed to find the cook engulfed in smoke from the toast, burning under the grill.

"What the hell is going on here?" I shouted; "are you trying to set the house on fire?"

"No, bwana," he replied tearfully; "I was waiting for the toast to jump out as it always does."

He was obviously of the opinion that, by some trick of 'white man's magic', I had organised the toast to defy the laws of nature.

Mind you, when you really consider things, it is pretty impressive to be able to turn a small knob on a wireless set and listen to people singing on the other side of the world, or to make a light by simply throwing a switch on a wall.

Whatever will they be able to do next?

CHAPTER 45
Our Wedding Anniversary

On the anniversary of our wedding we arranged a special trip to Tanga. We set out to spend the day on a long sandbank about four miles offshore. The tides were convenient, so that we could beach the boat without any trouble, let her dry out over the tide, and be ready to float off again in the late afternoon.

Our first priority was to make some shade, and this was done by the skilful deployment of a survey umbrella and the mainsail of the boat. The next job was to find a suitably cool place for the champagne, and after a few minutes prospecting we located a convenient hole in the rocks about ten feet down. This was fine, but just next door, in another convenient hole, there was a large moray eel, with its head projecting, taking careful note of what I was about.

Having heard tell of conflicting evidence on the habits of moray eels, that some were timid and others were pretty fierce – just like people, really – I reasoned that, if the next hole was currently empty, this creature would probably not object to having an inert neighbour for a couple of hours, so I carefully placed the bottle in the hole and retired slowly.

We spent some time swimming along the inshore edge of the sandbank, where the water was crystal clear. The bottom sloped gently away for about 50 feet from the shoreline, and then levelled out, and there were interesting coral rocks alive with small fish. Whilst following one larger specimen out towards the deeper water, I glanced up, and the whole field of view ahead was just a wall of fish, each about a foot long, swimming across my front, about 15 feet away.

While I was still wondering what might have caused them to gather up in this most impressive formation, as if at a command they all turned to face me, thousands of eyes staring directly towards me. I was well aware that piranha fish live in fresh water in South America, but what if some had escaped, and were now operating in the Indian Ocean?

Could I make it back to the beach?

Should I start? Now?

While these thoughts flashed through my mind, the shoal turned again, in unison, and resumed their procession. I remained stationary for some time, until the last of this shoal was out of sight, and then swam slowly back to the shore, with all my senses alert and my heart going like a steam hammer.

A short spell of sunbathing seemed called for, and then a quick dive down for the champagne, still guarded by my friendly moray eel.

After lunch, rested and refreshed, we went over to the seaward side of the sandbank, where the sand ran out very quickly, and the edge of the main coral reef was exposed. We swam over this edge and could see quite large fish well below us, cruising up and down, while, slightly further off, even larger shadowy shapes were visible. Having already been scared stiff in the shallower water, we retired gracefully and came out for another spot of sunbathing and some further refreshment, resting in the shade by the boat and watching the fiddler crabs waving their arms to one another as the tide came in.

When the boat floated off we set sail back to Tanga after a most memorable and enjoyable day. We had a good sail back to the yacht club, a quick shower, and then set out home again up the road to Pangani Falls, to our little house in the bush.

CHAPTER 46
The Tanga Road

We became quite familiar with the road between our house and Tanga, which was about 60 miles away. From the house we had to travel over rough bush tracks for about five miles before reaching the main road between Tanga and Korogwe. This was a proper metalled road and generally kept in good condition by the Public Works Department. During the rainy season it was sometimes closed for short periods due to flooding, or fallen trees, and one time, as we were driving into town, we came across a serious flood, extending for about 100 yards, right across the road.

A number of vehicles had stopped on either side, unable or unwilling to try their passage, and we walked up to see what was afoot. A pick-up truck was sitting in what once had been the middle of the road, with the water level up to the windows. Two or three people were standing on the roof of the cab, and a number of others were knee-deep, standing in the flat bed of the truck

Downstream, at the edge of the flood, some others were 'guddling' (a Scottish expression describing the process known elsewhere as 'tickling', usually for trout, whereby the guddler lies down on the river bank with one arm immersed up to the shoulder) about in the water, attempting to retrieve their possessions, which it appeared they had thrown over before abandoning the vehicle. People from the assembled lorries had also gathered, and were shouting advice and instructions to the marooned passengers, and directing the people who were recovering their belongings from the river.

By observing the tidemark of grass and twigs along the edge of the road, it was obvious that the flood was receding, and by dint of a lot of shouting, assisted by the general chorus of the spectators, we persuaded the people on the truck to wait a little until the water level had gone down a bit, when we would be able to come out and rescue them.

Once the level had fallen a bit more, we directed a volunteer to wade out, carrying a measuring stick as in crossing Tanganyika bridges, and then gave him our long tow rope – which was part of the kit for travelling in the bush – hooked him up to the heaviest lorry in the queue and pulled them out, to much cheering and rejoicing all round.

While we were waiting, a truck came along from the opposite direction, and tried to ford the flood, keeping to the higher side of the curve of the road. He seemed to be doing quite well until he was about halfway across, with the water almost over his wheels, when he gave a bit of a lurch, and then submerged. With amazing agility the driver climbed out of his cab, and was then able to stand on the roof, still up to his knees in the water. As he seemed unlikely to come to any further grief we turned back to the power station and sent a message back to town to advise them that the road was blocked.

Next day we tried again. The flood had subsided, the road was clear, and resurfacing work was in progress. There was evidence of a large hole by the roadside, where the oncoming truck had gone in, but no signs of the vehicle.

What had happened was that a culvert had become blocked, the water rose until it was coming over the road, and it then eroded the downstream bank right back to the head wall of the culvert on the other side. What had been a raging torrent yesterday was now a small stream you could step across.

On another occasion I was travelling back to the site with Christopher, my Nyasalander driver. We were on the tarmac section of the road, and going quite fast as we approached a curve in the road, with a steep bank straight ahead.

"Slow down a bit, Christopher," I said, becoming slightly apprehensive.

"It won't go any slower!" he replied.

I leant over, switched off the ignition, and made a grab for the handbrake, which, on this vehicle, was under the driver's seat. One good thing about this, it got my head below the level of the windscreen.

We bumped up the bank and came to a halt almost into the forest.

After a couple of minutes spent in relief and thanksgiving, Christopher started the engine again, and it went screaming up to peak revs immediately. This called for some investigation, and after a brief inspection the trouble was located at the throttle return spring, which had somehow become inverted, and instead of closing the throttle when the accelerator pedal was released allowed the throttle to remain fully open and quite out of control. By a bit of judicious fiddling with some pliers, we managed to restore some measure of control over the works, and drove very carefully for the remainder of the journey.

Following this incident, I arranged for the Land-Rover to go into Tanga for a major service, and was issued an old Chevrolet pick-up truck as a replacement.

After the low profile of the Land-Rover, this vehicle gave a splendid view of the surrounding countryside, although, to begin with, the vast expanse of the bonnet rather gave the impression of driving an aircraft carrier.

A few days after collecting this monster I had to go down to the town to collect the pay, and after a comfortable ride down and a visit to the bank, where I collected all the money in a canvas bag, we set out back up the road in the late afternoon. Some 30 miles on I noticed that the temperature gauge was showing a bit high, so we stopped to investigate at a convenient place, near where a stream crossed the road.

Clouds of steam enveloped us as we stopped, and on lifting the bonnet the cause of the problem became instantly evident. The support brackets carrying the radiator had both broken, allowing the radiator to fall down. The fan had cut through the upper hose and all the water had run down over the hot exhaust pipe.

Temporary repairs were in order, and in a bit of a hurry too, as I was not happy about the thought of spending the night in a broken-down truck, with all that money and no food, being eaten to death by insects as the least ghastly fate. Or, more probably, run into by some inattentive driver.

We chopped down a small tree and set it across the wings of the truck under the bonnet, and tied it in position with some fencing wire. We then wired up the radiator onto this beam somewhere near its original position, and then made a temporary repair to the water pipe, using part of the tail of my shirt as a bandage.

By this time it was quite dark and I was less inclined than ever to remain stationary on the roadside, so we filled up all the possible water-retaining articles aboard, including my wellington boots, and fired up the engine. Apart from a small leak in the hose through the shirt-tail bandage, everything else seemed to be in order, so we set off once again, at much reduced speed, stopping at frequent intervals to top up the water supply, and continued on our way.

Back at camp, Jean had become a bit anxious about our non-return, and a search party had been arranged by Ernest, who had set out with a well-equipped rescue vehicle, ready to pull us out of a ditch, or put out a fire, or whatever. We met them at a corner about five miles from home and came back in company without further mishap.

I wrote out a long report the following day about the need for serviceable vehicles in the bush, and listed out the faults in the old Chev.

As well as the brackets and the hose for the radiator, I mentioned that the wings were somewhat dented and the bonnet was not fitting very well. Some immediate repairs were carried out by the station mechanics, the brackets were welded up, some new hoses were found and fitted, and the vehicle declared roadworthy again. At least it was better than walking.

A few days later my Land-Rover was repaired, and we made our last trip in the old Chev back to Tanga. It took some time to become used to the limited visibility and lack of power in the Land-Rover, but I felt it was unlikely that the radiator would fall off.

CHAPTER 47
A new Recruit

At this time another young engineer, John, was seconded to the site staff, and he came complete with his new wife, Margaret, and a new Land-Rover. They moved into another house similar to ours, nearby, and we had a welcoming and house-warming party, which went down very well. The poor girl was a bit overwhelmed by our somewhat primitive conditions (she had not yet stayed overnight in Korogwe), what with no glass in the windows, not a shop for miles, and being surrounded by Africa.

About a week after their arrival I had a message delivered by cleft stick and runner (?) to say that John had broken down about five miles away, and asking if I could come out and bring him home. I loaded all the equipment I thought I might need – ropes, ground anchors, planks, cross-cut saw, spare water, extra spare wheel, fuel, oil, and some food and drink – and with a couple of men from my working party set off to the rescue.

When we arrived at the scene, John was sitting on a bank at the side of the track, watching a column of ants, and the vehicle was parked nearby, apparently undamaged.

Never, never sit down near ants. These marching columns usually sent out outriders to ensure that their way was clear, and also to prospect for food. . Almost anything animate was fair game to them, and immediate flight was the only protection. So far he had escaped detection, but we moved away to a safe distance to discuss the problem with the vehicle.

He said there was an irregular clanking noise coming from the engine compartment, and he thought it wise to stop and call for help.

"Quite right," I agreed; "now, let's have a look under the bonnet."

At first sight there did not seem to be anything out of position or obviously broken. There was water in the radiator, oil in the sump and fuel in the tank.

When the engine was started there was indeed a lot of clanking noise. Further investigation disclosed that a spare oil filler cap had been fitted to replace the original one, which had fallen down the side of the engine. This must have been dislodged by the bumpy roads, and was rattling about between the fan and the engine

140

compartment sides. When the engine was stopped, the cap fell back onto the fan, and was difficult to see. This was soon fixed, and I asked him for his tool kit, to adjust the fan belt.

He replied that he had left all his tools at home.

"What happens if you have a puncture? How do you get the wheel off?"

"Oh dear, I suppose I should have brought along the jack as well," was his somewhat contrite reply

"This is not the North Circular Road around London. There are no garages or service stations within 50 miles, and you must be able to look after minor repairs on your own. I suggest you get a book on elementary car maintenance and study it carefully before going more than five miles out. However, you did the right thing by stopping and not running any further with all that funny noise going on under the bonnet. It could well have been something serious."

The following day I noticed that his driver was busy collecting a similar set of equipment to that which I normally carried about, so the message must have got through.

What kind of engineers are they turning out nowadays?

CHAPTER 48
The Tanga Shop

The main shop in Tanga was owned by a Greek family and carried a variety of food, drink and assorted hardware. Most of their business was conducted on a credit account basis, with the bills being posted out at the end of each month.

There were often a few customers who departed from the territory without paying their bills, and the store had a novel method of recovering this money. A note was displayed in the shop window with the creditor's name and the amount outstanding, and if this had not been paid off after a further month all the goods in the store had their prices increased by one penny per item until the account had been recovered.

I never heard of the prices being reduced if some late departed customer coughed up, but most shoppers accepted the system with only minor protests. The alternative was to shop elsewhere, but this was rather a matter of Hobson's choice. The only other way was to pay your bills shortly before the mailboat left each month.

The Greeks must have had a word for it.

CHAPTER 49
A borrowed Car

My Land-Rover was in the garage for a service, and I had borrowed a car from another engineer who was working in the town office. It was an Opel saloon, and, compared to my Land-Rover, of very flimsy construction – totally inadequate for operations on the site roads, but still better than walking.

I was driving pretty fast along one of the site access road in the rain and came round a sharp corner to find a Land-Rover approaching from the opposite direction.

There was no chance of stopping on the muddy surface, so I took to the country on my left, and went up a steep bank until we had passed one another.

When I stopped, back on the road, I discovered that the little cleat which holds the canvas on the back of the Land-Rover had connected with the corner of the wing of the Opel, and opened it up, just as if it had been cut with a tin-opener. The only things holding the wing on the car were a small area of metal near the headlight and about six inches of panel down the door hinges.

Apart from a small smudge of paint on the cleat, the Land-Rover was undamaged, but the driver was most surprised to see me still on the road behind him. We exchanged documents, and I drove on a bit more slowly, hoping that the wing would remain attached until I reached Tanga.

When I arrived at the office it was still before eight o'clock, so I parked the car outside and went up to fill in an accident report and to see if my own vehicle was ready for collection. Some time later I came down to find that someone had placed a pole with a large 'No Parking' sign on it directly in front of the car. I drove off to the garage to collect my Land-Rover, completed my other business in town, and went off back up the road.

There was a bit of long-distance palaver about the dent in the Opel, which I managed to explain away, but there followed a summons from the police concerning the parking offence. I wrote back saying that the sign had not been in position when I had parked the vehicle, and asking how I, an out-of-town visitor, could be expected to know where the local constabulary would erect this temporary sign.

Would it be possible to plead only half guilty to this charge?

A second summons was the reply, with a handwritten note to say that I could only plead guilty or not guilty. I wrote back again, saying that as only half the car was covered by the sign, and I had only been there for half the time, in truth I could only plead one quarter guilty.

I thought that this had worked the trick, as nothing further was heard for a couple of months, but then came a further summons to attend about one month later.

I wrote back once again explaining my difficulties in answering the charge truthfully, and, in the meantime, orders came through from London to cease work on the project, due to some lack of confidence in the future financial situation. This was probably because of the impending legislation concerning the establishment of an independent government, and some doubts about the capacity of such a government to manage the economy.

I was instructed to make the arrangements to pay off my men, close down the site, complete all the outstanding drawings and records and leave them with the station manager, clear the petty cash, close the bank account, and arrange our passage home to London and call in to the office sometime during my home leave.

By a special piece of good luck, I was able to arrange our departure date about three weeks before the hearing of the traffic case. I wrote another long letter to the court explaining my difficulty in being unable to plead guilty to the charge, but was prepared to admit that the car had been parked where the policeman had subsequently set up his pole with the 'No Parking' notice. I also pointed out that I had been in the office directly above the car, and would have come down and moved it had I been so requested.

This must have raised some doubts about the case, but as it took them so long to consider a reply we were long gone before they had the hearing – and, so far, I have not been pursued.

I think the statute of limitation might now apply.

CHAPTER 50
Going Home

The next few weeks were very busy, winding up the work, completing all the drawings, leaving directions for the location of survey stations and paying off my men. We were very sad to be leaving, as we had settled in comfortably, the garden was growing nicely, and we had made a number of good friends.

We sold our remaining half of the boat to our friends in Tanga, left the dog with Ernest, and said goodbye to our little house in the bush.

As we passed through the camp we were touched by the display of affection shown to us by the men who had worked for me, and it was with a moist eye that we set off on our final trip down the road to Tanga.

We stayed in the old hotel for a couple of nights, then boarded the steamer in the harbour and sailed for Mombasa, where the ship berthed at the quayside, and stayed over for a few days.

In Mombasa we visited all our friends in the town and had a final sail from the yacht club – before leaving east Africa, bound for Aden, Suez, Marseilles, Gibraltar, and finally London.

And so ended our first spell in Africa.

PART TWO – TIREE, 1958-1959
General map of the United Kingdom

UNITED KINGDOM
IRELAND

NORTHERN
IRELAND

SCOTLAND

ENGLAND

WALES

IRELAND

© Copyright Bruce Jones Design Inc. 2003

147

CHAPTER 51
Travelling Arrangements

We sailed from Mombasa in late October, passing the Horn of Africa one very calm morning, about three miles offshore. The land to the west appeared as a long, dark sandstone cliff surmounted by a white lighthouse, marking Cape Guardafui.

As we rounded the cape, the cliff receded, and was replaced by a sloping sand dune, completely barren of vegetation, which ran down to the shore, where there was a collection of huts. All in all, a pretty dismal sort of place. There were a couple of dhows moored close inshore, probably waiting for the wind to get up before making their passage round the cape, and down eventually to Zanzibar and Dar es Salaam.

Our first port of call was in Aden, and we entered the busy harbour attended by tugs and pilot launches, with a fleet of floating emporiums manoeuvring for the best trading positions alongside. Before we were secured men were diving for coins thrown over by indulgent passengers, and other traders had established contact with potential customers.

The ship's rails were crowded with interested spectators, and soon a roaring trade had developed, lines had been thrown up, and goods hauled up in baskets, examined and rejected. Further bids were made and arguments ensued. Other passengers then took an interest, and the whole situation became very confused, with a sort of auction going on, passengers from different decks shouting bids and counter-bids down to the traders, and – it seemed – a good time was being had by one and all.

We decided to go ashore to stretch our legs in Asia, and from Steamer Point we took a taxi to the main centre of Crater Town. Here was another manifestation of the demon barefoot driver, and, after a short prayer, we were off with the mandatory blast of the horn, up an incredible road through the cliffs to the town.

This road had been cut through a considerable hill of red rock, and was in a cutting similar to the canal at Corinth, about half a mile long and some 200 feet below the crest of the ridge.

The shopping centre was laid out in a square pattern, with the shops, mostly single-storeyed structures, with large canopies out over the pavements, set up as tourist traps. Inside the most tumbledown-looking buildings were stores selling duty-free cameras, binoculars, cigarette boxes, lighters, perfume, tobacco, carpets, rugs, beads, ornaments, relics, oil lamps, etc., etc.

In the doorways, plausible gentlemen invited us inside, to have tea or coffee, and inspect their wares, all offering delivery by the next ship of the bulkier items direct to addresses in east Africa or Europe, and pressing cards with addresses of their agents all round the world. Defeated by this expert salesmanship, we bought a pair of binoculars, a camera, half a dozen small ornamental cigarette lighters, and returned to the street.

This was an extraordinary scene: there were people on foot, others pushing handcarts or pulling barrows, a few horse-drawn taxis, and the odd cart being pulled by a camel. Others were riding on donkeys, or on bicycles, and no great attention seemed to be being paid to the rules of the road. The men with the camel carts seemed to take precedence, probably because their camels were a bit taller than most other means of transport, and more likely to come off best in any argument in the street. There were also a very few motor cars and small vans, but these were accorded no rights of passage, and had to proceed behind a barrage of noise caused by revving their engines and hooting their horns.

After a couple of hours of this we found another taxi and returned to the comparative peace of Steamer Point, where a cool breeze was blowing off the sea, and things seemed to be a bit more organised, with customs officers, port health and immigration officers and other forms of bureaucracy operating from high cool offices – and taking their time about it.

We sailed off again into the sunset and proceeded up the Red Sea, passing numerous tankers and many other ships making their way to or from Suez, where the canal had recently been reopened. The Red Sea belied its name, as far as colour was concerned, but the heat was most oppressive. Perhaps the name derived from the impression that everything was red-hot.

The pilots came aboard and we steamed slowly through the canal, past the many wrecks and abandoned ships which had been caught up in the war resulting from the nationalisation of the Suez Canal Company, and berthed in Port Said, where a few passengers disembarked. We were advised not to go ashore, and stood at the rails watching the antics of the bumboat men and the traders, using similar techniques for conducting their business as those in Aden.

Somehow, despite the ship's security system, one gully-gully man (a type of fairground busker, whose main trick was known worldwide as 'find the lady') managed to get aboard, and he was the centre of a crowd of admirers as he performed his vanishing nut trick with his three small cups.

I suppose, if he could work that act, it was not surprising that he got aboard.

We sailed off in the evening into the Med, heading in the general direction of the Straits of Messina, and shipboard life resumed its routine. Part of the entertainment was horse racing. Everyone bought tickets in a draw for horses, and another for jockeys, and, by some fluke of chance, between us we drew four horses and three jockeys. The system was that the owner could choose his jockey, who shared in the loot if they won, and the course was set up in the dining saloon.

Each race had about ten horses – wooden models set on flat skids, which were pulled across the floor by the jockey cranking a small handle on a little winch. We had great success, and won about £25 in prizes and bets, which was immediately consumed in strong drink. (Hard work, this shouting and winding.)

Other activities and entertainments were: deck quoits, requiring skill and determination; table tennis, which was made more interesting by the roll of the ship, and the intermittent opening of a door onto the deck, allowing a 20-knot wind into the table tennis enclosure; whist and bridge sessions; and the regular calls to meals.

The weather was fine, but the wind was blowing quite hard, and the ship was rolling gently in the swell as we continued through the Straits of Messina, past a fine view of Stromboli in eruption – like a gigantic fireworks display. We sailed close in along the coast of Italy, where we could see the houses perched on the waterside at the foot of towering wooded hills, and on to Marseilles.

We went ashore here for a few hours, walking about in the pleasantly warm sunshine, admiring the yachts in the harbour, and selecting a suitable restaurant for our lunch. We settled on a large café on the harbour side that seemed reasonably full of local customers, and sat down to order. While we were becoming accustomed to the gloom an order was delivered to a neighbouring table, which was occupied by a large family, one of whom was a priest, who seemed to be in charge of the party. The waiter delivered a huge plate of moules, piled high, and they were at them as if they had never eaten before. As soon as the pile was half finished, more and yet more again were brought in, and this was only the first course. We settled for a more reasonable diet of soup and grilled sole, and when we left some time later the moules party was still in full swing.

Back aboard again we sailed for Gibraltar, and arrived early in the morning, just as dawn was breaking, and came into the harbour as the sun came over the side of 'the Rock', warming the whole scene. We went ashore at once, and did the tour of the caves, and the trip up the mountain while it was still pleasantly cool. Our bus driver had cautioned people that the apes were very smart, and quite unafraid, and that ladies were advised not to take their handbags out of the bus. We were sitting on the parapet of the road, admiring the view and the scene of bright beds of narcissus, when there was a cry of anguish. Some foolish woman had ignored the driver's advice, and an ape had snatched her handbag, and was leaping away down the slope with it. Some people won't listen. Back in the town again, we paraded through the shopping streets, where the same range of goods and prices as we saw in Aden were displayed, in slightly cleaner and more pleasant surroundings, but with the same promises for deliveries throughout the world.

We sailed later in the evening, out into the Atlantic. The weather continued fine as we rounded Cape St Vincent and turned north, following the coast of Spain and then Portugal, about five miles offshore, until we cleared the north of Spain and set out across the Bay of Biscay.

Some new passengers had joined the ship at Gibraltar, among them a Spanish couple, who sat at the next table in the dining saloon. It appeared that they intended to get their money's worth out of the trip, as they were always the first customers at every meal, and ate their way through every choice and every course on the extensive menu. I think they must have had elastic seams in their clothing.

The passage across the bay was fairly smooth, and we closed in on the coast of England on a calm morning, off Plymouth, where the pilot came aboard with copies of all the English newspapers, dated that day. We sailed on up the channel in calm but somewhat foggy conditions, and eventually anchored off Southend, as the visibility was too poor to allow us to proceed upriver to Tilbury, where we should have docked in the afternoon.

We spent a further two days at anchor off Southend, and this made things a bit difficult, as all our spare clothes had been packed away in readiness for unloading, and it was damp, cold and foggy on deck. Eventually the weather cleared a bit, and we steamed slowly up the Thames to the dock at Tilbury, where we disembarked.

The passport and port health formalities were soon completed, and we went ashore to clear our baggage in the customs hall. There was a bit of an inquisition following our declaration of a couple of cameras, a movie projector and the binoculars bought in Aden, and when asked if that was all, I suddenly remembered the cigarette lighters.

"Oh, yes," I said, "we have six table lighters – as little presents, you know."

The officer raised his eyebrows.

"And how much did you pay for these?" he enquired.

"Well, I think they were about four shillings each."

This rather floored him. What sort of people were these, bringing four-shilling lighters as gifts? What kinds of friends must they have?

With a sigh of disbelief he scrawled his mark on our belongings, and we were clear to go, and were soon on the train into London, where we had booked into a hotel for a couple of nights – our first time ashore since Gibraltar. We made contact with all our friends and relations, and arranged a grand tour for ourselves over the Christmas period, and next day set out on a great shopping spree to kit ourselves out for the winter. After a couple of years in the tropics the weather at home was not very pleasant, and having to wrap up in coats and mufflers was all a bit of a bind.

We started out on our progression: first to Essex, where we met part of the family, and bought a small car; then to Devon, where we spent Christmas; then to Scotland for the New Year, meeting more family and friends; and then off for a couple of weeks in the north before returning to Devon, where we established a temporary base.

I was summoned to London and asked if I would go out to Iraq on a single basis for a year, and declined, and said I would wait out my leave and see what else might turn up in the meantime.

After another month of home improvements in Devon, I replied to an advertisement for a job on the island of Tiree, rebuilding the pier there, and went up to Glasgow to see what it was all about. After a very pleasant interview I was offered the job, and accepted, providing I could find suitable accommodation on the island; as agreed, I set off by train to Oban, and then by the steamer to Tiree the next day, to see how the land lay.

I had already made some enquiries about the housing market on the island, and had been given a couple of addresses of likely places, and when I arrived I made some further investigations from the pier master, and at the post office, and collected details of another two or three possible places. I arranged for a taxi to take me around, and had a rapid tour of the island, visiting all the potential places, with a commentary delivered by the driver on the owners, neighbours and the possible advantages or drawbacks of each establishment.

One of the problems was that the majority of available houses were owned by people living in Glasgow, and were used only as holiday homes for a few weeks in the summertime, and in consequence the owners were not available for face-to-face negotiations with regard to periods of occupancy and the other facilities available.

There were several suitable places, however, and I returned to Glasgow to confirm that I would take the job, and phoned Jean to let her know of the situation before going back to Devon.

The next few days were spent in letter writing and telephone calls to firm up on a suitable house, and then to make the final arrangements for the move. I would drive up to Oban and take the steamer across, settle in the house, and sort out the various domestic arrangements, such as getting the coal in and fixing a gate on the garden to keep the sheep and cattle out, and sorting out the postal deliveries and the shopping delivery routine, and where to collect milk and eggs, etc. At the same time, Jean would pack up our kit and have it sent off by carrier to Glasgow, for onward shipping on the biweekly cargo boat from the Kingston Dock.

She would then come on by train and steamer, to arrive after the delivery of all our worldly goods.

The steamer left very early in the morning on Mondays, Wednesdays and Fridays, and it was sensible and convenient to arrive in the early evening in Oban, have a meal, and then retire to

153

your cabin on the steamer. The passage time from Oban to Tiree was about seven hours, including stops at Tobermory and Coll, and the arrival time at Tiree was usually about noon. From Tiree, the steamer went on to Barra and stopped overnight there, then returned to Oban the following day, stopping at Tiree, Coll, and Tobermory.

The start of Jean's journey had to be worked in with the due arrival date of our baggage, and the three possible days of the steamer passage, and had to include a train journey from Devon to Glasgow, changing stations, and then again on to Oban – during which time communications would be very difficult, especially since I had no telephone on the island, and had to rely on the assistance of the post office to take messages and pass the word otherwise.

When all these arrangements were in order, I set off by car from Devon to Oban.

And so to Tiree…

CHAPTER 52
Tiree

I stopped off in Glasgow to complete the formalities of my contract and collect the new company procedures on pay, union arrangements, office routine and reporting, the contract drawings, a letter to the bank, and a whole lot of other guff, which I set aside to read at a more convenient time; I spent the evening with my parents in Glasgow, and set off in the morning for a leisurely drive to Oban.

I arrived in the afternoon, and later saw the car loaded onto the steamer's deck by a pair of wire slings suspended from the ship's derrick, without appearing to do it any harm. The ship was not due to sail until early the following morning, so I had a walk round the town, had some dinner, and returned to my cabin.

When I awoke it was about five o'clock in the morning and there was a great roaring of engines, clattering of winches and violent vibrations as we left the quay.

The noise subsided a bit and I got up and had an excellent breakfast as we sailed up the Sound of Mull to Tobermory – a place of fond memories from when I had been there previously, sailing on a small yacht.

After leaving Tobermory we went out past Ardnamurchan and set a course for Coll, where we stopped in the bay while passengers and goods were transferred to a launch to ferry them ashore. The passengers were easy enough, but there were a number of calves bundled up in sacks, which were not nearly so keen to go ashore.

The next stop was at the old pier at Tiree, where the car was safely unloaded, and I set out to dump my bags at the hotel, collect the keys for the house, meet the pier master, see the postmaster about our new address, and put the word about that I would be looking for staff the following morning. I collected the house keys, and went along to see what I would need to allow me to move in as soon as possible.

The house was of fairly recent construction, with two rooms downstairs and a small hall and a porch. The kitchen was a little annexe at the rear, and there were two bedrooms and a bathroom upstairs. The house was wired for electricity, being fitted with lights and power points and a cooker, while the main heating was from a large range in the sitting room, and there was a small supply of coal outside in the yard. The water supply was from a well in the garden, with a semi-rotary handpump in a cupboard under the stairs. With all the taps shut, it was possible to fill the storage tank in the roof sufficient for one day's supply, including a bath, with about 700 strokes of the pump. With one tap open it was not possible.

The house was owned by a native of the island who was then living in Glasgow, and we had agreed that he would come out for his two weeks' holiday in July, and we would go off at the same time. The neighbours were very kind and helpful, and I had managed to move in within a couple of days with the basic equipment supplied with the house supplemented by a few local purchases. Our heavy luggage arrived a few days later, and the place was reasonably fitted out before Jean arrived.

The travelling arrangements worked out successfully and Jean arrived in good time after a calm crossing. We set out for our new house, about three miles away on the western side of the island. I had been a bit vague in my description of the house, merely saying that it was quaint, rather nice and well out in the country, and when she disembarked and we were travelling across the island I pointed out some of the traditional 'black houses' as examples of the local architecture. These houses were fairly common, and were about 50 feet long by 20 feet wide, with five-foot-thick walls, small windows and roofed with thick thatch, generally covered by a black-tarred canvas. Very few of them were in regular occupation, and they did not look all that inviting.

Some relief was expressed when we came to our house, which had a conventional appearance, the fire burning in the grate, and everything set out neat and tidy. We had a general inspection, including a demonstration of the water supply system, with special emphasis on the need for taps to be turned off, and an agreement on how to rearrange all the furniture, and an explanation concerning the workings of the range.

Meanwhile, work continued at the pier. I had engaged half a dozen local men, the foreman and the crane-driver had arrived from another site, and we established the office in an old railway van, complete with coke stove, at the end of the pier. A similar van was set up ashore for the men's bothy, and an old shed taken over as a workshop.

Jock, the foreman, hailed from Dundee and had a very strong accent; so much so that it was often difficult to understand his speech. One of the local men came to me saying that he wished to resign, as he could not make out what Jock was saying. I told him that I was having the same problem, but after a week or two it would probably get better. I certainly hoped so. Anyway, I persuaded the man to stay on, and had a quiet word with Jock to speak more slowly. Maybe he was having the same trouble with me.

It was a great joy to work with the local people. Most of them had been to sea at some time and all were natural handymen, and could tie knots in ropes, understood the basic principles behind moving weights, and the art of keeping their fingers out of moving parts of machinery. They also had an excellent sense of the importance of things; on one occasion they came in to advise me that they would not be in to work the following morning. Somewhat surprised, as we had no previous history of labour troubles, I asked them why not.

"Well," they said, "there is a good tide tomorrow morning, and the herring are in the bay, but we will come in later and do the day's work when we have finished the fishing." This seemed a very reasonable proposition, and I asked if I could join them.

"But of course you can. We will see you at the harbour at five in the morning; make sure you bring warm clothes and your oilskins," was their immediate reply.

After a few days on the island I discovered that the local vet had been at school with me, and we renewed our friendship. He had a small open motor launch, and we used to fish lobsters as an additional source of nourishment, and as a means of paying for the insurance and maintenance of the boat. We went out most days during the season, and could expect a plentiful supply of crabs and a couple of lobsters each week for home consumption, and a reasonable number to store in a big wooden chest floating just inside the pier. This was cleared about every four weeks, and sent off to market, for which we received the large sum of six shillings per pound weight. As well as lobster fishing, during the long evenings in the summertime we went out after cod.

The local bank manager was a noted fisherman. It was rumoured that, if he fell overboard, he would come up with his pockets full of fish. Whatever it was, he had it, and people took bearings on his positions, while noting the state of the tide and the wind, and frequently fished alongside him, though never achieving his success rate.

One evening we went out in the boat, six of us, and took station not far from the banker, who was hauling them in as usual. We were in about 40 fathoms, using hand lines, with worm as bait. The vet and I were being quite successful, the three wives were catching the odd fish, but the last member of our party was having no luck at all. We exchanged lines, altered our positions in the boat and tried to synchronise our techniques, but all to no avail; Bob was still unlucky.

The fish we were catching were about five or six pounds average weight, and soon we had about 60 fish aboard. One of the last ones to come up managed to jump off the hook as it rose up, and was swimming in a distressed manner on the surface.

Probably it was suffering from the bends, but the vet was determined to have it aboard, and leant right over the side to catch it in his bare hands. As this made the score 65 we reckoned we had done enough, and set off back to the harbour, where we shared out the catch between us, and then set off to try to dispose of it around the island.

There were very few houses with refrigerators, and we could persuade each house to take no more than one or two fish, and by this time it was getting late.

On our way home we came across one of my staff on his bicycle, proceeding with some difficulty, and we stopped to offer him a fish. "Well, thank you very much," he replied; "I have just the very thing here for carrying it." So saying, he produced a long roll of heavy canvas from his pocket, and proceeded to tie the fish along the crossbar of his bicycle. Then, with profuse thanks, he mounted and rode off, still going very carefully. I think he must have been at the home-brewed whisky.

Most people had gone to bed by then, and we went home with half a dozen large cod still to dispose of. In the morning we handed them out to all our near neighbours, and any passing motorists who chanced by.

Our friend, the banker, was also the official counter of birds and seals on Tiree, and the other islands nearby, and we arranged visits to Dutchman's Cap, and Lunga, to see the seabirds at hatching time. These islands were uninhabited, part of the Treshnish Isles, lying between Mull and Tiree, across a fairly open stretch of water, and we had to choose a suitable day when the weather was settled and the sea calm for this expedition. We set out in the vet's launch, with two extra outboard motors – just in case – on a fine, clear morning.

Dutchman's Cap is a strange-shaped island, with cliffs all round about 100 feet high, generally with a flat top, but with a hump in the middle, standing up another 100 feet or so. It was not difficult to see how it had got its name.

The only landing place was at the south end, where there was a small cove in the cliffs, with a short gravel beach where we could land and secure the boat while we went ashore. We climbed up a rough track to the level of the plateau, where there were some pretty wild sheep, grazing on the short grass. They were probably only visited once a year, being landed in the spring and taken off again in the autumn, and they had no fear of people.

We started our counting and reporting. There were many gulls all over the short, rough grass, nesting, and just sitting there until you were right on top of them before they would fly off.

On the central hump there were hundreds of short burrows, and the banker thrust his hand into the first one we came across, and pulled out a puffin; very irate it was, too. He passed it across to me, head first, saying, "Hold onto this one, while I see if it has been ringed," and, all innocent, I held out my hands. The wretched animal took a firm grip on my hand between my thumb and forefinger, and I had to perform all manner of contortions before I could get it under control, and I see that I still carry the marks from this encounter to this day.

We went all round the hump, pulling puffins and Manx shearwaters out of the ground while the banker did his counting and ringing, and then moved off to the other end of the island, where there were thousands of razorbills and guillemots nesting on ledges on the cliffs, and numerous other seabirds – kittiwakes, petrels and fulmars – on every conceivable nook and cranny on the sheer face, all busy about their own affairs, and quite unconcerned by the presence of the visitors.

The banker made some estimate of the numbers and details of the various species in evidence and we went back to the boat and sailed for Lunga. This was a larger and much more inviting island, with a small beach of sand and some trees. It had been inhabited some time previously, and there were the ruins of farm buildings and shelters. There was a resident population of large, black longhorn cattle, and they followed us about in a menacing manner, always about ten feet away.

On the grassy surface there were many skylarks, a number of oystercatchers, some lapwings and many gulls, all with eggs or chicks, and we had to take care as to where we were walking for fear of standing on them. At the cliffs at the north end of the island the area was also alive with birds. There were some particularly smelly cormorants, which sat on their nests and squawked at us as we passed, then opened their beaks and exposed the bright yellow of their gullets, and spat a foul-smelling sort of curd at us.

At the very edge of the cliff there was a group of razorbills, and I took our camera and crouched down, then crept towards them for a real close-up snap. As I as getting within range, lying on the damp mossy ground, and getting wetter by the second, Jean came over, took the camera from me, and walked right up to the birds, who appeared completely unconcerned.

We had our lunch and continued our exploration of the island, still being followed by the cattle, until it was time to be setting off back to Tiree. So ended a very satisfactory visit to the Treshnish Isles.

Another of the banker's activities was to report on the seal population, and in the late autumn we set out in his company around the north end of Tiree, where there were a number of sheltered coves and rocky inlets where the seals came ashore to have their pups. The seals were fairly tame, and would allow you to approach to within about 30 feet before they lumbered off into the water, sometimes leaving their pups behind in little hollows in the rocks. Seal pups are quite the most appealing little animals, with their large brown eyes and soft silky-white fur, but within a few weeks they have changed into a dark roly-poly sack of blubber, and then they are ready to take to the water.

We also took the launch across the Sound of Gunna, between Coll and Tiree, and performed another headcount on the seals there. On these expeditions we usually trailed lines behind the boat for mackerel, and we considered ourselves unlucky if we did not catch a few on each trip. Apart from the eating, they were good bait for the lobster creels, being almost as attractive to lobsters as salted herring.

CHAPTER 53
The Steel-fixer

I was in need of a steel-fixer on the pier – the man who assembles and ties together all the steel reinforcing bars in the structure – and, as there was no local tradesman available, I asked the office in Glasgow to engage someone and send him out on the steamer. A few days later the man arrived, and set about his work in a very capable manner.

The system for pay on the site was that I sent into Glasgow the particulars of each man's hours every week, and they calculated the wages, allowing for tax and all that, then sent me a note, which I raised as a cheque on the bank, where they made up the cash, which I collected and then had to load into individual envelopes, and paid out each Thursday.

My steel-fixer had not been employed away from home before, and I overheard him muttering about having to wait at the post office to get a postal order to send back to his wife, and the inconvenience and expense of this exercise. I suggested that I could have the office in Glasgow make an allocation direct to his wife by post each week, and this would mean that the cash would be with her, regularly, on a Friday each week. A few days later he agreed, and the allocation was established.

The following Friday Willie came into my office, and asked to be allowed to go home over the weekend. This meant that he would not have the overtime for working over the weekend, as he was not returning until noon on the Monday, and I asked if there was anything I could do to assist. "No," he said; "I have had a telegram from my wife, asking me to come home at once." I replied that this would be in order, and if he had any need to stay home longer he should telephone me, or go into the office in Glasgow, and find someone there to assist.

He went off, and returned on the Monday, very choked off. When questioned he produced the compliments slip that the office had included in the allocation to his wife.

"Stupid bloody woman," he said; "the first thing she said to me when I got home was, 'What's this about complaints? What have you been doing? Does this mean you have been sacked?'"

Apparently it took some time before order was restored and the explanation accepted, and Willie had to work on a few hours' overtime to catch up on his fares and expenses over the lost weekend.

CHAPTER 54
The Cattle Sales

One of the conditions of the contract for the construction of the pier was to ensure that traffic would not be interrupted by the construction work, and that special provisions might be required to permit the loading of livestock over the pier at the times of the annual cattle and sheep markets. Traditionally, there were separate sheep and cattle sales held in the middle of the island during the summer, and the stock were driven across and held in a field near the end of the pier. Dedicated cattle boats had been scheduled, and the animals were loaded out over the end of the pier, and sailed off to Oban. The pier was constructed in a corner of Gott Bay, and there was a small complex of sheds and offices at the landward end, a long stone breakwater with a high wall on the seaward side, and an iron fence on the inside. This led to the actual pier head, which was open on three sides, with a fence along the back.

Our first experience was with the sheep.

We had a visit from the 'cruelty man' to inspect our arrangements, and to advise us of possible breaches to our barricades and fences. We had set up some large boxes and a wall of heavy timber baulks all along the exposed edges of the pier, and had to make up temporary defences along the frontage, which could be moved as required to accommodate the loading ports of the different ships, and declared ourselves ready for the fray.

Out of curiosity I went along to the field, and saw the various batches being sold and marked, and marshalled for their journey to our paddock. All seemed complete confusion, what with sheep rushing about, dogs barking, drovers cursing and whistling, and numerous spectators and participants exchanging greetings, comments, and the occasional dram.

Away from the market, things seemed to be a bit more under control, with flocks of about 100 sheep being taken along the unfenced roads, and straying onto the machair – the open grazing unfenced land, which, in the springtime, was a mass of yellow flowers – and so on down to the paddock by the pier.

The pier master asked if we would assist him with the loading, and as it was impossible to continue with the construction at this time we prepared to man the barricades. When the ship had come alongside we set up the gangway and adjusted our fences to provide a straight, clear passage, from a gate on the pier head directly to the gangway, and signalled to the drovers that we were ready. They then put a dog into the paddock, and chased about a dozen sheep at a time onto the long breakwater, and the dog followed them down to the pier head. So far so good. The sheep were not happy about leaving the shelter of the breakwater and going onto the open pier head, but with a bit of shouting and prodding, and some action from the dogs, we managed to get the first batch aboard without too much trouble.

Things settled down into a routine, and we felt we could reduce our forces at the barricades, when something went rather wrong. One of the dogs had jumped up onto the top of our defences, and half a dozen sheep suddenly changed direction and leapt over the fence. They went careering down the length of the pier head and jumped straight over the edge, into the sea, about ten feet below. We tried to get the dog to go in after them, but he seemed to have more sense, and swimming was probably not in his contract. He stood barking furiously while the sheep, now formed up into a tight formation, were heading off towards Mull, and a good long swim.

We had to launch the pier boat, which once had been a 'puffer's' lifeboat. (During the first half of the 20th century most of the trade between the islands and the mainland was carried out by small steam coasters, usually with a three-man crew. As these ships were registered with the Board of Trade they were obliged to conform with the relevant regulations concerning safety equipment, and each puffer carried a heavy lifeboat, equipped with buoyancy tanks, water cask, compass, etc. and weighed about a ton. Our boat had been released from official service, and was in use as a stable platform around the pier construction and repairs.)

Our intention was to row after these sheep, turn their heads towards the nearest beach, and urge them ashore.

One of the drovers and his dog had walked across the bay, and was waiting to collect them when they staggered ashore, and we rowed back to the pier in triumph, ready for another rescue operation if required.

There were a few more minor alarms, but nothing as desperate as before, and all the sheep were loaded out in good time. Next day we spent a lot of time washing down the surface and clearing away our fortifications.

Having experienced the wilfulness of mere sheep, we reinforced our barricades in readiness for the cattle sales. It is one thing to scare off a sheep from the security of a ship's gangway, but quite another to face up to a determined, large cattle beast, probably weighing the best part of a ton, already excited, and not at all keen to go down onto a steamer.

I went along to the sales again, and once more there appeared to be utter confusion. Cattle were paraded and auctioned off, then put into a marking enclosure and marked, then sorted out into consignments for different ships, and held in separate fields. The drovers then began to herd them along to the paddocks by the pier, where they were mustered in separate batches. The procedure was then to cut out about half a dozen animals from the paddock, and, with one drover on the high wall and his two dogs, drive them down onto the pier head, where another crew, including the 'cruelty man', encouraged them down the gangway into the ship.

We had tried to make our barricades funnel into the end of the gangway, so that there was a good chance of single-file loading, but such was the pressure of reluctant beasts that our defences slipped open from time to time, and we had a hard job keeping some measure of control.

Once again we settled down into routine, and I had gone into my office to try to get on with some paperwork when I was aroused by a lot of shouting and barking. In one batch of half a dozen cattle there was a huge red creature with long horns, which was most reluctant to enter the narrow funnel leading to the gangway and broke back at the last moment, with some danger to life and limb, causing confusion among the other animals that were behaving in a more orderly fashion. He was left to run back down the breakwater, to be brought up again with the next batch.

Right enough, he was collected and brought back again with the next group, and was greeted by shouts of welcome by the drovers on the pier head. "Look out, here comes that big red bugger again, keep well clear of him" – or words to that effect. Once again he refused to cooperate, and galloped back to join the next group. This performance was repeated several times. You would think he might tire of it.

The drovers in the paddock began to make comments about their colleagues on the pier head, and it seems that a challenge was issued that one of the shore-based men should come out and show the others how it should be done.

A suitable candidate was elected, and he climbed up on the high wall, and with his dog started to bring the creature out towards the pier head. He got him about halfway out, but was unable to move him any further, so he climbed down off the wall and approached the beast, which had now adopted the typical bull-fighting attitude – head down, foot stamping and pawing the ground, and with his tail straight up and a determined gleam in his eye.

The drover advanced, to calls of 'olé!' from the gallery, waving his stick and encouraging his dog to go in and turn him, but the beast stood his ground, and suddenly began his charge. With remarkable agility the drover vaulted over the fence on the inside of the pier, and dropped down onto a ramp leading to the beach.

One up for the old red bugger.

A revised technique was developed. A pair of farm gates were lashed together and set up across the approach, then manned by a whole crew of drovers and the 'cruelty man', who, shouting and waving their sticks, advanced this gate down towards the pier head.

We watched with interest from a safe position behind our heavy timber baulks on the pier head. This position of safety was only relative, as the only place we could go in the event of a breakout was over the side and into the sea, where we had prudently left the pier boat in the water.

Faced by this advancing noisy barricade, bristling with sticks and shouting men, the big red bugger retreated slowly towards us, and to shouts of encouragement was passed into our jurisdiction. Faced with a much smaller force spread over a larger area he went berserk, and rushed about, swinging his horns, and causing a general withdrawal to the safer sides of our defences. By the use of a spare gangway, which allowed for a short distance between the men and the beast, he was eventually persuaded to start down the gangway onto the ship, but stopped about halfway down, out of reach of the men on the pier.

There was talk of prodding him with the pier boat-hook – a great long pole with a hook and a sharp point on one end – but the 'cruelty man' ruled this out of order, and instead of this long-range policy elected to go out alongside on another spare gangway, and give him a bit of a prod with his electric prodder. There being no other obvious solution, we launched the spare gangway and the 'cruelty man' went down and gave him a jab with his machine.

WHAT A REACTION!

The beast gave a convulsive leap over the side of the gangway and landed on the main deck of the steamer, just opposite the ship's galley, and continued through it out onto the other side deck, leaving a mess of spilt food and a very excited cook in his train. He proceeded to rampage round the ship, terrorising members of the crew who had been taking their ease on the sunny side of the ship.

The mate took control of the situation from his safe position on the bridge and shouted instructions that no one was to go near this demon until they had set up a hurdle on the deck to divert him into the cattle hold, and then to keep out of the way, and let the drovers get on with it. While the hurdle was being set up, some discussion took place as to whose responsibility it was once the animal was actually on board, but this was settled amicably, and a couple of drovers and their dogs went down, and without any further trouble herded the beast into the hold.

After that it was pretty easy stuff, and, apart from clearing up the mess and removing our barricades, the pier was open for normal traffic, and we could get on with our work.

CHAPTER 55
The first Regatta

Each year there was a meeting of the local boats, and a sailing match was held from the harbour on a course set outside off the island. This event was open to all local and visiting boats, and was keenly contested. Most houses on the island owned a boat, although many of them had fallen into disrepair, but there was a profusion of old sails about, and one of the men suggested that we should enter the pier boat, and he could probably arrange to borrow a sail for us. When asked if he would like to sail with us he remarked that he had his own boat already. (I realised later that this was a reply allegedly attributed to the Macneil of Barra, when invited by Noah to join him in his Ark.)

The 'resident engineer' entered into the spirit of the event, and we collected the sails and tried them out on the pier boat. This vessel had originally been a puffer's lifeboat, and was still fitted out with eight oars, buoyancy tanks, water cask, etc., all as required by the Board of Trade. It also weighed about a ton, having been covered with a coat of thick tarry paint every year of its existence.

Undeterred by these handicaps we set out to find the best sails, and practise the operation of the dipping lugsail. In principle, the rig comprised a mast, which fitted through a hole in the forward thwart onto the keelson, and a sail attached to a long yard that was hoisted by a tackle attached to one side or other of the boat and served also as the support for the mast. The tack, or lower forward corner of the sail, was attached by a hook to a ring set into the bows. This is a very efficient rig, having few moving parts, and is self-regulating; if anything breaks the sail falls down. The snag is that to change tacks the sail has to be dropped, the yard passed round to the other side, the tackle unhooked and transferred to the other side, and the tack secured again before hoisting and resetting the sail. We practised this manoeuvre until we had it down to a fine art, as we were well aware that our boat was much heavier than the others, and only by smart drill could we expect to put up any kind of a show. Also, we needed a lot of wind.

Came the day. We had to sail round to the harbour for the inspection before the race, to see if we were carrying any illegal aids, and we sailed out to the start.

We had a good race, and managed not to finish last, which was a bit of a surprise to some of the locals. The event was such a success that a second meeting was arranged for the following weekend, and one elderly spectator had been so impressed by our performance that he offered us his old boat, which had been noted as a flyer in her day.

The following weekend the old boat was brought round to the harbour for the inspection, and after a couple of drinks we boarded and prepared to set out. Most of the other boats elected to row out to the start but, encouraged by our previous performance, and perhaps influenced by the drink, we decided to leave under sail, watched by a knowledgeable gallery.

We hoisted and set out at great speed and prepared to tack. Our previous experience with the pier boat was that we could put the helm down, run up the side of the boat, undo the tackle, unhook the tack of the sail, swing the yard through, then run down the other side to reconnect the tackle and hoist the sail. This is fine if the boat weighs a ton, but not appropriate on such a lightweight vessel. At the first jump the boat tipped over, and water came in over the side. We both then jumped up onto the other side, and more water came in there. The next move was to jump into the middle of the boat and let the violent rocking diminish, and then bail her out, while we worked out a better system of tacking.

Having decided on a revised drill, we hoisted the sail again, but we were much too enthusiastic for our elderly craft, and the ring in the bow pulled out, making the sail quite uncontrollable, and we had to retire back to the beach, explain away our failure to the owner, and drown our sorrows in the hotel bar.

Having participated in competitive sailing again, we resolved to have our own boat for next year's event, and began looking for a design that would be safe for the conditions off the island, and small enough to build in our garage. We settled on the Osprey, a 17-foot centreboarder, with plenty of built-in buoyancy, and ordered up the plans, fittings and materials to be ready for a start on the construction early in the spring.

CHAPTER 56
The Geese

One morning the mechanic came in a little bit late, and apologised, saying that the geese were in, and he had been shooting and got a bit carried away with the excitement. Further enquiries disclosed that shooting geese was a regular seasonal pastime, and provided a supplement to the diet of many of the population. I was asked if I would like to come along, and advised it would mean an early start, and some time standing about in the cold. Not unduly discouraged, I arranged to borrow a gun, bought some cartridges, and prepared for an early start the following morning.

This was still early in the season, and there had not been a regular pattern discerned as to where the geese would come in, so it was rather a matter of chance as to whether the expedition would be successful. But, anyway, we set out at about 4:30 on a cold and windy morning, still black dark, and parked the car in the lee of a farm building out of sight of the road, then walked quietly down to the beach, about half a mile away, and took station in the shelter of a dry stone dyke running along the edge of the beach.

After a short time had elapsed we became aware of others on each side, and quiet questions were being asked along the line. "Is that you, Donald?" "Aye, is that yourself, Hector?" Having established that the keeper was not in the immediate vicinity, we continued our vigil. We could hear the honking of the geese as they approached, and suddenly they were visible, flying fairly high, and just discernible in the faint light of the early dawn. This was encouraging; at least there were geese about, although that particular flock was out of range.

A few minutes later a second wave came in, much lower, and coming directly towards us, and suddenly the whole coast exploded. There must have been about 30 guns firing, and birds were tumbling out of the sky and falling in the field behind us. For a moment nobody moved, and then there was a rush to claim the fallen birds, and clear them from view before the next flight came in. By this time it had become much lighter, and we could see who our companions were, and then the next wave came in and we were all firing again.

It was then about seven o'clock, and as some of the men had to go to work the positions were abandoned, the spoils shared out in what was judged to be an equitable manner, and the parties dispersed.

I took home one large bird, and it was hung up in the rafters of the garage for a few days, so that I could pluck it at the weekend. On the Saturday morning I set to work and by lunchtime the work was completed. Unless you have actually tried it, you don't realise that pulling feathers off a goose is pretty hard work, and there are a lot more of them than you might at first imagine. The deed was done, however, so I tied the plucked carcass back up into the roof of the garage and went in for lunch.

The next thing we noticed was the farm dog walking down the road in front of the house with something black flapping from his mouth. I rushed out and saw that it was a part of the foot of a goose, and when I went into the garage there it was, as they say, gone. The wretched dog must have jumped up onto the bench, and from there leapt out at the goose. All that was left was a broken piece of string dangling from the rafters, and a sack full of feathers from my morning's work.

The shooting continued for some time, and we had a few more geese, but from then on I took them to the site, and had the feathers burnt off by the oxyacetylene burning and welding torch, and then they were taken home and kept in the house.

As a bit of a joke, one of the men had shot a cormorant, and had it plucked and its head chopped off, and presented it to me. All unsuspecting, I passed it on to Jean, who looked up some special recipe for roast duck in her cookery book, and spent a Saturday morning in the preparation of this delicacy. When served up in the evening it was quite awful, tasting and smelling of decomposed fish, and the whole meal had to be discarded. This caused further trouble, as we had neglected to stock up on any other food for the weekend, and as all the shops were shut by then, and would remain closed over the Sabbath, it looked like a hungry weekend ahead. We broke out the piggy bank and went out to dinner at the hotel, however, and managed to survive until the Monday.

CHAPTER 57
Visitors

Tiree is a very sunny place, and ideal for holidays, lying as it does well out in the Western Approaches, under the alleged influence of the Gulf Stream. During our period on the island we were visited by many friends and relations, and had some entertaining experiences.

Jean's parents visited in the summertime, and after an uneventful journey to Oban came across on one of the stormiest days we had seen. We were waiting for them on the pier in a whole gale of wind, and the steamer made three attempts to come alongside before finally succeeding. Later, Jean's father said he was prepared to jump if they hadn't made it the last time, rather than go on to Barra, and try again on the way back the next morning. We took them back to the house to settle down, and to explain about the water supply system.

A couple of days later I had gone off to work, Jean and her mother had gone shopping, and her father was left alone in the house. It was a fine sunny morning, so he took a deckchair out into the garden and settled down to read the paper.

Unfortunately, he had failed to shut the gate, and when he awoke some time later he found himself surrounded by large black cattle. Unused as he was to such large and heavy-breathing creatures in such close proximity, he sat petrified until someone came along and shooed them away. After that, he always made sure to close the gate properly.

We were troubled by these same beasts some time later, when we were giving a small party in the house. It was a very stormy night, with wild showers of heavy rain, and continuous low cloud. When the time came for our guests to depart, someone went out to bring their car up to the door, and came back in again as quickly as he had stepped out. He claimed that something large and black had rushed past him as he stood in the porch. We all thought this was very amusing, and theorised that it might have been a ghost, or perhaps a bit of black tarpaulin blowing past in the wind – but, anyway, he shouldn't be so silly. Reluctantly he went out again, this time with a torch in his hand. While this sortie was in progress, I had gone out of the back door to get some more coal for the fire, and disturbed a large black cow that had been sheltering in the lee of the house, and which went blundering round the side, to confront our friend.

You can well imagine how difficult it is to see a black cow when you have just come out of the light, and he was scared stiff once more as it went past him. Just to convince him that it had not been something supernatural, we went out of the back door and found another two cows sheltering from the storm, and managed to persuade them to walk round the house, past the front door, as exhibits B and C.

Amazing what drink can do for one's confidence.

During the snipe-shooting season there were a number of shoots that came specially to Tiree, where the snipe were reported to be particularly difficult, and these parties could be found all over the island, accompanied by the keeper and a party of beaters and pickers-up. One day I had come home to lunch to find a line of parked cars on the road, and a jolly party in progress in the garage, out of the rain. Jean informed me that the keeper had called and asked for permission for the gentry to make use of the garage for lunch, and she was a bit bothered about not having invited them into the house for a dram. As there were half a dozen of them plus their entourage, this was probably just as well, as they would soon have drunk us dry. They were soon off again, however, and left us not a bird.

There were always a number of Highland cattle wandering about on the machair, and as the majority of the roads were unfenced they often remained on the tarmac, enjoying the heat being radiated up from the surface. There was one particularly large animal called Sheba, who lived nearby and was often out on the roads in the summertime. She was very soft and gentle, but had an enormous spread of horns. I came along one day and found two young ladies with bicycles stopped some distance from Sheba, who, as usual, was standing in the middle of the road. Not wanting to get out of the car, I said that they should just give her a push and go past, or if they were really troubled make a slight diversion into the grass. This they declined to do, so I had to get out and move her myself. There are said to be special words for moving cattle, such as 'kush, kush', although I had never actually heard these words in use at the cattle sales. Anyway, Sheba responded to a gentle push, and soft murmurings such as "Get off the road, you great hairy monster," and the road was cleared like magic.

Tiree is a very mixed-up place geologically, and at the end of the spring term numerous students from the universities came to undertake their studies for their field papers in geology. As we were practically the first people they met on the pier they often enquired about the best places to report on, and the best lodgings nearby, so we became quite a students' directory before the summer was out.

One chap came as a student of folk singing, and came into our office while his recording equipment was being unloaded from the steamer. He told us that he was looking out for elderly people who could remember old songs and stories. The office boy, who was present during this conversation, was able to suggest a number of names for him. As soon as he had gone off, I suggested to Willie that he had better get around the island and warn the populace of this man's coming, so that they could arrange a suitable fee structure, and some means of passing him on to the next pensioner, etc., and he departed on his motorcycle to pass the word.

I met the student some time later and he complained that there seemed to be some sort of Mafia at work, with a fixed price of half a bottle of whisky as the standard fee per story. This might not have been too bad, he remarked, but it was often the same story, over and over again. It also appeared that there was some form of underground system for suggesting to the student where he might find other older inhabitants who would have additional songs and stories, usually with the added information that: "Old so-and-so likes his dram!"

Thirsty work, all this singing and talking.

CHAPTER 58
Various Diversions

Apart from the cattle sales, the only interruptions to the work on the pier were caused by the weather, either directly by heavy seas and high winds, or indirectly by delayed deliveries of materials and equipment. There was one period of about a week when the wind speed never dropped below about 50 knots, and this caused a bit of damage and alarm in general. For one thing, the island nearly ran out of drink, and bread was beginning to be a problem, with people reverting to home baking as long as the stocks of flour and yeast lasted.

Despite its lashings, the men's hut blew over, with the tea boy inside, brewing up on the coke stove. As luck would have it, he happened to be at the other end of the building as it rolled over, and said later that he had just walked round the walls and roof as they came past.

A complete hen house was blown away, chickens and all (probably Rhode Island Reds, being the most expensive); at least, that was what the insurance claim said. There were numerous sheets of corrugated iron roofing material blowing about, and these could be very dangerous. I came across one sheet, about ten feet long, blowing across the machair. I stopped the car and watched it come to rest about 50 yards away, and decided that this was my chance to get past, but as I started to move another gust lifted the sheet, and it went swooping across the road in front of me, and dug itself into the bank by the roadside.

When this storm subsided the whole island was as clean as a whistle. Everything that could have been shifted was gone, any loose ends of rope had been flayed into long tendrils of threads, and tarpaulins that had not been properly secured had flogged themselves to ribbons. For the next few days there was a lot of work in hand, fixing additional ropes around fragile structures, replacing tarpaulins, etc., but the weather moderated and their efforts were not immediately tested.

As the spring came on the work on the pier was progressing well, and other things were happening all round the island. There was a brand new crop of lambs in the fields, the machair was covered in small yellow flowers, and various plantations of bulbs were blooming. Birds were nesting and the geese had long since departed.

Work on the new boat was also going well in the garage (and in the house as well, as I had to make up plywood panels to a length of about 20 feet to cut the individual planks, and the garage was only 18 feet long. We glued them in the garden, clamped them, and then manhandled them through the porch and up the stairs to allow the glue to set at the recommended temperature.)

Towards the end of the shipbuilding we were invited to the wedding of my sister in Glasgow, and as part of the general preparations for this event I tried on my best suit, which had been hanging in the wardrobe since arriving on the island. It did not appear to have suffered any damage from the moth, but due to all my shipbuilding activities, and other strenuous outdoor pursuits, my shoulders had increased in size to such an extent that I was unable to get my arms into the sleeves, and a new suit was required. This was put down as an additional charge on the boat-building, which caused that budget to overrun.

During this period Jean had to go to Oban to visit the dentist, and on her return passage, on a fine sunny morning, she was sitting reading on the upper deck of the steamer as it was sailing out of the Sound of Mull towards Coll. A lady approached and said, "How can you sit there reading, when you are sailing through some of the most spectacular scenery in the world?" and was most put out when told that she had been there frequently over the past few weeks, and had seen it all before.

The lobster fishing season was with us again, and we went out nearly every day, either in the early morning or after work in the evenings. We seldom drew a blank, and the chest off the pier head was filling up nicely.

One of the directors came out to see how the work was going on, and I offered him a couple of lobsters to take back with him the following morning. I spoke to Jock, who went off to organise the packing, and next morning Mr B was presented with a plywood box as he was boarding the steamer. Next day he telephoned to thank us for the lobsters, and to say how surprised both he and his wife were when they opened the box and found them alive and clattering about. Next time, he asked, could we have them ready cooked for him?

On the occasion of his next visit he enquired as to what the funny noise was that was going on behind the office, and was a bit shattered to learn that it was Jock, boiling up a pair of lobsters in a bucket, using the oxyacetylene burning and welding torch.

CHAPTER 59
Communications

During our operations on the pier we were using our crane for certain critical tasks, and we had to be certain that the electricity supply was not likely to be disturbed. The Hydroelectric Board had a resident electrician on the island, and it was our custom to check with him before we started on such critical tasks. At that time there were no mobile telephones, nor private radio networks, so to find the electrician we called up the services of the post office, and asked if they knew where Mr Hutchinson was.

Sometimes he would have left messages as to where he was going, and, if so, we would be put through straightaway, but often he would not have left any notice, and we could hear the operator calling up people at strategic crossroads on the road network, and we never failed to track him down.

This was what is called an 'operator-friendly service'.

From time to time there were special events organised, mainly for local charities, and on one occasion we had attended a dance in the village hall. During a short pause in the proceedings, Willie, the tea boy, came over with a small problem. The steel-fixer had perhaps had a little too much to drink, and was sitting, fast asleep, outside on the edge of the pier approach. I was asked if I would authorise the use of one of the mechanical barrows from the pier to take him home, as he might fall into the water, or get lost, or something.

I said, "Yes, of course, but remember that these machines are not licensed for the roads, so be careful and keep to the grass; and also, as there are no springs, make sure his head is protected."

I came out some time later and both parties seemed to have disappeared, and I assumed that all was well and that Willie had collected his friend and taken him home.

In the morning there was a certain air of lethargy about the place, and I observed to Willie that the steel-fixer had not booked into work, and asked if he had taken him back to his lodgings last night.

"Oh dear," he replied; "when I brought the barrow down from the pier head he was not to be seen, and I thought he had walked off home under his own steam."

I thought about how soundly he had been out the previous evening, and became somewhat concerned about his welfare. "Willie, get on your bicycle and go round to his digs, and see if he is all right."

He came back shortly to announce that the steel-fixer had not been at home that night, and his bed had not been slept in.

"You had better go round all the houses nearby on the route between the hall and his digs, and see if he hasn't been taken in somewhere."

So off he went again, and came back, looking quite relieved. It seemed as if the steel-fixer had mistaken his direction, and had gone into another house, where he was found all wrapped in a big collie dog inside the porch.

He came in later and apologised to me for being late into work, but excused himself by saying that he had overheard the proposal to run him home on the barrow, and decided that he would be much safer walking on his own. "Most of these houses look the same to me, anyway."

Part of the work on the pier was to blow up a large rock on the seabed about 25 yards off the end of the new pier head, and we arranged for a diver to come to site to drill the holes and set the charges. This chap arrived with all his gear, and I explained exactly what was required, and agreed a set of signals with him.

We would go out in the boat and mark the positions of the holes with a long pipe, and he was to drill them, using the compressed air drills, and then mark them by stuffing rags down the holes. When all the holes had been drilled he would go out again and charge them with explosives, which we would pass down to him from the boat. He had a telephone in his helmet, but I thought it better for him to be in touch with the men operating his air pump, and we arranged to communicate by signals from the boat during this operation.

He completed the drilling fairly quickly, and had come in to stand under the pier before coming up his ladder (perhaps he thought it was raining upstairs?) when I noticed a few items of equipment that had been knocked off the end of the pier – a couple of hammers, some long bolts, etc. – and I went along to his telephone by the pump to get him to collect these items on his way past. No amount of shouting down the phone seemed to rouse him, and, in desperation, I took a long thin piece of reinforcing steel, stood over him, and tapped him gently on his helmet.

Well. What a commotion. I don't know what he thought it was, but it certainly produced some action. He shut his valve and blew himself up to the surface like a cork, and when he got out of his suit threatened all manner of violence to my person.

I explained, at a safe distance, that all I was trying to do was attract his attention as I was worried that he might have been in some sort of trouble, as he was not answering the phone. I suppose I should have written a note on a board, and lowered that down beside him, although perhaps the sight of a board coming down alongside would have given him just as much of a start as a gentle tap on his helmet.

Thereafter, relations were a bit strained between us.

After all this palaver we loaded the explosives and had an excellent bang, sending a column of water about 30 feet up into the air, and pretty well destroying the offending rock – but, despite this massive blast, not a single fish came up.

I reckon he was a pretty useless diver, anyway.

CHAPTER 60
The second Regatta

The work on the boat continued apace, and soon we were ready for the launching.

This entailed the construction of a trailer, first to convey the boat from the garden to the pier, and then down the beach into the water. After far-reaching searches around all the old car dumps on the island I located a pair of wheels and the front axle off an old Ford Seven, and, suitably modified, it served very well. We loaded the boat out from the garage and down to the beach without incident, and prepared for the ceremony.

The intention was to break a bottle of Babycham on the bow, and then push off for the first sail. This, we discovered, would mean that Jean would have to stand in water over her knees to get at the bow, and this was not approved.

We pulled the boat back a bit, and Jean swung the bottle. It failed to break on the bow, and left a nasty dent in my beautifully finished stem head. This would never do.

Next we tried to break it on the small metal fitting projecting through the deck for the forestay, and, again, the bottle failed to break (Babycham bottles are very strong). Eventually we rigged a piece of board with a rock tied on over the bow, and, this time, the bottle broke. There was the usual bit of fizzing and dripping, and the next problem was revealed. All the broken glass was under the boat, and liable to damage our feet – or, more likely, someone else's feet – so we had to clear up the mess before pushing off for our first sail.

We had a pleasant little trip round the bay without damaging anything, or scratching the varnish, and declared the boat fit for service, and ready for the regatta.

During the next few weeks we had a number of practice sails, and everything worked out well, nothing broke, and we lost only a minimal amount of skin and blood... There was a lot of local interest and always plenty of assistance during launching and hauling out, although some reluctance to come out sailing with us in what they considered to be one of those newfangled machines, needing a crew to sit out to keep it from turning over. (This was, of course, quite true.)

On the day of the regatta we sailed round into the harbour from the pier, and went ashore to collect the sailing instructions, and note the opposition. It was agreed that we would sail two rounds of the course for every one by the local boats – a reasonable handicap – and we prepared to go aboard again. There was a bit of side betting going on, and a few of the older hands were suggesting that we should load some large cobblestones as ballast, as they were against the principle of using the crew as a counterweight. We were able to decline politely, however, and set out for the start.

In the end it was not much contest, as our boat was so much faster and more manoeuvrable than the traditional vessels, and we finished, even over our double distance, well ahead of anyone else. Mind you, our boat was not much use for the fishing, and it would soon have worn out along the bottom if it had been pulled up and down the beach as often as some of the competition.

CHAPTER 61
The Doctor

During the construction of some sheds at the pier end I had the misfortune to drop a heavy hammer on my foot, and, despite wearing heavy boots, damaged my toe. I went along to the local doctor in the evening to see if he could do anything for me.

By then it had swollen to such an extent that I was unable to wear my shoe, and was forced to hop about with the aid of a stick. After a lengthy examination he pronounced that I had broken my toe, and would need to go to Oban to have it X-rayed. As the accident had happened on the Saturday, and the next boat out was not until Tuesday, I could see that I might have a problem.

"Apart from the X-ray, what else will they do for me in Oban?" I asked.

"Not much. They will bandage it up and tell you to come back in a couple of weeks," was his reply.

"It strikes me that this X-ray is all a bit of a nonsense, especially if I have to wait until Tuesday to go to Oban, if you can do the bandaging right here."

"Well, I suppose so," he said, and proceeded to bandage me up there and then.

"I thought you might like a couple of days in Oban."

CHAPTER 62
Closing down

As the work on the pier was drawing to a close, I made some enquiries as to my next possible assignment, and it appeared most likely that the next contract would be in Wick, but this was not due to start for a couple of months yet. Having spent the last two years under the benign influence of the Gulf Stream, the prospect of a long contract in Wick did not fill me with enthusiasm, and I began to look elsewhere to more pleasant climes.

I wrote away for a job in west Africa for the construction of a new bridge in Nigeria, and was invited down for an interview, which ended up by me being offered the job, provided I could pass the medical and satisfy some other assurance conditions.

I went along to the doctor in Harley Street, and met yet another specialist in tropical medicine, who was also an expert on snipe and wildfowl, and after rather a perfunctory examination and a long chat about Tiree he passed me out, gave me some further injections and inoculations against some obscure diseases, and gave me a bunch of marked-up certificates.

I returned to the island and began closing out all the work on the pier, making arrangements for returning to the mainland, shipping off all the plant and machinery, paying off the staff, clearing the office, closing the bank account, etc., and leaving our little house.

We were sad to be leaving, as we had had a wonderful time and met many kind and interesting people during our stay, but such is construction work, and you have to move on when the work is finished.

We stayed in Glasgow for a couple of weeks, completing all the documentation for the contract and saying goodbye to our friends in the office, then left for a couple of weeks' holiday before taking up work in London, prior to going out to Nigeria.

Our trip down to Devon was rather a slow one, as we had to tow the boat on its trailer, but it did make an additional place to carry the luggage.

The intention was that I should go off to Nigeria within a couple of weeks, and should stay in a hotel in London in the meantime, while Jean would remain in Devon until the accommodation was established in Nigeria.

PART THREE –
THE PERSIAN PERIOD, 1959-1961
General map of Iran

IRAN

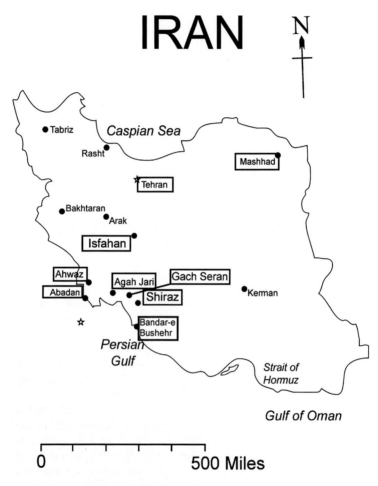

N

Tabriz

Caspian Sea

Rasht

Mashhad

Tehran

Bakhtaran

Arak

Isfahan

Ahwaz

Agah Jari

Gach Seran

Abadan

Shiraz

Kerman

Bandar-e
Bushehr

*Persian
Gulf*

*Strait of
Hormuz*

Gulf of Oman

0   500 Miles

CHAPTER 63
The London Scene

When I joined the company it was with the intention of going out to Nigeria to assist with the construction of a new bridge over the River Niger, at a place called Onitsa. I expected to be in the London office for a couple of weeks to study the drawings, comment on the construction methods, and then to assist with the selection and ordering of special items of equipment, before going out to site to make a start on setting up the temporary works.

On the day I arrived, however, there was a change in the arrangements. Another contractor had submitted an alternative design, and we were also asked to produce a design and build proposal. This, as far as I was concerned, looked like a six-month delay before I would be required on site, and I made arrangements for Jean to come up to London, where we would try to rent a flat for the interim period.

We visited many estate agents' offices and came away dismayed. What we had estimated as our maximum rent would have provided a single-room dungeon flat, probably with an outside lavatory, and we had to elevate our proposed minimum by about 100%. We went around again, and saw several more reasonable places, but as we were only looking for a short lease these were withdrawn when we made more formal enquiries.

We did find a place quite near the office in Ealing, and after protracted discussion with the agents, and a meeting with the landlord, agreed on a tenancy for a limited period.

Before moving in, we had to have the agreement registered at Somerset House, and had a 'marching in' inspection, attended by the landlord and an inventory clerk. During this inspection we noted that some of the furniture was screwed down to the floor. (The landlord observed that he did not like his tenants to be moving the stuff around.)

We also were interested in the discussion between him and the clerk as to whether the decoration on the top of the wardrobe was Greek Key or Chinese Fret pattern, but managed to assure him that we would not make off with it, and passed on to the bathroom. This had a gas geyser on the wall, with a set of handwritten operating instructions pinned up beside it (included in the inventory), and a bath along one wall, which was boarded in at the end and side (noted as mahogany trim).

In the kitchen, after we had counted all the cutlery and cooking utensils, the clerk discovered an enamelled tea tray, advertising a well-known brand of cigarettes, which, after discussion, was listed as 'tea tray, seagull pattern'.

We settled in and enjoyed living in London, with all the shops and cinemas very handy, and the office within reasonable walking distance – quite a change from Tiree.

A couple of weeks after we had arrived there was a ring at our doorbell, and presumably on all the other bells on the outside door, and I went down to see two young men from the Jehovah's Witnesses, obviously in the mood for a long discussion, and thinking they had landed a likely lad. I put them right off their stride by informing them that I had been saved already, thank you very much, and politely closed the door on them. The landlord, who lived in the ground floor and the garden flat, appeared just after I had terminated the interview and said, "Thank God they have gone. The last time they called they had me on the doorstep for half an hour. What did you say to them?" When I told him he was much amused, and said he would try the same ploy if ever they called again.

Meantime, things had changed a bit in the office. Another job had come in for a site visit for a contract to build a road between Qom and Persepolis in Iran, as a mark of respect to the Shah, who, it was hoped, would drive down, more or less along the same route as previous illustrious leaders had taken in the distant past.

It just happened that the series of inoculations I had recently undergone were exactly what was required for persons visiting Iran, and I was asked to go out, in company with another young engineer, to assess the material and supply situation that would be needed during the construction. The other chap had previously been in Iran, and went out immediately, and I followed a few days later, when all the permits etc. had been cleared through the embassy.

This was to be a joint venture bid, and I met up with another engineer from our partners, who was to form part of the route survey party, and who would study the problems of various river crossings and the general alignment. Before leaving I was given a large wad of Iranian banknotes, and had to sign my life away with the Exchange Control Office, and had my passport stamped all over, but eventually made it to the airport and made contact with my travelling companion.

We flew out, stopping at Rome (which had not improved much since my previous visit), Beirut, Damascus and then Tehran, where we were met and conducted to our hotel.

CHAPTER 64
Tehran and the Survey

The hotel was a fairly new building, full of chromium-plated, uncomfortable furniture, and with a bar that was like the inside of a coal mine. I think even the lights were black.

My friend had a message left in his room to take a taxi to an address in the town, where some other staff from his company were living, and was all set to go when he realised that he had no local money.

"I seem to have plenty," I remarked, brandishing my thick wad. "Let me lend you some of this." It was only then that I discovered that the notes were all designated in Arabic script, so we had a problem, first in deciding how much would be adequate, and also how he should describe it on my receipt, e.g. three green ones with a heart shape, upside down, etc. Help was at hand from the telephone, which had the sense to have a dual dial, and from a protracted exercise on these numbers began our education in Arabic numerals. (The numbers in most Arab countries are written in Arabic script. It is a curious arrangement that correspondence is written from right to left, but numbers are written from left to right. The numbers are based on a scale of ten units, or multiples thereof; just as well, too, or how else could you perform the simple calculations of adding, subtracting, multiplying or dividing? It does make things difficult for the typist.)

Next morning I had to attend in the office of the Immigration Department, and hand over 20 signed photographs of myself, to be distributed to all manner of officials dealing with the different provinces of the country where we would be operating.

I could not think what they might do with all these pictures, but I was reminded of a story told by the chief estimator in London concerning a visit to Liberia.

Apparently the visitors came to a provincial border, where officials demanded additional photographs before the visitors could cross, and, of course, there were some people present who had run out of their supply of pictures. The client's engineer in charge of the visit had a contingency plan ready for this, and produced a heap of photographs of earlier visitors, saying: "Here is one chap with glasses – you can have that one – and here is another of someone with no hair – how about you?" At the end of the visit, guests were invited to leave behind any spare photographs for the use of future callers.

Back to Tehran. Having sorted out the permit situation, we determined a split of responsibilities, and both set off by Iran Air for Esfahan, flying in a Vickers Viscount aircraft.

By this time I had acquired a small phrase book, produced by the Anglo Iranian Oil Company, and together with my elementary knowledge of the numerals at least I could find my hotel room, and, if I was speaking to English-speaking Iranians, managed to get by.

My colleague had started on the part of the study concerned with local plant and equipment, and relations with local contractors; I had to go down the route and find suitable sources of water (both for concrete and general services), rock quarries, fill material, possible sites for construction camps, access from existing services, as well as looking out for possible problem areas, such as flood plains, or narrow gorges where snow and rain might restrict the programme.

Armed with my phrase book, I went down to a local garage, where the hotel had suggested that I might be able to hire a car and driver, and after a protracted discussion came to terms, on the understanding that the vehicle was a Land-Rover and the driver could speak English. What I had failed to agree was the extent of his English.

Next morning the driver reported to the hotel, and said, very clearly, "Good morning, sahib. My name is Mohamed, and I am your driver." So far so good.

It turned out that this was practically the full extent of his linguistic ability, but the car was in pretty good condition, and probably one of the very few suitable vehicles in town, so I collected my bags and we set off.

Esfahan is a very old city, at an altitude of about 3,000 feet, with wide, sandy deserts all around. There is quite a large river through the town, and a very famous multi-arched bridge, which I stopped to admire. This was a mistake. Within a very few minutes I was the centre of a throng of beggars, and the police came along too, enquiring what I was about. After distributing some baksheesh, and smiling my way past the police, we were on our way again, out onto the open desert. As we went along I made cryptic notes in my log concerning gravel banks and possible water point locations, and we made frequent diversions into the hills on either side of the road.

The water supply to the villages in the plains was very interesting. From each village there were lines of small mounds of earth, each line going towards the hills, sometimes several miles distant. On examination, these mounds were the arisings from a series of vertical shafts and connecting tunnels, and the hill water was running, sometimes at a depth of as much as 20 feet below the general surface. This system, known as quanats, had been in use for centuries, and in most cases was still operational, though rather lacking in maintenance.

There were two small cities between Esfahan and Abadeh, where I hoped to spend the first night of the tour, and we stopped in an interesting sort of café in Sharia for lunch. I gave the driver some money and told him to be back in half an hour, while I had lunch, and when he returned he brought along some potential passengers for the next stage. They were a pretty fierce-looking bunch, dressed in dirty turbans and long dusty coats, and I tried to explain that we would be making frequent diversions into the hills, and it would not be convenient to have them along. The owner of the café seemed to get my message, and told them to clear off.

The next stage was very similar: long, rocky desert plains, with a few scrub bushes from time to time, little villages with broken-down mud walls around the fields, and generally a depressing air of abject poverty and hopelessness. We stopped at the little town of Abadeh and booked into the hotel (?).

This was a large café-cum-tea house at the end of the town, in quite a pleasant street, with shops and some trees. It was quite pleasant, that is, until traffic came past, when the dust blotted everything out, and then slowly settled all around, only to be disturbed by the next vehicle. They probably kept at sufficient distance apart from one another to be driving in clear air.

The café part was quite attractive, with plenty of local wine available, and not bad local beer, but the accommodation was something else.

There were two large rooms at the rear, in the garden, each containing about a dozen beds. These were string charpoys – a sort of loose net thing, covered by a coloured blanket, with nominal mosquito netting over each bed. The washing and toilet facilities comprised a standpipe in another shed, and a couple of holes in the floor. Definitely 'bring your own paper' variety.

After what I thought had been a reasonable dinner I retired to my bed, and woke up in the middle of the night surrounded by sleeping and snoring bodies, and with the most violent stomach pains, and the running habjabs. I made my painful way out to the hole in the ground, and squatted there for a long time, regardless of the insect population, before crawling back to my bed.

I was no better in the morning, but managed to swallow some special bung-up pills, and went back to bed, where I remained for the next day and a half – without food, and only small sips of wine. I was up, but certainly not fit for the road for another two days, and then made a miraculous recovery, called out the driver, and we set off again. For all I knew, he had been running a private taxi service in the town during my sickness, but at least he was still there, and the car appeared undamaged.

On the next leg of the journey we met up with one of the line survey parties, who were living in an army fort – a square blockhouse, with beds on the roof – and I spent a night with them, sleeping under the stars after a good supper, washed down with some of the local brandy. Next day was a holiday, and we had a donkey derby in a field about two miles away, on donkeys on hire to the survey party. All they wanted to do was go home, and it was very difficult trying to keep them headed round the course. This spectacle drew a large crowd from the local village, and towards the end of the performance most of the riders had been bumped off, and the donkeys were trotting back to their homes, leaving us to walk back. There was a bit of a knees-up in the village that evening, with a three-note flute and a drum, but somehow none of us managed to get the hang of it, and we retired to our fort for an early night.

Next day we moved on towards Shiraz, the next main town on the route. The ground was a bit higher on this section, with snow visible on the tops of the mountains, and the road running in steep, narrow gullies through the hills. The vegetation was still pretty sparse, and we could see the parties of the migrating tribes preparing to move down to the coastal plains for the winter.

We travelled on, and came into Shiraz in the early evening, and booked into a local hotel. This time it was much more opulent than

Abadeh, with proper individual bedrooms with integral baths and toilets. What luxury.

In the hotel garden there was a large ornamental pool, with many carp swimming about. The drawback was that there was an army of cats perched all round the edge of the pool, yowling and mewing as they waited for a fish to come within reach – a noise that continued all night.

There was quite a lot of work in progress in Shiraz being undertaken by European contractors, and some of the staff came into the hotel for a drink. Recognising a fellow traveller, they invited me to join them, and after a meal we returned to their billet in an old house in the town. When they heard that I would be going down to Bushehr, a small town on the coast, they suggested that I should look up their office there, and see if they could be of any assistance. At the end of the evening I was loaded into a horse-drawn barouche and taken home to the hotel, where the cats were still at it.

Next morning we were off again on the road to Bushehr, and after travelling a couple of miles out of the town we were stopped by a soldier carrying a rifle. He announced that he would be joining us, and we would take him to a small village about 20 miles out of our way, and climbed into the back. It all seemed a bit organised, and I suspected the driver had been persuaded to fall in with his wishes as the result of some previous deal, but I did not think it was a good idea to argue, and we continued our journey in a sort of strained silence.

The road descended from the mountains in a fairly easy grade, and then suddenly plunged over an escarpment. At one place, we could see seven levels of road below us, as it twisted and looped down the slope.

We had been passing groups of tribesmen on the road, travelling on foot and on horseback, and driving their sheep and goats before them, and when it came to the escarpment they gave no heed to the road but came straight down the hill, creating great clouds of dust, and a bit of a hazard from falling rocks. We had to stop several times to allow family groups to go past, with children and sometimes dogs and chickens mounted on their ponies. We still had our rifleman in attendance in the back, and he and the driver shouted out encouragement or admiration to the passing hordes.

At the foot of the escarpment the soldier indicated that we should turn off the main road (?) and take a dusty and ill-marked track across the foothills, and by a series of questions made up from my phrase book, and a bit of sign language, together with my growing comprehension of the Arabic numerals, I made out that it was about 15 miles off our route; but, when we arrived, I would be expected to have tea with his parents.

Some miles later, and still on the bumpy and dusty track, we arrived at his village, and caused quite a stir. He stood up in the back of the vehicle and waved his rifle over his head, and soon we were at the head of a cavalcade of local children, and drew to a halt outside a small tea shop. His parents came out and were obviously surprised and pleased to see him, and we were ushered into a shady courtyard, where a small meal was rapidly prepared of fruit and salad, together with a bottle of wine and some coffee.

Conversation was difficult, but it was all very friendly, and from somewhere in the village an old man was produced who could speak English, and we had a few words of formal welcome before we made our excuses and set off back to join our original route.

By the time we arrived in Bushehr it was getting dark. I made enquiries about a hotel, and was directed to a pretty miserable-looking building near the harbour. As it was the only place available, I went in to book a room, and met with an Indian engineer, a colleague of the people I had met in Shiraz. They had sent a signal down to their office in Bushehr to look out for me, and took me off to their mess in a large house on the outskirts of the town, much to my relief, as I must confess that I did not fancy the hotel at all.

I gave my driver some money and sent him back to town, with instructions to come back at 7:30 in the morning (or, at least, that was what I hoped I had said), and settled into my room in the mess.

The house was a large, square block, set in spacious gardens with a high wall all round, and built into a courtyard. On the ground floor were the public rooms, kitchens, and offices, while upstairs each room had large open verandas on each side – one to the outside of the building, the other to the courtyard. After a bath in a king-sized bath I went down to dinner and met the other residents. They were supervising the unloading of materials for their contract, and arranging for the overland transport back to Shiraz, and had been there about a year. The mess was well run, and they were a pleasant

197

company, and after dinner we went up and sat on the roof under the stars. What a pleasant end to a long and dusty day.

In the morning I was awoken with a pot of tea, and went down to breakfast much refreshed. My driver reported at 7:30 as ordered, and we went down to the harbour, to continue my investigations. The harbour was really only a landing place for small flat-bottomed barges, and all the ships unloaded about two miles away, with the cargoes being lightered to the shore.

As this system had been used with some success for our friends in Shiraz, unloading all their heavy equipment there, I thought that we could manage in the same way.

The town itself was a strange mixture of Arab and 19th-century Western commercial architecture. Round the harbour were the usual offices for the customs, and various shipping agents. Most of these were in tall buildings with open courtyards within, and many shuttered windows. All had magnificent timber doors, with intricate carving – mostly, alas, in a bad state of repair.

On the other side of the harbour were a number of run-down warehouses, most of them out of use and falling down from neglect, and all round them were miserable fishermen's cottages, and the homes of poor dockers and other local workers.

Farther back from the waterfront there were numerous fine houses in walled gardens – the residences of wealthy merchants and traders, and some of the local officials.

I spent a couple of days sounding out the possibilities of the harbour being improved, and spoke to the various officials and agents concerning tariffs and capacities of their organisations, then took some time to write up my report on the facilities available (and also to civilise my report on the state of the roads in general) and the overall impressions of potential sites for camps, services, quarries and sources of local material along the proposed route.

With that work completed I arranged to fly back to Shiraz, where I met up with my colleague, and we set out together to survey the last part of the route, to Persepolis, about 30 miles away.

Persepolis itself is an archaeological site of some considerable interest, and there was quite a good hotel there, with a small town of artisans and excavators nearby.

The monuments had been sacked at the time of Alexander the Great, but they were of such massive construction that most of the enormous columns were still intact, although many of them had been knocked down by a series of earthquakes, and lay scattered on the ground. The whole place was kept in a tidy and clean condition, and the resident curator and his staff were very obliging, and most instructive.

When we told them that we were surveying for a road to bring the Shah down from Qom on the 2,000th anniversary of the founding of the Persian Empire, he was most impressed, and said he would need to improve the conditions in the hotel before that event. We did not disillusion him with the prospect of a workman's camp on his site for a couple of years before the formal opening.

Our next appointment was with our joint venture colleagues, and we flew off to Esfahan for a meeting there, and then returned to Tehran, where we had further meetings before flying back to London.

We had a bit of a farewell dinner before leaving, and I was put aboard an American Airlines aircraft in the very early morning. Within a very few minutes I was fast asleep, and was rudely awakened by the hostess some time later, saying we were about to land at Damascus, and enquiring if I would like some coffee and a huge wad of chocolate cake. There was no reasonable way to say 'no' to that question.

We landed at Damascus, Beirut, Rome, Frankfurt and eventually London, with meals offered in flight and on the ground, and because of the time shift as we flew west we were quite disorientated by the time we landed.

Back in the office next day we tried to sort out all our impressions, and spent three weeks making out our formal presentation, with cost and programme implications.

During this exercise I was developing the programme for excavating equipment for the quarries, and telephoned various companies asking for information, delivery dates and prices of their equipment. Many of the people I spoke to were not very enthusiastic, and quoted figures directly from their sales sheets, but one chap, who took a bit more interest when advised that we would be looking for about 20 large diggers, replied: "Don't move from your office, I am on my way over," and appeared within the hour, loaded down with booklets, performance tables, factory production schedules and his managing director. These were the machines we selected for our proposals.

Meanwhile the design and planning work continued on the Nigerian bridge, and I rejoined the department dealing with this proposal.

During this time we were awarded a contract for the construction of an airfield at Agah Jari, in south Iran, and as I was fully inoculated and vaccinated, and also was now recognised as an expert on Iran (I still had my phrase book), I was asked to go out and take charge of part of this contract.

CHAPTER 65
The second Visit

For about two weeks I slaved in the London office, sorting out drawings, making up programmes, listing out equipment, setting up the shipping programmes, clearing with the embassy the necessary permits to work, collecting letters of credit for the bank, and all the hundred and one things needed to get the show on the road, and eventually had a long weekend leave, cleared the flat, took Jean down to Devon, sold the car, and travelled back to London on the train.

I called in at the office on the way to the airfield, and they gave me a heap of papers and said there was a parcel waiting for me in the lobby. It turned out that this parcel was a dozen spare starter motors for our transport, which had somehow been mislaid, and which weighed about 20 pounds each. I had quite a problem getting them all into a taxi, and could foresee all manner of bother passing them through the customs, at both ends. As time was short, however, there was no trouble at Heathrow, and the boxes were loaded as passenger baggage.

This time I flew to Abadan, and arrived in the early evening. Since Abadan was one of the major oil ports of entry, and as our contract was with the oil company, the sight of incoming passengers bearing strange parcels of a technical nature was not uncommon, and after stamping everything in sight I was allowed through with no excess to pay.

I had been booked into the oil company rest house, and took a taxi from the airport entrance. The driver looked askance at my load of starters but loaded the lot, and we drove round to the rest house. This was another demon driver; I always seem to pick them. At least he was wearing shoes, but before setting off he indulged in a few words of prayer, passing his beads rapidly through his hands. Then, turning the radio to full blast, we set off at great speed through the suburbs of the town.

The main feature of Abadan was the smell from the refinery. It was all-pervasive and most unpleasant. I am sure it would be considered a health hazard in these more enlightened times, but after a few hours it did not seem too bad (a bit of overkill to the sense organs in the nose, perhaps).

The rest house was a fine, low building set in pleasant gardens, with many flowering shrubs and small trees. The accommodation was pretty good by any standards, with cool rooms and shaded verandas, and comfortable public rooms. There was a message waiting for me in reception to the effect that I would be collected the following morning, and could expect to drive out to the site at Agah Jari, across the desert – a distance of about 100 miles.

As a result of my previous experience of desert driving, I dressed in my working clothes the following day, and was picked up as arranged, and we set off.

Except in the rainy season, access across the desert is fairly simple. There are a number of well-defined tracks, going in the generally correct direction, which open out and converge again between pipeline crossings. There were a series of oil pipelines running across our route about every five or six miles, and the crossings provided were simple mounds of earth built up over the pipe.

Individual vehicles were making their way across the desert in both directions, and it was a matter of some chance as to who got to the crossing first. This was quite important, as if you fell in behind another vehicle you had to travel in his dust cloud until you could chance overtaking him, or branch out onto another track.

Travelling in a dust cloud made it difficult to see if there was any approaching traffic, and nobody liked to be overtaken, first, as a matter of pride, and second, they would then be in the dust. Another small problem was that the pipelines obscured the field of view ahead, and it could be alarming, to say the least, to meet a large lorry coming the other way over the crossing having had no previous knowledge of its existence.

Despite all this excitement, we reached Agah Jari and drove up to our temporary accommodation, in one of the oil company's senior staff bungalows, on a slight escarpment overlooking the plain to the Gulf coast, about 40 miles away. These houses were very comfortably furnished, and we were provided with resident servants, who looked after us quite well until we could move into our own camp accommodation on the plain below.

When I arrived a start had been made on the construction of our compound; the floors were already laid, and the prefabricated sections of the houses were on site.

These houses were essentially square boxes, with varying arrangements of the rooms and facilities, and the addition of a corrugated iron upper roof as a means of providing some shade, and a roof space for the breeze to blow through. When completed, they were quite comfortable, if a bit austere.

Meanwhile the offices were being assembled, power and water laid on, and a start made on the stores and workshops buildings out on the actual site, about half a mile out into the desert.

During this early phase of the construction I was approached by a wizened old man, who craved my indulgence and begged permission to set up what he referred to as 'a small hotel' at the end of our access road. He would provide tea and drinks for the men, and would also erect another building for our staff.

This seemed a good idea, and showed commendable enterprise on his behalf, so I agreed, and next day we found a thatched roof structure on bare poles on one side of the road, and another, with dwarf walls about a foot high, on the other.

He had some sort of boiler going, and was offering tea to all the staff for free, but charging the men some extortionate rate for the same beverage. I shudder to think where he drew the water from, and I must say I confined my patronage to bottled drinks, but, even then, how did he wash the utensils? We came to some arrangement about paying for the staff drinks, a matter that caused him some surprise, but it was felt that we might have some better control over his activities if we were actually paying for our own drinks. No doubt he was having to pay someone else to continue his franchise there, anyway.

He continued his operations there throughout the duration of the contract, and we never had any trouble from him, or his clients, nor was anyone taken off sick as a result of his elementary hygiene.

The work on our houses was soon completed, and we moved down. As expected, we were overrun by applicants as cooks, houseboys, garden boys, and others looking for an easy number. Each man had a packet of testimonials, written in bazaar English (sometimes bizarre English, too), extolling his virtues and skills, and each household had to make their own selection.

It had been arranged that there would be married accommodation for seven families, and the remainder of the expatriate staff would live in a communal mess, with individual rooms in a building nearby. The whole campsite was laid out along the line of a road, with the generator at one end and the blocks for the mess and the houses at the other. There was a small front garden for each building, and at the rear the desert had been marked off about 20 yards out, with posts cut from old boiler pipes, and heavy industrial wire mesh. I believe the hope was that later occupants of these buildings might eventually make reasonable gardens there.

In other oil company houses there was normally a small plot behind each house, and the residents did grow a few vegetables, or at least provide a bit of shade with shrubs and bushes.

Part of our construction programme was to sow grass seed and encourage watering in these plots – mainly, I think, as a sort of trap for the ever-present blowing sand from the desert outside the fence.

As the work was progressing well, and the time for moving into the accommodation was rapidly approaching, I was asked to go off and see about the furniture and equipment for the camp. The nearest large town was Ahvaz, and I arranged an expedition with an English-speaking clerk and my Land-Rover driver to try to set up a series of contracts for the supply and delivery of these items.

CHAPTER 66
Setting up Camp at Agah Jari

We set out for Ahvaz in the early morning, driving first on oil-bound roads, and then across the desert. Ahvaz was a large market town, and was enjoying something of a boom in the construction business. There was work in hand on a brickworks, new offices and factories were being erected, and there was considerable work in progress for the oil company.

We arrived at about lunchtime after a fairly uneventful journey, and stopped at the Swan Hotel, a place with some delusions of grandeur, although the food was quite passable.

After the obligatory siesta we set out into the artisans' area to seek out suitable furniture-builders (there were no department-type stores in town – anything you wanted you had to have made to order), and we visited a number of workshops before making a fuller investigation of one particular place.

We were shown various booklets of sketches of typical units of furniture, and even a few photographs of completed items, and settled on a middle range, somewhere between crude and overdeveloped, and then began to discuss the proposed delivery.

The man had no idea of the problems of producing a complete range for seven houses and a bachelor's mess furniture within a limited period, and was waffling on about one house a fortnight, but even that would be very difficult, etc., etc. I then asked if I could see his workshop and factory, to see if I could suggest some method of improving his production output, and we went in through the back of his shop into a large shed, with half a dozen men squatting on the floor, sawing and planing away at assorted bits of stick, while round the walls stood some antiquated pieces of woodworking equipment, all belt-driven from a series of shafts hanging from the ceiling, and a very elderly electric motor, covered with dust and cobwebs, in one corner.

As this had been the most promising shop in our investigations I began to despair of our ever getting any furniture, but, as a result of some harsh words from the clerk, we managed to produce some action, and the machines were all cleared of their clutter of odd lengths of wood and demonstrated to be in running order, and it also appeared that at least two of his staff knew something about operating them.

After some delicate negotiations I was able to suggest that I could assist in rearranging his shop, so that there would be some semblance of continuity of work, and a reasonable flow of products from one process to another.

For one example, I pointed out that his saw was situated against a wall, and in order to cut long lengths of timber into more convenient sizes his men had to go outside the building and poke the timber through a hole in the wall, at the same time blocking access into another shop next door.

We gave him a provisional order for one set of house furniture complete, to be ready in a week's time, and he agreed that he would alter the arrangement of his machinery as I had suggested, and would then guarantee to complete the remainder of the order in a further four weeks. There was no hope of such a promise being fulfilled, but I reckoned that with regular provocation we would be able to cope with the arrival of families, and the completion of the camp buildings.

We visited a number of other shops and stores in the town for such items as crockery and cutlery, utensils, linen, blankets and all that sort of stuff, and I rather made my name with the clerk when we were bargaining for some sheepskin rugs, which I thought would be a comfort when stepping out of bed onto a cold concrete floor.

The currency in Iran was the rial, of which there were about 500 to the pound, and also, to reduce the number of zeros attached to each price, there was another unit of a tuman, equal to ten rials. The negotiations had gone on for some time, and I agreed a price in rials which I thought to be not unreasonable, but when I produced the wad of notes, and counted them out, it transpired that we had been at cross-purposes and that the price he had been quoting was in tumans, i.e. ten times what I had been counting. I told the clerk to tell the merchant that, for that price, I would be wanting the meat and the horns as well.

This gained me several points in the esteem of my driver and the clerk, but we did not get any further trade in that shop.

By this time it was quite late, and as we had been advised against travelling back in the dark we returned to the Swan Hotel, and slept in comparative comfort. In the morning we paid another visit to our furniture-maker and gave him a few prods, and then, collecting such items as we could carry, returned to camp.

I made weekly visits to Ahvaz to maintain pressure on our suppliers, and became well known on the ferry across the river at Khalafabad. The river was about 200 feet wide, even in the dry season, and the ferry was run by the local authority, but maintained by the oil company. It was a large barge, which hung on two heavy wire cables suspended across the river and was pulled back and forth by an engine mounted on the side of the barge.

We came down one morning to find a long queue of vehicles waiting for the ferry, and the driver and I walked down to the river's edge to see what the trouble was.

The ferry was on the other side of the river, and it appeared that the gearbox for the winding engine was out of service. My driver explained that I was a most important person, and needed to get across immediately, and somehow conned someone to take us across on a small rowing boat to see what the problem was.

After the now standard explanation of my status, we were invited into the engine room, to see for ourselves what the damage was. The cover for the gear lever box was loose, and the actual lever itself was broken, but I could see the end of it inside, and was sure we could bodge it up somehow. After some negotiation with the captain, he agreed that he would take us across if we could get him going, provided that we would take him back to the other side afterwards, as that was whence the mechanic would come (?).

We were ferried back in the rowing boat, brought the Land-Rover down to the head of the queue, and took the toolbox across in the rowing boat again. After a few minutes poking about we managed to select a gear, and chugged across to the other side, where we loaded the Land-Rover and a few lorries and set off back again, using the jack handle as an operating lever.

When we got ashore on the far bank, I told my driver to go back across and see them safely landed on the other bank, and return, with the jack handle, in the rowing boat. This message must have been misunderstood, as no sooner had he set the ferry in motion than he came leaping out onto the ramp and jumped ashore, waving the jack handle in a sort of salute. I had visions of the barge pulling itself out of the water on the other side, still stuck in gear, but as there was not much we could do about it then we continued on our journey into the town.

When we returned in the evening, I could see that the ferry was operating as usual, and that there was another captain in charge. The story must have gone around, as we were greeted by shouts of "Dashti Jack, Sahib" (Mr Jack Handle), and I was known by this name at the ferry for some time to come. It appeared that the captain had stopped the engine while still in gear when he had arrived at the other bank, and no harm had been done.

CHAPTER 67
The Sandstorm

The ladies had arrived, and most people had settled into their houses with the rather strange furniture and odd selection of utensils and crockery. Even the gardens began to indicate some response to the tender loving care lavished upon them, with distinct impressions of green grass growing through the sand.

The houses faced out onto the desert, where, about a mile away, it was just possible to see our machines working, and the arisings from the perimeter ditch stood out, almost like mountains, against the flat horizon.

There was a football match in progress some way out between the gardens and the site, and I suppose there were about 50 or so spectators, standing about and cheering on their respective teams.

From the rear of the house we could see the sky becoming blacker and blacker, and a sort of grey curtain closing in on the match. The players also became aware of this, and the game was abandoned, and everybody fled back towards the camp.

We could then hear the noise of the wind, although it was still flat calm in the gardens, and there was a concerted rush to get inside and shut all the doors and windows.

Within a few minutes the storm struck. The wind shrieked through the roof space and all the crannies of the buildings, everything went dark, the generator shut down, and the inside of the house became full of fine blowing sand. We lay down on the bed and spread towels over us like a sort of tent, and tried to filter as much dust out of our breathing as possible. After about an hour of this misery the worst of the storm passed over, and we were able to get up and move about to take stock of the damage.

The air was still full of fine dust particles, and we still had to wear handkerchiefs over our mouths, but at least we could see across the room.

Slowly the dust settled, and the camp came back to life. The first job was to restore power at the generator, where we found that a section of the roof had collapsed onto the fuse box; this was quickly sorted out, and we had light again.

Inside the house the sand and dust had got literally everywhere: inside the fridge, inside the cooker, inside the toilet cabinet in the bathroom, inside all the drawers, suitcases and boxes, and all the loose food was covered in a layer of sand.

On top of all the doors there was a small ridge of sand, which blew down when the door was opened, and generally the whole place was a mess. We worked away until dark, trying to clear the place up, and then took off to the oil company club, which had not been in the way of the storm, and had an extended supper. Then back to the camp to shake out some more dust, and eventually get to bed.

In the morning we found that the paint had been stripped completely off the sides of two lorries out on the site, and the tearooms of the small hotel had been destroyed.

A good thing this kind of storm is fairly infrequent.

CHAPTER 68
The Quarry Site

One of the first jobs required during the actual construction of the airfield was the location and development of a quarry for the stone surfacing material, and I was sent out prospecting for suitable sites. The site of the airfield was on the plain below a small escarpment, and the rain falling on the higher ground had eroded many steep-sided gullies and exposed suitable surfaces of stone. I selected one such site about five miles away, and, after clearing with the oil company the subject of the ownership of the land, began to plan the development of the site.

The valley was about half a mile long and about 100 yards wide, with steep sides rising to about 300 feet above the floor. I decided to establish a platform about 30 feet up, cut into one side of the valley, so that material could be loaded by gravity through chutes to lorries waiting below, and by some mountaineering activity set out some marks on the side to indicate where the machines would need to start levelling out to make the platform.

Next day the foreman arrived, with his large bulldozer. I pointed out the marks on the cliff, and began to discuss how he would go about getting his machine up there to cut the ledge. By this time the machine had been unloaded, and the driver set off breaking out the side of the cliff and making a ramp up to the required level. It might have been part of an act to impress me, but he worked away furiously, and very rapidly climbed his machine up to the required elevation and began to level out the platform. The whole operation looked extremely dangerous, and as I was not anxious to be called as a witness at the inquest I decided to leave things to the foreman, who seemed to know what it was all about, and left them to get on with it.

When I came back later in the afternoon the machine was sitting securely on a wide platform, and there was a substantial pile of material ready waiting for the chutes to be installed, and the following day, when they were delivered, we were in business. As some of the rock brought down by the bulldozer was in large lumps we required a compressor to work drills and pneumatic hammers, and a suitable machine was brought along and set up on the valley floor.

After a few days we went along to see how the work was progressing, and found the compressor missing. This was quite serious, as it was an expensive piece of kit, and its absence was causing some problems to the production. I called out the watchman, who had been on duty the previous night, and he was unable to assist in our enquiries, but he did remark that there had been a heavy shower of rain during the night and a bit of thunder and lightning, and he had remained inside his tent up on the elevated platform, beside the bulldozer.

This gave us a bit of a clue, and we examined the walls of the valley more carefully, and found that there was a sort of tidemark of dead grass and odd branches etc. about eight feet up the sides, which we could not remember seeing before. We followed these marks out of the gully, and they grew less marked and lower as we came out into the plain, and eventually disappeared. We continued our search out into the plain, which was cut up with many shallow wadis and covered in scrub thorn bushes, and there, in one of the wadis, we found our compressor.

This machine weighed about two tons, and had been displaced about 200 yards, and was sitting, apparently undamaged, still on its wheels, in this small depression, quite invisible from the general level of the valley floor.

We whistled down the bulldozer, made an access ramp, and pulled the machine out in quick time. Apart from some grass around the corners it seemed undamaged, but I thought it best to have it taken back to the workshops and have the oil levels and things checked, as it must have been partly submerged overnight.

Later enquiries disclosed that such floods were not very uncommon. A severe storm in the nearby mountains, where there was very little vegetation and practically no topsoil, would result in such flash floods. Maybe the watchman knew what he was about when he pitched his tent well up from the floor of the valley, but he must have been a sound sleeper not to be awoken by the raging torrent that carried our compressor away.

Anyway, he said that he knew nothing.

CHAPTER 69
The Hunting Expedition

We employed a number of local contractors to supply labour and equipment on the site, and on the occasion of some local festival they clubbed together and arranged a hunting expedition for some of the staff. We had two long-wheelbase Land-Rovers with open backs, and a series of folding chairs were set up in each vehicle, two facing one way and three the other. The driver and a guide sat in front, and two shooters stood looking forwards over the cabin. The contractors' men were all armed with long knives and revolvers stuck in their belts, and they also carried some pretty fearsome shotguns and rifles, with bandoliers of cartridges slung across their chests. They had borrowed a number of 12-bore shotguns, which appeared to be in reasonable order, and these were handed out to the remainder of our party.

There was a short briefing before we set out; the intention was to go out into the desert in line ahead, about 50 yards apart, where we might come across some wild pigs, or a flock of turkey buzzards, and – very unlikely – some deer. We would then stop for lunch at a small oasis, and after a suitable rest, and refreshment, we would come back again. We had an elementary check to see if our arms were provided with safety catches, and that they were on, and set out.

The desert there was mainly sand and pebbles, pretty flat in general level, but with frequent gullies and wadis about six to ten feet deep that were difficult to see until it was almost too late. We trundled along quite sedately for a couple of miles without disturbing anything significant, and then came on a fox.

Very unsportingly he was greeted by a barrage of gunfire, and, undamaged, he set off at a good pace, keeping just out of range but still being fired at from the two shooters over the cabin. (The remainder of us were too busy hanging on to do any firing.)

The second vehicle came up and joined the fray, again firing over the front, and then, with great cunning, the fox doubled back, and passed between us, causing some consternation, but giving each car the chance of a broadside. He still kept nicely out of range and led us to the edge of a deep gully, which he jumped into and disappeared from view, to emerge – still out of range – on the far side, encouraging us to follow him. We tracked along the top of the wadi for some distance, but were unable to find a safe way down, so resumed our search across the plain.

We started a few partridges, but they were much too small and quick for us, although they were subjected to the odd fusillade, and we continued until we came upon some turkey buzzards. These are quite large birds, about the size of a turkey, with a wingspan of about six feet. They could run, in a straight line, at about 30 miles per hour with some wing assistance, dodging through the bushes, and we had a hard job keeping up with them. The ground was so bumpy that we only managed the odd shot over the cabin at them, and they were really pretty safe. They also had the ability to stop and change direction through 90 degrees and be off again, still wing assisted, and at such manoeuvres we were no match for them. We continued the chase, however, until they too went over the edge of a wadi and ran on out of sight.

We regrouped, changed around the firing positions, and set off towards the oasis, which we could see some distance off – a few date palm trees and a small thatched house beside a well in the desert. When we arrived we were astonished to see the spread that had been prepared for us. There was a display of fruit, dates, oranges, melons, grapes, and nuts that could have graced the Harrods food counter, and trays of rice and various meats, including the proverbial sheep's eyes.

For drink there was whisky, gin, local brandy, local wines, imported beer, lemonade and other soft drinks, all set out on little low tables, on a series of magnificent carpets, in the shade of the palm trees. The only things missing were the dancing maidens, but perhaps, we thought, they might be along later.

No such luck.

We sat there eating and drinking, and taking our ease for about a couple of hours, and then decided it was time to continue the shikar. Once more we set off across the desert, this time proceeding with a bit more decorum – for a start, anyway – and set up a family of wild pigs. To hell with the decorum, then. We chased these animals for about half an hour, without really being anywhere near enough for a shot, although this did not deter some people.

As the family ran, they kept making violent alterations of course, and enthusiastic shooters were obliged to change from one side of the vehicle to the other, causing some stability problems as we bounced over the rough ground, until eventually, like the turkey buzzards and the fox, the quarry dived down into a wadi and were lost to view.

On our way back to camp we set up a few more partridge, again without success, and as we were nearing the site we saw skeins of geese flying in overhead, making for an area out well beyond our boundary of the airfield. When we arrived back at camp, without any game to show for our day out, we invited the contractors into the mess for supper, and tried to repay some of their hospitality from our stocks.

During this period we found out that there was some shooting out on the desert, and they would arrange an evening shoot for us, warning that it was cold and damp out there, and the geese were very smart and extremely difficult to approach. The tactic was to go out before dusk and dig yourself into a shallow trench, and hope the birds landed within range. They did not seem at all bothered about sitting birds. I declined to go out on such expeditions, but some people went off, and returned damp, cramped, and empty-handed.

Served them right, too, I said.

CHAPTER 70
A Bear Party

When all our accommodation was reasonably complete, and most people had settled in, we decided that this was the excuse for a party. By this time we had about 30 expatriate staff and dependants on the site, and local staff employed in the stores and offices totalled another 15 or thereabouts, so we felt there was a critical mass, and we had sufficient room in the mess to be able to cater for about 60 persons. Invitations were sent out to various local dignitaries and officials in the oil company, and serious preparations made for the festivities.

Arrangements were made that individual houses would loan items of crockery and cutlery, and a committee of the ladies would supervise the catering in the mess, as well as providing individual delicacies from their own kitchens; there was a great sweeping out and clearing up of gardens, and the whole place looked quite presentable.

One of the ladies had been presented with a Persian lamb, which lived mainly in her garden, but was quite agile and often escaped along the line of houses, and was frequently found foraging in the dustbins and generally scattering rubbish all over the place. Suggestions were made that this would be a fine party piece for the supper table. Such ideas caused some pain to the owner, and private arrangements were made to ship the animal out to another family living in the oil company town at the top of the hill.

There were some travelling tribespeople moving through the area to their winter grazing, and someone prevailed on two camel-owners to bring their animals along and provide camel rides around the compound. This eventually went off very well, with impromptu races round the buildings. Care had to be taken not to fall off onto the hard surface of the road along the frontage. As anyone who has ever been on a camel must know, they proceed with a rocking movement, even at slow speed, and there is not a lot to hang on to. Also, it is a hell of a long way down.

Another entertainment was provided by a travelling bear, called Beirut. The bear and his two attendants arrived in town at the local bazaar one afternoon by bus. The attendants had travelled inside, like ordinary passengers, but Beirut had come on top of the bus, sitting amongst all the assorted parcels and bundles that seem to be part of the travelling Iranian's equipage, on the extensive roof-rack.

I arrived in the bazaar just after the arrival of the bus, and watched with interest as the bear was enticed down off his perch, where he seemed to be quite comfortable, with a good view of the passing scene. He came down quickly enough when bribed with a small piece of honeycomb, and sat down to await further orders. There was some argument going on about the fare for the bear, and possible claims for compensation by passengers whose luggage had been squashed, but when the attendants thought that this discussion had gone far enough they had Beirut stand up to his full height, about six feet six inches, and lean against the back window of the bus.

This demonstration settled all the arguments, and they went on their way down to the end of the town.

As well as the camel racing we thought it would add a bit of interest to have a performing bear at the party, and one of the young engineers was instructed to complete the arrangements for his attendance. He came back to report that all was set up, and the bear and his attendants would appear at about eight o'clock in the evening.

While the final preparations were being made, and people were dashing about with plates of food, cakes, etc. to the mess, I heard a bit of hammering going on in my front garden. I went out to see what it was all about, and found the attendants hammering a long stake into the garden, as they were proposing leaving the bear there while they went off for some food in the town. Not relishing the thought of my wife coming out of the house and running into a large bear tethered in the garden, I asked them to take it away and tie it up by the mess, and they obligingly moved off. They probably reckoned that they would get a lot more food there, anyway, and would be able to claim authorisation for their presence.

The party got under way with lots of food and drink, and the performing bear was a great success. He sort of shuffled about (dancing, I believe it is called) and both he and his attendants seemed to be enjoying the party. Invitations were issued to be his partner for the next waltz, but there were no takers, and the attendants continued their performance. After each episode the bear was given a drink from a coke bottle, as were the attendants, and I fear there was more than coke in the bottle. They seemed to be enjoying it, however, and kept coming back for more.

During a pause in the music someone asked if there were any other tricks Beirut could perform, and the attendants, now very willing to oblige, made him lie down, stand up straight, and hop on one leg, and, as a final pièce de résistance, they made him climb up on the low mesh fence round the garden. This he managed without too much trouble, but, alas, the fence had not been designed to withstand a super-load of one large bear, and it collapsed gently, allowing Beirut to step down in a very controlled fashion – rather like the ballet, really. Following a round of applause, and another drink, the attendants were thanked for their performance, collected a fair sum of rials for their trouble, and set off back into town.

It was difficult to make out who was in charge of whom as they ambled down the road. The party continued well into the night, without any other scheduled innovations, and was generally thought to have been a good night.

In the morning I had to go out to the site fairly early, and as I was going past a roundabout at the end of our road I could see the bear and his two assistants curled up, fast asleep, inside the perimeter. Having some idea of the amount of spiked coke they had been drinking, I thought it best to leave well alone. This was probably going to be a bear with a really sore head.

I heard later that they had departed on the bus in the afternoon, and there had been a bit of excitement, as Beirut was finding it difficult to climb up onto the roof-rack.

CHAPTER 71
Shopping Expeditions

The small town of Agah Jari had been a centre of operations for the oil company for a long time, and apart from the usual local shops and stores, the company had set up a sort of supermarket for its employees, where the normal run of Western foods was available. This shop was in the old town, at the top of the escarpment, about five miles from the camp, and it was the normal source of supply for food and drink for our staff.

A new store had been opened in Bandar-e Shahpur, which was a port on the coast used for shipping out bulk oil, and it was also a pilot station for ships going further upriver, and was being further developed for general cargo.

I had been to the port to try to sort out a small problem with the customs and shipping agents concerning a bulldozer, which had been sent out from England for discharge at Abadan. For some reason, the machine had been loaded below deck, but the blade had been deck cargo, and, as it had been in the way of one of the hatches, had been lifted off at Bandar-e Shahpur and inadvertently left there, documentless, in the port area.

The tractor part had been unloaded at Abadan, but some bright spark had realised that the machine was incomplete – i.e., no blade – and technically was not a bulldozer but a tractor, so he was unable to release it without contravening all sorts of diktats. I had visited the customs at Shahpur, who denied all knowledge of the blade, even though it was sitting there, in plain view from their office.

The shipping agent was not much more use. He agreed that there was a blade lying in the dock, which had the same number as our tractor in Abadan, but the number in his book referred to a bulldozer, which was obviously not here, and he could not process a blade without proper documentation – and, anyway, it was not officially here. How daft can you get?

After a lot more palaver, both at Abadan and Shahpur, we managed to persuade a ship to take the blade to Abadan, where it was reunited with its tractor and eventually released – at great expense – and carted to the site, where it began to earn its keep.

During my visits to Shahpur I had discovered the oil company store, which was vastly superior in scope to the old store in Agah

Jari, and on each visit I brought back some goodies. The news of this soon leaked out, and an expedition was arranged for some of the families to go down one weekend for a family shopping spree. We arranged to borrow the site manager's car, a huge, pink Chevrolet, which we thought might be a bit more comfortable than the Land-Rover, and set out early in the morning.

The road ran dead straight for about 42 miles, with very slight variations in grade over the desert. It was a bitumen-surfaced road, officially 'all-weather', and relatively dust-free. This encouraged excessive speeds, and we wound up the old Chev, foot flat on the floor, to see how fast it would go. The limit was not the availability of power but the alarm caused by the signs of wrinkling along the wings, as a consequence of disturbed airflow. On consideration, this indicated very thin metal, which a rapid inspection confirmed, and we reduced speed thereafter, keeping the wings free from wrinkling rather than watching the speedometer.

The bitumen surface of the road meant that there was not the normal cloud of dust following each vehicle, but this also meant that approaching vehicles, or vehicles being overtaken, were not visible from afar, and therefore came up on one rather surprisingly.

Another problem arose, caused by sections of the roadway being covered to varying depths of sand drift, not unlike snowdrifts in more temperate climes. These could cause some alarm, and, after the first experience, tended to temper the speed at which one progressed.

The story goes that the road was set out by someone starting off from the back of Bandar-e Shahpur, and going straight for the loom of the flares on the heights above Agah Jari, and this was probably the longest bit of straight road in the world. The desert in between was quite featureless, except that, from time to time, the ditches alongside the road were half full of water. We did think of towing a water-skier along behind the car, but were unable to persuade any volunteers for this trick.

The novelty of the new shop soon wore off, especially as it entailed about three hours confined to the back of a bouncing car, and the main shopping expeditions were resumed to Agah Jari, with the very occasional visit to Ahvaz.

CHAPTER 72
Gach Saran

Work on the airfield had fallen into a routine, and was progressing well, when we were awarded another group of contracts by the oil company in the small town of Gach Saran, about 80 miles off to the north, and in the mountains. I was asked to go to Gach and take charge of the operations there, which were the completion of the new town services and the construction of a number of oil rig drilling sites and access roads within the main oilfield area.

Jean and I went up to visit the area over a weekend, and were very pleased at the chance of a change. The hot weather was pretty oppressive at Agah Jari, with temperatures in the sun of over 120°F, and relatively high humidity, whereas the daytime temperatures in Gach were more like about 100°F, but at least it was much cooler at nights. The small camp that had been established for earlier work was of three houses – similar to our present premises, and again built in a line, but this time on a slight rise, with a splendid view out onto the mountains, which were about ten miles away.

The old town was a staging post on the route between the coastal plain and the uplands around Shiraz , and there were a number of interesting old buildings around the market place.

There were a number of successful oilfields in the area, which had been producing for about 30 years, and the town had become the administrative centre for this activity. With the discovery of many additional oilfields in the adjacent area a new town had been built, with roads, water, drainage and sewage, telephones, and all the other benefits of modern civilisation, together with extensive new construction for the oilfield production and maintenance activities.

The road between Agah Jari and Gach Saran went straight up into the mountains; the bitumen surfacing stopped at the end of the town, and from there on it varied between a rough stony track, barely wide enough for a large lorry, and open sandy stretches across the high level plains. One of the hazards on this road – and, indeed, on other narrow mountain roads in the region – was the sudden appearance of large blocks of rock lying in the roadway. These, we discovered, were chocks that had been wedged behind the rear wheels of lorries that had encountered breakdown or other mechanical failure, to prevent them from running backwards down the hill.

When a more powerful vehicle came along, or the original machine had rested for a bit, they would set off, leaving the chocks behind. Some of the more elderly vehicles we noticed had special wooden chocks attached above the mudguards, and these could be deployed by the driver's assistant, so when the vehicle got under way again they trailed behind on their chains, ready to be used, until that particular hill had been surmounted. This was both a friendlier and more efficient method of dealing with the problem, as it was not uncommon to come upon some unfortunate vehicle that had run into some other person's chocks and done itself a bit of mischief.

Frequently, as a result of this damage, a second set of chocks was required for the damaged vehicle.

We made a second visit to Gach with the remainder of our household equipment, and settled into our new house.

My office was in a small caravan outside in the garden, and we also had a large marquee tent where the local staff conducted their business. The caravan was often very hot and stuffy, but reasonably sandproof, while the tent was generally cooler and, apart from the sand between the pages of all the paperwork, not a bad place to be.

I had a staff of two other expatriates, and seven local foremen, and we recruited labour from a number of local contractors in the area.

The work in the town was residual items of completing electrical connections, plumbing houses, laying footpaths and minor roads, and the operation of a small quarry. Out of town we always had two rig sites under construction, and sometimes three or four access roads to and between these sites, which were scattered in all directions all round the town at distances of up to ten miles.

Each of us tried to visit at least two different sites each day, meeting in the office at lunchtime to review the situation and settle on the next priority of work, men and equipment, then back out again to the next visit, with another conference in the evening. We each carried a gallon Thermos flask, filled with ice and lime juice, and tried not to drink any of it before eleven o'clock so that we would have enough to last us back to camp by lunchtime, and a similar load was carried in the afternoons. At each location we visited we left messages for the following engineer, saying what was required, and where we would be going to next, and this system worked very well. Any time any of us was delayed by punctures or other breakdown, assistance was never far behind.

There were a number of other European contractors working in the area, each with a small staff and living compound, and cooperation was generally very good between us. Frequent functions were held by each contractor, on the occasion of birthdays, or persons leaving on completion of their tour, or just for the hell of it, and these evenings enlivened the scene all round.

One contractor whose mess was quite close to us gave a party one evening that ended in a bit of uproar. One of the staff had taken in a bear cub some months earlier, and was looking after it very well. It always appeared clean and spruce, and full of fun, and it was given the run of his house (needless to say, his wife had gone home). The cub followed the man everywhere, and generally sat quietly in a corner, even while energetic singing and dancing was in progress, and on this particular occasion the cub had gone into the bathroom and was sitting quietly on the floor, causing no harm to anyone.

A new employee, just out to the site, unaware of the situation and obviously with some drink taken, went into the bathroom, and was confronted by what he took to be a wild animal (which it was, of course). He rushed out, slammed the door, and began the general evacuation of the gathering. The cub, hearing all this commotion, broke down the door and came bouncing out to join in the fun, and some degree of panic ensued. Order was soon restored, the unfortunate new man sent out in disgrace, and the party continued, with the cub sitting in a prominent position so that everybody could keep their eye on him.

Other entertainment was available in the old bazaar, where, from time to time, itinerant storytellers would set up their stall and collect an audience. One man comes to mind. He had a sort of large easel, which he set up with a series of pictures arranged in a roll on the top, and when a sufficient crowd had gathered he would begin by announcing the title of the particular story he was about to recite. He would then appoint members of the audience to take the parts of animals or features of the story, and rehearse them before he began. Thus he had people barking like dogs, mewing like cats, making the noise of a waterfall, etc., and when he was satisfied he would begin.

The first scene would be unrolled, and as the story unfolded he would bring in the audience by pointing to the item with his stick, and conducting the performers with his other arm. Most of the stories were traditional, but he always managed to introduce some topical items, and persons of local note. Officialdom was not often favourably described, and generally took a poor view of the performance. On some days there were jugglers and acrobats performing in the bazaar, and it was always a very lively place to visit.

The new club, which was opened shortly after our arrival, was also well patronised, with its swimming pool and tennis courts and a good English library.

On the occasion when it was officially opened by the field superintendent, God's local representative, there was a marvellous party, attended by all the oil company and local officials, all in their best boots and spurs, being served drinks and small chop (a selection of fruit, nuts, crisps, etc.) on the lawn around the pool.

As most of the local staff were Muslims they were not expected to partake of the strong drink, but someone had managed to rig a refrigeration plant to freeze vodka, and this was being handed out to selected patrons as ice for their orange squash.

Some of this so-called ice went astray, and was being given to some of the local staff ladies, who, probably, had not experienced such fine orange juice before and became very unsteady on their feet, until someone realised what had happened and organised a swift dispersal of the affected ladies.

There was a bit of an inquest into this affair the next day, but solidarity won and the culprit was never officially discovered, although many fingers were pointed.

There was an open-air cinema in town, and regular English and Indian film shows were given. Very often, the sound was so bad that it was difficult to tell whether the dialogue had been dubbed into Farsi or was in the original Hindi or English, but as the projectionist frequently put the reels on out of sequence such finer points rarely made much difference.

The screen was a large surface held up on poles, and at the slightest wind it swayed and bulged out of shape. Paying patrons sat on chairs on one side of the screen, and the town sat on the other side of the screen, receiving a slightly distorted image.

CHAPTER 73
A Plague of Locusts

During our travelling around the sites we came across great flocks of locusts, generally still in their non-flying mode. They came across the desert in great hordes up to half a mile wide and probably three or four miles long. When we encountered these swarms we drove back and forth across them, killing thousands at each pass, but still they came on, a great, green moving carpet, devouring all before them. We reported them back to town and they advised the Locust Control Office somewhere further up the line, and no doubt plans were made, and action taken somewhere, but still the plague came on.

We had gone to the cinema one evening, and were sitting in the semi-dark, when some woman in front jumped up, clutching her head and shouting and dancing about. It seemed that a flying locust had landed on her head. Immediately after this the beam of light from the projector was seen to be full of these locusts, each about three inches long, milling about in the beam, and probably looking for someone else's head to land on. A rapid evacuation followed, and we drove carefully back to our house through swarms of flying insects, almost like snow.

When we arrived home we seemed to have run out of the main stream, but there were still a few flying around so I went out and woke up the watchmen, who lived in a small tent at the end of the garden, and gave them strict instructions that all locusts were to be kept off our garden, especially the Indian corn, which was almost ready to eat. They did not protest too much at this unreasonable instruction, but went about flicking all our corn and stamping on the ground. I wondered about leaving the car lights on, directed at one of our less friendly neighbours, in the faint hope that the locusts might be distracted from our garden, but quelled this unkind intention and, with a further warning to the watchmen, went quickly into the house and turned all the lights off.

Actually, we were only on the very edge of the swarm, and suffered only slight damage to our garden; the houses were pretty well fly-screened and insects came in only when doors were left open, so we had no trouble inside. The watchmen were still rushing about when I came out in the morning, and displayed a heap of dead

226

and damaged locusts, and only a few ears of corn had been devoured. I told them they had done a good job, and promised them some extra cash in their pay on the next pay day.

In the town the majority of the gardens had been ruined, with practically all the green material devoured, and wide brown patches left on the roads where vehicles had run over the swarms that had landed. Of the flying insects there was no sign left; they had all gone off to attack some other place. We still saw the walking-mode locusts on our travels, but they did not come back to town again – they had obviously cleared the place out of suitable food.

CHAPTER 74
Garden Works

We had moved into the house in the middle of the hot weather, and inherited an old gardener from the previous occupants. In the front of the house there was a sort of open patch, with many withered sticks projecting from the ground, and I was often at the old man to dig them out and make the place look a bit more attended. Each time we spoke on this matter he became quite excited, and kept telling me that these were not just sticks but fine flowers, and it would be a mortal sin to dig them up.

"Just wait and see what fine flowers they are when they come out at the end of the hot weather."

Meanwhile he continued with his other work, which was to cultivate our crop of maize. These plants had been planted about four deep all round the side and rear of the garden, more as a hedge than a food supply, and we were informed that this was no way to grow maize, as it needed to be planted in blocks to permit proper fertilisation.

Not being much of a gardener myself I told him that that was an old wives' tale, and it was up to him to see that the heads were fertilised, either naturally or by taking a feather to them. He accepted this as a bit of a challenge, but went about it somehow, and eventually the corn was ripe.

His other job was to see to the grass, which had been sown earlier all round the rear of the house, and this was his pride and joy. Each day he would water it carefully and remove any signs of weed or foreign growth, and he kept it well cut by means of a knife blade fixed to the end of a walking stick. He took all the cuttings of the lawn back home each evening, and fed them to his chickens, I believe.

In one small corner of the garden the previous owner had planted some horseradish, and in the course of conversation with my chief clerk one day he was admiring this plant and said that the local people used the leaves as a kind of salad. As most green produce in Iran is eaten as soon as it appears I was not surprised, but told him that we normally used the root as flavouring, and I had never heard of anyone eating the leaves. Just to show willing, I pulled a bit of it out of the ground and said, "I will taste the leaves, and see how it is." Actually it tasted rather like peppermint, and was quite pleasant.

Not to be outdone, Sulimuni broke off a large piece of the root and put it into his mouth. He was much too polite to spit it out, but coughed and spluttered, with tears streaming down his cheeks, and commented: "That must be very hot sauce."

I explained that it was intended to be used sparingly and not eaten raw, but added that perhaps he could start a fashion in this mode, and said he had better come in for a long drink.

Sometime later in the year we had to go off to Tehran and Mashhad on a survey, and we were away for about three weeks. During our absence the miracle occurred, and when we returned our front garden was a mass of white chrysanthemums, and the old gardener was so pleased with himself.

"Look," he said; "like I said, these sticks were flowers all the time."

I agreed that he had done a miraculous thing, and had been quite right to talk me out of digging them up, and gave him a bonus in his pay.

During the remainder of the cool weather he brought flowers to the house every day. I never questioned where he collected them from, but there were a lot of marigolds growing round the oil company club that bore a marked similarity to the ones he brought in.

CHAPTER 75
The Carpet Men

Our evenings were occasionally enlivened by a visit from one or other of the carpet men.

Shortly after we arrived in Agah Jari we had our first experience with one of these gentlemen. Our houseboy came in to announce that there was a person to see us at the door, and, somewhat intrigued, I went to see who he was. A very plausible Iranian was standing on the doorstep, who introduced himself as a carpet merchant from Esfahan, and wondered if I would like to see his wares. As we had nothing arranged for the evening we invited him in, and immediately he shouted out to his companion to bring in the first three rugs.

He had a large pick-up truck parked outside, stacked to the roof with carpets and rugs, and his man brought a selection to the door while the merchant himself displayed them on the floor, pointing to the qualities of each one, and giving a potted history of the maker and the finer points of the designs and figures on each rug. We examined about 20 carpets all told – and then began the bargaining.

The technique was to express a mild interest in one particular specimen, but really to prefer another, some way down the heap, which was now about ten deep on the floor, the merchant having already taken out those he knew would be too expensive at this stage, and some others which were too large or otherwise discarded. Some time later we ended up buying two rugs for what I took to be a reasonable sum, and his man started to remove the remainder from our floor.

He returned with a pair of new rugs, which we had not seen before, and after a long peroration requested that we should keep these two, a matched pair, in use until he returned on his next visit. This struck me as the beginning of some involved underhand dealing, but I was unable to see how it would work out to his advantage and eventually agreed, saying we would look after them for him, and keep them rolled up in a cupboard until he returned.

This was not what he had in mind, and he explained that we should make use of them, as carpets improved with being spread out and having light traffic on them, especially on tiled floors such as ours. Not very reluctantly we agreed, as it would be a great comfort to step out of bed in the morning onto a carpet instead of the cold tiled floor, and duly spread them out in the bedroom.

About a month later another carpet merchant arrived, and a similar procedure was adopted. He immediately identified the two rugs we had down in the living room, and correctly named the previous merchant. He offered to exchange either or both for different rugs he was carrying, and brought in a whole pile of others for our selection. We eventually said we would take one of his in exchange for one of ours, and buy one other, and this seemed satisfactory on both sides.

Having concluded the deal I thought to ask him about the two rugs we were holding for the other man, and brought them in for his inspection.

"This is a very good pair of rugs," he informed us, "but still too new for trading," and gave his opinion of the price they would make when they were run in, as it were.

"These two merchants must be in cahoots, and they are bumping the price up by agreement, hoping for some sucker to snap them up," I thought.

He offered to take them off my hands at about half as much again as our earlier supplier had suggested they might make, but as I was uncertain of the ethics of trading in goods without proof of ownership, and without a licence, etc., etc., I declined and the man departed.

Sometime later we moved to Gach Saran, still with the two new rugs being run in. I wondered if there was some kind of secret police that would keep tags on all carpets and ensure that householders in temporary charge of dealers' rugs would be prevented from leaving the country. Shortly after we moved we had a return visit from our first merchant, who took away the two carpets, but left another two in our hands for safe keeping. He also recognised the carpet bought from the other dealer, and admired it wholeheartedly. He reckoned that we had got a bargain.

Some months later we had a visit from a third merchant, this time on a Friday evening. He came in and went through the official procedure, carpets were displayed and discussed, our current crop was recognised and commented upon, and another carpet displayed that he thought would be just right for us. After a fair degree of bargaining we agreed a price, and a general discussion followed.

When chided about trading on a Friday, the Muslim equivalent of the Christian Sunday, he declared that he was a Christian Jew and not bothered about all that stuff about Fridays. He also told us there was an agreed scale of prices for carpets, depending on the nationality of the potential customer. Americans were charged a considerable excess, English a bit less, and Dutch were rock-bottom (generally recognised as being a pretty miserable lot). There did not appear to be a special rate for Scotsmen, but this was negotiable.

These travellers carried a stock of about 100 carpets in their vehicles, each probably worth about £30 or £40, and seemed to travel freely about the country, without any fear of bandits or thieves. They were generally pretty short of actual cash, and went to great efforts to ensure a sale of some kind on every visit. They also appeared to be remarkably well thought of by older residents in the country, and they generally advised that it would be perfectly safe to pay for goods and have them dispatched to addresses at home.

I must say that, when we consulted Iranian friends about the rugs we had purchased, they all assured us that we had done pretty well, and would always be able to sell them back at a profit should we so desire. They also confirmed that value was added by wear, and that some people put their rugs out in the street, to be run in by passers-by. (Perhaps they nailed them down.)

CHAPTER 76
A small Earthquake

From time to time we had visits from the main construction people at Agah Jari. About once a month the accountant came up, and every fortnight the plant manager came up to see that we were not abusing his equipment. Most of the heavy equipment was hired out by the oil company, and they called in individual items for periodic inspections and overhaul to their main workshops in Gach Saran.

I had gone in with the plant manager to argue about the costs of some repair allegedly undertaken on a tractor, and as we were standing talking I felt a bit of a jar underfoot. Most of the native population dropped whatever they were doing and rushed outside.

"What the hell was that?" the plant manager exclaimed. "Why are they all running out?"

I really thought that someone had backed a lorry into one of the columns of the building, but made a flippant remark that it was probably only a small earthquake, and found myself in splendid isolation.

It turned out that my flippant remark was correct. Thinking about it later, I was probably in the safest place in the country, standing on a reinforced concrete floor inside a steel-framed building, with a steel-sheeted roof; perhaps a thunderbolt could have got at me, but I was reasonably proof against earthquakes.

I went home quickly to see if there had been any damage done in the house, but nobody had even noticed the tremor!

CHAPTER 77
Some Problems with Machinery

Part of the oil company's programme was to drill holes, exploring for new oilfields and confirming and developing existing fields. They had a number of contractors, mainly American, who operated the drilling equipment, and as work finished on one hole the company directed the rig to the next hole site on the programme. When drilling was completed, and the rig ready to close down, a separate team took over the operation. These people were experts at dismantling, loading onto lorries, moving, reassembling, and setting up ready to drill at the next location, and there were unofficial races for each move.

When work stopped on the drilling and the hole was closed down the moving crew would move in. The equipment was taken down and laid out on the site in a very orderly manner, with marks set out for each item, so that lorries could drive in and be loaded with the least delay, while at the new site similar marks would be set out so that each vehicle would know exactly where to stop and which way to be facing when it arrived.

By the nature of the structure, the last item to be taken down and prepared for transport was the first item to be assembled on the new site, and the layout of the sections delivered earlier at the new site needed to be carefully planned to ensure that there were no delays.

It was an unwritten law that rig-moving traffic took priority on the roads (the only prior precedence was in case of fire, when everything stopped and kept clear). The average rig, with all its attendant equipment, probably amounted to about 300 tons, and generally between 15 and 20 lorries and specialist carriers were involved, some of them making several journeys. Our work was to prepare the rig sites, and to build and maintain the access roads.

The country around Gach Saran was quite hilly, and there was a considerable range of mountains to the north and west, with snow-covered peaks lasting well into the hot weather season. In the springtime the whole place was covered with grass, and there were many wild flowers blooming; poppies grew in profusion on the plains, and in the rocky gullies the sides were yellow with jasmine. During this period the melting snow on the mountains produced a steady flow in the rivers, and the frequent heavy showers and local storms caused flash flooding.

We had been building a road out to a rig site about 15 miles from town, and had put a couple of large pipes in a wadi and filled in over the top with rocks and stones to bring the road level up to match with the sides of the gully. One of these flash floods had washed out our culvert overnight, and our grader had come along and dived into this gap, which had not been there the previous evening.

When I arrived at the scene, about ten minutes after the accident, I was glad to see that no one had been hurt, but the machine was standing on its nose, with the back wheels clear of the ground, and the front looked a bit out of alignment. It was also blocking the road.

For those who are not technically minded, a grader is a long, six-wheeled tractor with four wheels at the rear, a long bridge section, which carries a swivelling blade, and the two steering wheels at the front.

When we pulled our machine out, it was seen to be quite badly bent (not surprising, really – it had just fallen down about 15 feet, and landed on its nose); the bridge was cracked near the front axle, and the whole front section was decidedly squint.

Here was a problem. It was obviously quite seriously damaged, but looked as if we could mend it on site. If we took it back to the oil company there would be all manner of inquests, allocation of blame, arguments about costs, possible write-off and total replacement, and it would all take weeks if not months before any repairs were started. We needed the machine back at work in order to complete the road for a rig move in the following week, and the thought of all the hassle with the workshops and accountants made up our minds.

We took our welding set out to the location, with a squad of mechanics, and started right away. Fortunately it was only the structure on the machine which seemed to be damaged, and this repair was a relatively quick operation as there was no waiting for spares, bearings, connections, etc., so we managed to get the machine back on the road without anyone noticing within three days. Then a quick dab of yellow paint, and a lot of dust, and no one would ever know.

During the rig move that followed the base workshops engineer happened to be visiting, and I remarked on the skill of his staff, having repaired this obviously serious damage at some time in the past

"Yes," he said, "they are pretty good at that sort of thing in my workshops."

Nothing like a bit of flannel to divert some people's attention.

We had hired a new air compressor for our work on a small quarry, and it was the very latest model, specially built for desert operations. One of its unusual features was an inertia starter, as opposed to the normal battery-operated electric starter. This machine had been delivered straight from the workshops, untouched by human hand, as it were, and was taken out and put to work. Our foreman fitter, an Armenian (a very skilled man), was fascinated by this starter, and decided that he would need to know how it worked.

When I arrived on my normal morning visit the machine was running quietly, and Pannus was sitting on a blanket, surrounded with wheels and springs. He did have the decency to confess that his curiosity had got the better of him, but assured me that he knew exactly where everything had come from, and said that I was not to bother as it would be all reassembled before nightfall. I remarked that it had better be, otherwise he would be spending the night making sure that the engine did not stop for want of fuel, and I asked how I was going to explain *that* away to the oil company workshops?

On one of the rig roads we were setting up a small quarry for stone on the side of a steep hill, and when I had visited the previous day the bulldozer was cutting a ledge into the hill, to allow for a place for the crushing machine to be established. On my way out the following morning I met Pannus on the road, on his way back to camp. After the usual greetings etc., he suggested that I should go back to town with him, and perhaps come out later in the day.

It transpired that the bulldozer had slipped off the ledge, and was sitting, the right way up, with its tail on a large boulder, and was unable to move, firstly because the tracks were off the ground and second because the driver had gone off – whether from fright, or from fear of what Pannus might do to him, was not yet established. We managed to stabilise the situation with wires from the winch in the tractor, and some ground anchors that we dug into the ledge, and when Pannus came back he brought news of another tractor coming in from another site, with a 'sensible' driver (his words).

We decided that we should try to shift the big rock that was holding the first machine, and then try to lower it down on its own winch until the slope ran out a bit, and it could come down under its own steam thereafter. Pannus suggested that I might have better things to do elsewhere, and I took the hint and went off back to town. He came into the office later in the evening to advise that all was well: the machine was back working, and the other machine had been returned whence it had come. The only signs of damage were lots of places without paint on the machine, and some large yellow marks on the side of the hill.

He suggested that if we ever had problems with machines falling off ledges we could probably put them down to small earth tremors, which were fairly frequent, and thus very difficult to disprove.

CHAPTER 78
A Local Wedding or Two

One of our contractors supplying labour and equipment was closely related to a nearby khan, a tribal landowner who was also a member of the Majlis, the parliament in Tehran. This fellow lived in a large castle about 20 miles away, with a small village nearby where his retainers and local agricultural workers lived.

We had daily dealings with the contractor's agent, and we were invited to a wedding at the castle, where his son was to be married to the daughter of another local landowner. We took advice as to the protocol to be observed by way of dress, and wedding gifts and such things, and made our preparations.

Our general manager and his wife had also been invited from Tehran, and they were due to stay with us for a few days over the wedding period, and also to visit the various sites and construction work in progress.

On the day of the wedding we set out in the early afternoon in a saloon car to the castle, and were welcomed by the khan's steward, or farash. Having parked the car we were escorted inside the fortress, through a gatehouse in the walls, which were about 12 feet thick, and then to a large room on ground level. This room was furnished with small tables and chairs set out all along the walls, and on each table was a selection of bottles and a number of glasses.

The farash showed us to a table, and poured out large measures of spirits for each of us, and then produced a tray with fruit and nuts, and small sweetmeats.

The room itself was about 50 feet long, and there were a number of large carpets spread on top of one another all across the floor, which made the tables a bit unsteady, and the progress of the farash even more difficult. He was wearing a long sort of cloak over his normal baggy trousers, and on his head he sported a sort of tam-o'-shanter bonnet, with enormous overhangs all round, seriously impairing his vision.

More guests arrived, and introductions were made and small talk exchanged, the procedure being that incoming visitors made a sort of circular tour of those already seated and then sat down at the next vacant table. The bridegroom and his party then arrived, and took their place at a table at one end of the room, and we all went past them, offering our congratulations and best wishes, before resuming our seats.

The next event was the entry of the bride's family, and they took their place at another table, and once again we all trooped past. During each of our absences from our seats, the farash had been going round filling up all the available glasses, and after each progression we had a few toasts and some more nuts.

The next item on the programme was dancing by the village ladies, outside in one of the courtyards, and we moved out onto the roof of one of the outhouses of the castle, while the band, comprising a three-note flute, a two-stringed fiddle and a drum, mustered on another roof on the opposite side.

The music started and the dancers formed up. The first dance was for ladies only, and about 20 of them, dressed in all their finery of long dark skirts and brightly coloured blouses, coloured shawls over their heads and shoulders, and rings and bangles on all moving parts, began to dance.

They had formed up in a sort of crescent, with their arms on each others' shoulders, and the dance went rather like this: three steps to the left, hop; two steps back to the right, hop; one step forward, hop; one step backwards, hop; and then continue, moving slowly round in a circle to the left, dancing the same steps over and over again, until they had made a complete circuit of the courtyard and were back at square one again.

After a mild round of applause, a second dance was started – this time for the men.

This was not so much a dance as a means of extracting retribution from neighbours for previous slights and insults. What happened was the band played a rather slow tune, if you could call it that, and one man took the centre of the yard with a stick, rather like a Boy Scout's pole. He then pranced about at a slow hop, enticing some competition.

After some further formal movements, a second participant joined in, also armed with a stick. The tempo increased, and they both hopped about, executing fancy turns and swinging their sticks about like drum majors, but keeping a wary eye on one another.

Again the tempo increased, and the first performer selected a position in the yard, and put one end of his pole on the ground and stood behind it, still dancing in time but keeping the stick between him and his partner. The other man, also still dancing in time, tried to bash his partner on the legs with his stick, and this continued until he succeeded, whereupon the first competitor retired (hurt) and the second man took his place. There must have been some means of selecting the sequence of the subsequent contenders, but the logic of it was beyond me.

Meantime, while this formal performance was in progress, all the small boys present were knocking hell out of each other all round the perimeter, and other actions were in progress by their mothers, trying to protect their offspring and keep the neighbours at bay.

The music changed again, and once more the ladies formed up and danced around in a circle, this time with handkerchiefs waving every four or five steps.

Before the end of this performance we were taken down to another room below the courtyard, where supper had been set out.

This was another large room, with triple carpets throughout, a large table in the centre and a similar arrangement of small tables and four chairs all round the walls, again liberally decorated with bottles and glasses, and the farash in his funny hat going round filling everything up regardless.

The khan made a short speech, followed by a few words from the young lady's father, and we were invited to eat. The food on the centre table was a splendid display. There were several large round trays of rice, heaped about two feet high, with floral decorations all round the base, plates of fruit and nuts, large tureens of lamb stew, various unleavened breads, jars of assorted sauces, and a number of unidentifiable objects, which still seemed to be going like hot cakes.

It was quite difficult, as we had not eaten since breakfast and were pretty hungry, but as soon as you put your glass down to go and collect some food the farash immediately filled it up again, with whatever bottle happened to be nearest, or what was still in his hand. This was all right the first couple of times, but a certain degree of restraint was called for, as we had to drive back to town at the end of all this, and the farash was obviously carrying out instructions.

When half the food had been demolished we went out onto the roof again, where the dancing was still in progress, and saw a repeat performance of the earlier entertainment, and then we were ushered back to the lower hall, where some order had been restored, and we sat about listening to various speeches and drinking further toasts to one and all. The pace slackened, and we took the opportunity to depart, but first had to go and meet the fathers of the couple for a farewell toast.

Somehow, we found our way out of the castle, located the car, and then endured a long, bumpy ride back to town. Next morning we were barely open for business, but by the evening had revived a bit and went off to the club for dinner.

Some month after this episode our accountant, based at Agah Jari, announced that he was about to marry one of the schoolteachers in the English School in Gach Saran.

We asked if we could assist in the logistics, and set about making the preparations for a party and finding accommodation for some of the guests. There was not much problem with the bachelors, as they could all bunk up in the spare rooms in our other two houses, and also in some rooms we could borrow from other contractors nearby.

The bride's mother would be coming out from England, and we arranged for her to stay with us, as there was no room in the schoolteachers' premises, and other arrangements were made for the priest, who was going to come from Tehran.

Due to the normal chaotic transport arrangements the priest failed to arrive on time (you might think he could have done better than that, with a few words upstairs), but the civil ceremony was conducted, and we had a splendid party afterwards. The bride's mother was not convinced that they were really married, but eventually accepted the fact, and when the priest arrived a couple of days later, and spoke his few words, all seemed to be well and everyone departed content.

This was the occasion for another party, and once more all the available space was full of visitors.

One factor we had not legislated for was the sudden demand for salt tablets. We always had a reasonable supply available, but, probably because of our large intake of water, never seemed to need them, but most of our visitors soon became dehydrated and salt tablets were in great demand. Perhaps they thought they were aspirins.

CHAPTER 79
A Call to Tehran

We had a visit from the manager at Agah Jari one day, who informed me that I had been asked by London to make a site visit to Mashhad to prepare an estimate for the construction of a new steam power station there, and should go off to Tehran as soon as possible to collect the documents and make local arrangements to visit the site.

I advised my staff that I would be away for a couple of weeks, and made arrangements to drive down to Agah Jari, pick up our local representative, and go on to Ahvaz, where he had already booked us onto the night sleeper to Tehran.

We drove down in the early morning and had lunch at Agah Jari, but when we came to set off again my car had a flat tyre, so we were offered the manager's car, a large Chevrolet. We went off across the desert and after about 20 miles the car stopped.

We tried all the usual tricks, but everything seemed to be in order – except that it would not start. The lights came on, the hooter worked, and all the gauges were registering, but when we lifted the bonnet we discovered that there was no fuel coming through.

Again, we tried all the usual tricks, of blowing and sucking, but still no fuel. In desperation, we rocked the car to try to see if we could hear the fuel swishing about in the tank; not a whisper. Despite the fuel gauge reading half full, the tank was empty.

Our colleague hailed a passing lorry, and was taken on to the next village. He returned some time later with a can full of fuel, and we were soon on our way again, but about two hours later than we had planned.

We drove furiously into Ahvaz just in time to see the train pull out, and enquired if we could catch it at the next station. This was not much help, as the next station was a long way up the line, and we would be just as well to drive the whole way to Tehran.

We then enquired about flights from Abadan to Tehran, and worked out that, if we hurried, we could catch the last flight out in the evening, so we set off again at great speed out of Ahvaz to Abadan.

The road to Abadan was across the end of the desert, just below the foothills, and was dead straight for about 20 miles, then there was a gentle bend, then another 25-mile straight, then a rather sharp bend, and straight again into Abadan for another 50 miles.

As soon as we left the town our friend in the rear seat began to advise me of a nasty corner ahead, and no sooner had we managed to negotiate that one than he began about the next corner. This one was marked by the presence of a large tree. How it had survived is a complete mystery. Why it had never been eaten by goats and camels before the advent of motor cars, or burnt as fuel by the wandering Arabs, or destroyed by vehicles that had failed to get round the corner I cannot guess, but there it stood, like a monument in the desert, with Ali in the back repeating over and over again, "There is a very bad corner at that tree ahead."

No sooner were we round that corner safely than he began about the narrow causeway that leads over some of the channels of the delta of the Shatt al Arab, at least 50 miles away. I had not been on that stretch of road before, and it was as well he had warned me, for suddenly the road narrowed to a single vehicle track, thus requiring a dramatic reduction in speed. We were still cutting things very fine, and I pushed on as fast as I dared, into the airfield, asked Ali to park the car, and jumped out with Jean to try to get tickets and book onto the last flight.

"Yes, sir," the man said, "two seats to Tehran. We regret that there has been some delay, and the flight will not leave for another two hours; please take a seat in the lounge."

With that Ali staggered in, carrying our luggage. We all sat down and had a few drinks to recover from the journey, and, for us, in preparation for the next stage. Ali had decided to remain in Abadan overnight, and had booked into the rest house.

Our plane took off two hours later, and we had a quiet flight to Tehran, and no problems at the airport, where we were met by our general manager and taken to his flat in the city. He was just going off on leave, and showed us round the place, giving us careful instructions about the heating and hot water systems, and then departed.

Our first meeting concerning the contract was not scheduled for another two days, so we spent a quiet time in the flat, finding out how things worked. The flat was on the first floor of a block of fairly old buildings, and it had a small iron veranda along the front, overlooking a large roundabout on the road. The streets in Tehran in that district were fairly wide, with deep ditches, lined with concrete slabs on each side, called jubes. The water from the mountains to the north flowed down towards the town, and was intercepted and directed into the jubes, where it served as the main source of public washing and waste disposal for the outer inhabitants.

We had been out shopping, and amongst other items had bought some cherries, so I decided that I would have a restful morning at ease on the veranda with a bowl of cherries and watch the traffic go by. The roundabout was at the intersection of five roads, and the traffic was quite alarming to watch. Within a few minutes I had seen a number of near misses, and two minor collisions. The near misses called for shouting and screaming by both parties, but when a collision occurred both vehicles were pushed down a side street and the argument continued out of sight. It seemed that the last thing you wanted to do was to involve the police.

Another thing very noticeable in Tehran was the profusion of electric power cables along every street. I heard later that there were seven different power supply companies operating in the city, each with its own network of distribution cables, mounted on its own poles. Some clever individuals had discovered that they could obtain a free supply of electricity by throwing a weighted wire out of a window onto the public supply, and another wire for the return. Very often they had connected to wires belonging to different supply companies, and as a result, when repairs to the line were being undertaken, what should have been a dead line as shown from the layout drawings was in fact very live when the householder lit his light or put on his electric kettle, and, on average, one electricity worker was killed every two or three weeks.

The traffic in the city was like Russian roulette, and although the office car was available we did all our travelling around in taxis. These vehicles could be identified by their white-painted wings. Most of them were in reasonable condition, and the drivers very helpful. Almost invariably they had a quick word of prayer before starting up, and then turned the radio up to full blast and set off at great speed.

There was one famous junction where each road had seven lanes of traffic, and as well as traffic lights there was a small island on a pedestal in the centre, where a white-helmeted policeman tried to settle the arguments between drivers in the wrong lane trying to cross in front of drivers in another wrong lane attempting to go in the other direction. Sensible drivers avoided this junction.

The taxi fare anywhere in the city was 15 rials, and, at nearly 500 to the pound, pretty good value. I was looking for a chemist for some aspirin tablets and jumped into a taxi outside the flat, and said to him in my best Farsi: "Take me to a pharmacy, please." The driver nodded, went through his take-off procedure, and we were away. We drove aimlessly around for some time, until I noticed a large chemist's shop on a corner, with a huge illuminated sign on the wall saying 'PHARMACIE'.

"Look!" I cried out; "pharmacy, over there on the right."

"Oh yes," he replied; "pharmacy," and swung the car round despite the following traffic and pulled up with a flourish outside the shop.

The steward, or farash, who came with the flat was a very willing chap, and was always busy about the place. Generally we made out a list of shopping from the bazaar, gave him some money and left him to get on with it. One day we decided that we should have some prawns with our dinner, and looked up the book for the Farsi word. No luck.

Not to be defeated I tried to draw a prawn, and illustrated its movements by making swimming motions. This was not really fair, as the poor chap had lived in Tehran all his life and the largest volume of water he had ever seen was the fountain in the roundabout outside the flat, but suddenly the drawing clicked for him and he set out, mouthing the correct Farsi word, and waving the drawing in triumph. When he had gone I was not sure he had got the right animal, as a wingless locust and a prawn could both be inferred from my sketch. He probably had the same idea, as when he returned he said the market was right out of prawns (or locusts).

That little incident reminded me of a tale I had heard from Bandar-e Shahpur.

A Dutch company had set up a prawn-freezing plant on the coast, and were making a good thing out of freighting frozen prawns by train to Tehran, where they were in great demand.

During the period of Dr Mossadeq the Dutch company had been taken over by some local entrepreneurs, who seemed to be managing very well, but they discovered that there was a very good market for ice in the town, where a block of about six inches square by 18 inches long was worth selling. In the cool weather, they thought they could get away with sending their frozen prawns to Tehran with a token block of ice in each consignment, but on the first hot night all the prawns went bad, news got back to Bandar-e Shahpur, and the smart operators pulled out, back to India, with their pockets full of cash, while the legal eagles made a lot of hay over claims for food poisoning and all manner of other criminal acts.

Meanwhile I was planning our trip to Mashhad, right up in the north-east corner, where Russia and Afghanistan both border Iran, and meeting with various suppliers of goods and equipment to try and obtain costs for shipping goods into the country, then moving them by rail to Tehran and then on to Mashhad, which was the last station on the line.

We booked onto an Iran Air flight that left early in the morning, just in case we would be unable to stay overnight in Mashhad, in which case we could still catch the evening plane back if required, and went back to the flat.

The domestic arrangements in the flat were unusual, you might say. There was a large lounge with an oil-burning stove in the centre. This monster used up about 20 gallons of fuel per night, and gave out a fierce heat for a distance of about ten feet. The remainder of the room cooled off rapidly as you moved away from the stove, and by the door it was about freezing. The temperature in Tehran varies greatly, from over 110°F in the summer down to about -30°F overnight in the winter. We were there in late October, and it was pretty cold at nights then.

The bathroom was also very unusual. To heat the water you had to go outside onto a small veranda at the rear, where there was a brick firebox, rather like a coal bunker.

Alongside this structure there was a bicycle pump, with the hose leading into the firebox. Nearby there was a fuel tank, and the procedure was as follows:

Open fuel tank valve.

Pump up about 20 quick strokes on the handpump.

Open the small glass door on the front of the firebox and throw a lighted match inside and slam the door shut.

If you were lucky there was a satisfying 'brumph' sort of noise, followed by the roaring of the burner, and you were in business.

If you were unlucky there was no comforting noise, but the fuel was still pouring in, and on the next attempt, after the mandatory 20 strokes of the pump, when the match was thrown in, there was a significant explosion, followed by the roar of the burner lighting up.

After a decent interval the next hazard was the bath itself.

The bathroom was a large room about 15 feet square, with a cold terrazzo-type floor, and the bath set midway along one wall. There was a shower fitting set up over the middle of the bath, and this was the only method of getting water out of the system and into the bath. Inexperienced operators turned on the shower taps, and were drenched in cold water. While recovering from the shock the hot water had come through, but it was almost impossible to reach the tap through the descending stream of red-hot water without the protection of a towel round the arm. (All this while you could feel your feet freezing to the floor.)

Having stabilised the water to the correct temperature, an enjoyable bath was then possible, but this was not the end of your trials. When the plug was pulled, the water discharged in a flood across the floor, probably carrying essential parts of clothing with it, and vanished down a four-inch diameter hole in one corner of the room.

Experienced flat-dwellers could gauge the progress of newcomers bathing by the series of shouts and screams of anguish and despair.

Fortunately, the four-inch pipe discharged into a large tank outside, and it was normally possible to recover any flooded garments before the watchmen found them.

Living in Tehran was different – you could say that.

In the morning we set off for Mashhad in an old DC-3. Mashhad is a place of pilgrimage for Muslims, and we were the only Europeans aboard. All the others were elderly turbanned gentlemen, dressed in long white robes, on their haj, or pilgrimage. All good Mohammedans are expected to try to make a pilgrimage to Mecca, Samarkand, and Mashhad, and when they do so can expect to go straight to heaven when they die. They can also wear a green hat, and have seven wives – or so I was led to believe – and can be addressed by the title of haji.

CHAPTER 80
Mashhad

We took off and soon were heading east, and into the mountains. The snow lay thick all round, and with some alarm I noted that the tops of the mountains were considerably higher than the aeroplane. From time to time the clouds cleared, and we suddenly had a marvellous view of the high peaks of the Elburz Mountains, all shining in the snow.

Inside the aircraft it was very cold, and I called the stewardess to ask her to turn the heating up a bit. It was then that I noticed some light coming in through the floor at my feet, and when I put my hand down there was a great blast of icy air blowing in, through what looked like a bullet hole. The girl came back with a couple of heavy blankets and the situation improved. We came down from our great height over a long, open plain, and could just make out the city ahead of us.

All round the city were extensive plantations of peach and apricot trees, the leaves of which were shining a bright red against the dull green and brown of the cultivated landscape, and in the centre of the city the gold dome of the principal mosque shone dully over the brilliant white-painted walls set above the emerald green of the lawns surrounding the building. We landed and were met by the manager of the power station, who had booked us into the best hotel in town, and had arranged a car and driver to take us about, and shortly we were ensconced in our room, with a good view over the roof of the town bathhouse.

Non-Muslims in Mashhad were as scarce as hen's teeth, and it appeared that we were some of the very few Europeans who had ever visited the town on a non- religious pilgrimage. We were advised to keep out of sight near the mosque, and Jean had to have her head covered at all times; in fact, she was told she should remain inside the hotel, or at least inside the car, if ever we went out in the city.

On the first morning of our visit I went down to the existing power station to see the present equipment, and find out about projected electrical loads, fuel supply situation, etc.

Everything seemed to be well under control as we sat talking with the manager, and then the lights flickered and went out. I was just about to ask if they really needed a new station, but thought better of it, and the manager said, "This happens all the time; the system is grossly overloaded, and my phone never stops ringing with people making complaints."

More or less convinced of the need for a new station, I went off round the town to the proposed site. This location was on the town midden, a large muddy area where all manner of items had been dumped, and in fact were still being dumped, from household refuse to the sweepings of some nearby camel stables. It might have been of some interest to an archaeologist, but as the foundations for a power station it left a lot to be desired.

When I returned to the manager I asked him where the new town refuse dump was to be, and the answer was that no formal location had been determined, but, doubtless, Allah would find a suitable place. Building power stations is difficult anywhere, but having persons dumping rubbish into your excavations was a problem I had not previously encountered, and I made a note for my report to the effect that special provisions should be noted in the contract.

The next place I visited was the railway station and marshalling yards, with a view to seeing what sort of area and service would be available for moving and storing heavy equipment, plant, etc. The stationmaster was a model of efficiency, and produced drawings, layouts, timetables, freight rates and passenger fares as we walked round his establishment, and I was impressed. The main traffic into the town was pilgrims, and regular services were run on the recently opened line from Tehran. The facilities were all new, and the stationmaster full of enthusiasm.

The buildings in the town were very old around the mosque, but further out there was an industrial estate being developed, and pleasant rows of shops and small houses were evident. In some contrast to the south of Iran, there was an air of bustle and activity about the place; every second shop was a workshop, where agricultural hardware was being produced, small three-wheeled tractors being repaired or converted, or furniture being built. The other shops were full of clothing and food, and seemed to be doing a roaring trade.

We went out of town to look for suitable quarry sites, and found a large new reservoir being constructed, with its own quarry already operational, and were welcomed into the manager's office, where we were soon discussing costs and rates of production.

All round the town were orchards of plum and apricot trees, and we drove down fantastic avenues of scarlet leaves to well-established farms and processing plants, with tidy gardens.

Our hotel was worth a mention. I suppose there were about 30 rooms on two floors, and we had managed to obtain the end room on the top floor. The furnishings were adequate, but the heating system was not. To the front we could see over the town to the dome of the golden mosque, a most impressive sight. Unfortunately, we were in direct line of sight to the loudspeaker in one of the minarets, which blasted off at regular intervals, calling the faithful to prayer and repentance or whatever. To the rear we overlooked the roof of one of the many bathhouses. Since most travellers had come to Mashhad on foot, or by horse or camel before the advent of the railway, I suppose public baths were a good idea, and there were many such establishments in the town. The roof was flat, over an area of about 100 feet square, with a small parapet all round, and there were numerous domes sticking up at seemingly random centres all over the surface. I was unable to obtain an explanation for these bumps, and never plucked up enough courage to enter the place and see for myself.

The dining room was in the basement, and was dominated by a huge, blue, enamelled stove, about six feet tall. It gave out great heat to those nearby, but around the edges of the room it was very chilly. As we were special guests, we were given pride of place, next to the stove – what you could call a warm welcome.

We thought we would take a run out to the Afghan or Russian border, but were turned back some 20 miles off and were unable to get any nearer. The border region was full of soldiers, but otherwise uninhabited.

Having found out all I could in Mashhad we booked our flight back to Tehran, and returned by the same leaky aircraft. Once again we had a fine view of the mountains to the north, with the peak of Damavand soaring up into the sunset. Then it was back to the flat to sort ourselves out, and start on the report.

There was a holiday weekend during our stay, and we arranged a trip to the Caspian Sea and set out early in the morning.

The road out of Tehran ran along the foothills for about 50 miles of good tarmac, and then turned into the mountains and became a rutted track most of the way.

At first we were in a wide valley, with crops growing along the edges of a wide river, and as we progressed the valley narrowed, the sides became steeper, and the course of the river was marked by thick clumps of trees and bushes. Soon we were in a narrow gorge, with the road clinging to one side, steep rocky cliffs above us, and a long scree slope down to the rushing river below. Except in the gullies and along the riverside the ground was bare, and there were no signs of cultivation.

As we climbed still higher we came out onto the side of another hill, with snow patches all round us, and at the foot of the ravine we could see a number of wrecked vehicles, which we presumed must have been carried down by an avalanche. Our driver informed us that what had happened was that during the winter, when the road was under the control of the military, a one-way system was operated, with army signallers at each end of the restricted section. Someone, full of his own importance, had overruled the signalman and had met the convoy coming in the opposite direction, with an officer in charge in the leading vehicle. He had called up his man at the other end, found that the oncoming vehicles had failed to follow his instructions, and without further ado unloaded the vehicles and had them pushed over the edge, to fall down about 300 or 400 feet. I suppose the occupants were fortunate that he had allowed them to disembark.

At the end of this long section of road we came out to the portal of a tunnel, with snow all around about two feet deep, and drove into the darkness. When we came out on the other side, it was like a different world.

The hill faced to the south-west, the snow line was hundreds of feet higher, there were trees and shrubs, and small streams ran gently down the mountainside. Even the houses looked different, with stone walls, stone-slabbed roofs and small gardens.

As we descended the air grew warmer, the slope of the road eased, and we were in a forest, with large fir trees and shady glades, and further down we came into orange orchards, with a splendid view of the Caspian Sea – and a tarmacadam road, too. We stopped at a small hotel in Rasht – rather a lovely building, with wide colonnades, in a beautiful garden, full of roses and other exotic plants. This area was the summer holiday playground for the wealthy citizens of Tehran, and there were elegant mansions set in the foothills all along the coast.

Next day we drove along the shore of the Caspian Sea, and came on an enormous harbour, about the same size as Dover, with only one small rowing boat to be seen afloat. This was one of the ports for the sturgeon fishing, and at some periods of the year the harbour was full of boats; but not today. We continued along the coast for some miles, still in delightful country, to Sharia, where we stopped for lunch, again in a well-appointed hotel, and then set off back to Tehran, and had a pretty uneventful journey.

After a few more days completing my report, we set out to return to Gach Saran, via Abadan, by Iran Air. We checked in at the terminal, and were quickly into the transit lounge, when there was an announcement that there would be some delay. Not unduly worried, we sat drinking coffee, and noticed a man in a white boiler suit with a long screwdriver in his hand go up into our aeroplane.

Some time later he came out again, and we were invited to board. We settled down, and first one engine, then the second, and then the third fired up, but there was no action out of the remaining one. We disembarked again, and once more the man in the white suit went aboard. We had had enough coffee by then, and had started on the gin.

Once more we went out, and, again, three engines fired up without trouble, but a different engine failed to start this time. Out again. More gin, return of the screwdriver man. More gin.

Called out again, gin beginning to take effect.

Still only three engines. Announcement from the captain. "We seem to be having some trouble with our starting sequence; please remain seated and we will try again."

Some fuddled calculation was going on in my head: if it takes four hours with four engines, how much longer will it take with three?

Return of the man in the white suit bearing his long screwdriver, who disappears into the depths of the floor of the cockpit. Sudden roar as the fourth engine fires up.

Round of applause for the mechanic. (Perhaps he should stay aboard?) Captain announces that all connecting flights have been missed, and accommodation booked for all hands in the rest house in Abadan, and we are off into the evening sky.

When we arrived in Abadan, cars were waiting to take us to the rest house, and free drinks were available in the bar. Good old Iran Air.

We had booked onward flights on the oil company's de Havilland Dove the following morning, and trooped out to the airfield and climbed aboard. The pilot went through his pre-flight procedure, and tried to start the first engine. Not even a grunt out of it. He turned to his passengers and said, "It often doesn't want to start these days; just hold on and I will give it a bit of a tweak," and so saying he climbed out and went over to the propeller, and hand turned it about half a turn. When he tried the starter again, it went off like a bird, and we were soon on our way back to Gach Saran.

CHAPTER 81
Further Events in Gach Saran

The oil company aeroplane, a Dove, flew from Abadan to the main oilfield towns of Masjed Soleyman and Gach Saran every day, but in a different order.

This meant that every other day the plane came straight from Abadan, which was the main port of entry for passengers for the fields, to Gach Saran.

One day we were expecting a new member of staff, and had heard he was in Abadan, and coming into Gach on the afternoon plane. There was some delay in Masjed Soleyman that day, and we were advised that the plane would stop over in MS, and would come on the following morning.

When we met the plane the next day there was a bit of an argument going on with one of the passengers (not our man, I hasten to add).

When he stepped out of the aircraft and saw the barren mountains all round, and the half-completed airfield, he declared to all and sundry that he had not contracted to work on the surface of the moon and was going straight back to Abadan, and then home. No amount of persuasion from his site manager was effective, and by the look of the man the place was the better for his absence, so he was bundled back aboard and flew out.

Our chap, who appeared to be very subdued, handed me a letter from the field superintendent at MS (God's local representative), stating that this man needed watching, and if he ever came into MS again he would be on the next flight out.

We collected his baggage and drove back to our camp, and after he had settled in I called him into my office for an explanation.

It appeared that when he arrived in MS, and had to stay overnight, he had been allocated a bed in one of the contractors' camps, and there had been a bit of a party going on to celebrate someone's departure (or release, some people might have said).

There had been a lot of drink about, and the party had become pretty wild, what with singing ribald songs about the field management, and one thing and another, and our man had joined in enthusiastically (thinking that, perhaps, if this sort of event took place often, Iran might be a very entertaining place to stay).

Due to some complaints from the neighbours, the party was broken up, and the FS, whose dignity had been besmirched, took what action remained to him. He could not do much about the main participant, who was leaving anyway, but all others present had a severe warning letter to the effect that they had better behave themselves in future, and any further transgressions would lead to instant banishment.

Our chap turned out to be a willing and cheerful member of the staff, whose main accomplishment off duty was his skill with an airgun. He could hit a beer can after it had been thrown up into the ceiling fan three times out of four.

Another of my young engineers had the misfortune to break his spectacles, and sent home for a spare pair. In the meantime he went down to the oil company hospital in Abadan for an eye test. The consulting room was in a large hall, with many people wandering about between the patient and the target board and interrupting the test. At the end of it all, the optician had said it would take a couple of months to have the spectacles ready, but he could arrange for a white stick if he liked. When he returned to Gach his spectacles had arrived in the post, but had been damaged beyond repair; the glass was all there, but in about 1,000 pieces. We got a message back to his mother, via someone who was going home, that she should send another pair to the office in London, and they would arrange for them to be brought out by the next traveller. In the meantime we went out to the bazaar, and found some ancient pair that were of some use, and so he did not need the white stick after all.

Communications with London were very difficult. The post was often opened and pages removed, any articles that were breakable were invariably broken, and the delays were immense. Telegrams were all censored, and almost always garbled beyond deciphering, and the international telephones were impossible. All our correspondence was sent off hopefully, but back-up copies were always sent by hand of people returning to the UK, and brought out by new arrivals. We frequently had a competition as to what a particular signal had been trying to tell us, but were as often wrong as right.

We were in Gach Saran on the day the son of the Shah was born, and the whole town was celebrating. Shops and offices were decked out with palm leaves, it was declared a public holiday, and all vehicles were decorated with leaves and flags.

We had been having a lot of extra work to complete a road ready for a rig move, and a day off was most welcome. In the evening we were sitting in the garden enjoying a quiet drink, watching the sun go down, when we heard a lot of shouting coming from next door.

Our neighbour was not then very fluent in Farsi, and the conversation was all about what the farash had been doing. Here it was, half past six, and he had not been woken, nor was his breakfast ready. The farash tried to explain that it was half past six in the evening, and the sahib had slept right through the day, but was not making himself very clear. I called over the fence that we were having a sundowner, and asked if our colleague would like to join us, and there was a stunned silence before he replied that he had better get himself washed and shaved first, but would be right round directly.

The drilling in the field was all undertaken by drilling contractors, nearly all American, and their staff lived in caravans in the town. They generally worked 12-hour shifts, and sometimes slept in cabins by the rig, but all their gear was in their caravan. Some drillers had their wives with them, and we were frequently invited over for tea, or a drink. One weekend we were visiting one of the caravans, which was shaped like an aircraft fuselage, and extremely well fitted out, when we heard a lot of noise coming from the adjacent caravan. We went out to see what all the noise was about, and found the occupant in a frenzy.

The system of pay for them was that all their money was credited to their accounts in the States, but any cash they required could be drawn from their local office on site, against signed chitties, and this was paid in rials. This chap had been set on going back to America on leave, and had called in the office to have them make the arrangements. He was a bit upset when they told him that he had spent all his pay locally, and there was not enough left in his account to pay for his fare.

Apparently, it had been his custom to draw his rials for immediate spending, and he had put all his unspent money and change in a number of shoe boxes, and other such containers, and he was just then trying to see how much money he had in the caravan, and if there were any other boxes he had not yet found. This to the accompaniment of much cursing and swearing. It turned out that he had a great pile of rials, but still not enough to keep him on leave, and he had to arrange to work on for another couple of months before going on leave. When he tried to exchange his rials for dollars the exchange rate had altered, and he was of the opinion that everybody was against him.

He was not a very bright individual, and even his brother drillers thought that he was not quite the full shilling.

In the middle of the dry season we had a sudden rainstorm in the town. It had not rained during the dry season in living memory, we were told, and this shower caused a number of problems. On one stretch of oil-bound road about ten miles out of town the surface became suddenly very slippery, and at either end of this section half a dozen lorries came off and ended up on their sides in the verges, together with three that were already on the road when the storm struck. One of our vehicles was stopped by this scene, and came back to get help, and we went off with the Land-Rover and some gear and called up our grader, which was not far away. Some of the oil company vehicles also arrived, complete with lifting booms and winches, and the vehicles were all recovered by midnight, mostly slightly bent but with no serious injuries to the occupants.

The storm also filled a wadi that ran through the town, and in the dry weather this was usually full of sheep pens, rickety huts, and a lot of the town rubbish. When the storm came, the wadi, which was about 20 feet wide and ten feet deep, filled up with roaring water, and swept the place clean – again, mercifully, without loss of life.

After the rain stopped the sun came out again, and the whole placed steamed. There was the most pungent smell of sheep dung, which had been lying about in little balls, all dried up, and which swelled up in the wet, and ponged. There was a sudden plague of flies. Normally we were little troubled by flies, with the odd one or two buzzing about in a desoultory fashion in the heat, but after this sudden rain they came out by the thousands. We had a washing line just outside the house, and it was completely covered by them, and looked as if it was about half an inch thick. The small house lizards had a proper beanfeast. It was pretty awful for the poor watchmen in their tent, but they had lit their small fire inside and were quietly kippering themselves.

That same day we had a camel train pass through, just behind the house. The camels were very frisky and trotted past in fine style, with their bells clonking, and the riders in their flowing robes urging them on.

Some time after this storm we went out one weekend to look at the line of a new road, about 15 miles out in the hills, and took a picnic. The weather was not too hot, and after the rain the grass was beginning to grow and some of the streams were running at the foot of the hills. We drove along the line and stopped for lunch, which we spread out on a blanket – jug of wine, loaf of bread, and all that – in the wilderness, and lay back afterwards, enjoying the warm sunshine.

We went on a short drive around the top of a small hill, and suddenly the front wheel of the Land-Rover fell into a hole in the ground. When we got out to see what had happened, we found that there was a hole in the surface, about two feet wide, which opened out below into an underground cave. We tried dropping small stones in to see if we could find out how deep it was, so that perhaps we could fill it up with nearby material, but the bottom was well beyond our reach. Next we tried to place rocks under the other wheels, to give some traction, as they were by then all clear of the ground, with the chassis resting on the top of the hole.

This required a lot of jacking up, and the fitting of selected rocks, and there was always the chance of the top of the hole collapsing, and the whole machine going in.

We worked away carefully for about an hour before we dared to try to move the car, and it came off very gently. We then had a good look around, and discovered that the top of the little hill was full of similar holes, and realised that we must have been very lucky not to fall into one earlier. We then proceeded with extreme caution, with one person walking ahead, scouting out the best route, until we were back onto the valley floor again.

The oil company had been doing some geological exploration some miles south of the town, and we met with the geologists in the club. They went on tour into the hills for about six weeks at a time, with mule trains and tents, and carried out their preliminary surveys literally on the hoof. They came back into Gach Saran to write out their first report, then went off to Tehran for some leave, and then returned to the field to continue their investigations.

This particular party was supported by helicopter, and they ran seismic surveys and other investigative work, and located a sizeable new prospect. Their first location for a trial hole was about three miles away from one of our sites, across some really dreadful country, and a special team of road-builders was called in from America. They brought along their own caravans, and within a week of their arrival they had a good camp set up, with free ice cream, a good mess, and even regular film shows, out in the mountains.

They had imported the biggest and best equipment, and were tearing into their road-building, blasting down whole sections of the hill and smashing their way through to the proposed rig site without a care for what went on in the river bed below.

They finished their operations and the rig arrived and was set up within their target time, the road-builders went off, the drilling started, and this turned out to be a bumper oilfield.

By this time we had been in Iran for about 18 months, and must have been showing the strain. The normal tour was about a year, and as our section of the work was nearing completion we began to close the site down, and left for Abadan, leaving a number of new friends behind and some happy memories of our stay.

When we arrived in Abadan we took our baggage to the rest house, and went on a short tour of the town. There were some interesting shops and we bought a few odds and ends, and then I thought it would be a good idea to exchange all our rials for English pounds, so we inspected a number of money changers and their rates, and settled on the least dishonest looking of them, and began to bargain.

When we had agreed on a rate he produced a handful of new £5 notes, which had been introduced during our stay in Iran, and was quite hurt when I rejected them and demanded the proper large white fivers.

He had none of them in his shop, however, so we renegotiated the rate in US dollars and settled on them. Come to think of it, he could just as well have made them in his back shop for all the skill I had to detect forged dollar notes.

The smell of oil in Abadan was just as bad as ever, but we seemed to be out of it a bit in the rest house, and after a comfortable night set off to the airfield.

We were due to fly out in a de Havilland Comet, coming on from Karachi or somewhere east, and arrived in the airport buildings with our baggage in plenty of time. A delay was announced, and we were invited to have coffee. A further delay was announced, and it was suggested that we should return to the rest house for lunch. We had nearly finished a large lunch when we were called back to the airfield, in time to see the aircraft land, and were soon aboard and fastened in.

The Comet take-off was most spectacular; we ran along the runway pretty fast, and then seemed to stand on our tail and go straight up. We levelled out and the crew came round with drinks and another lunch, and soon we were landing in Beirut.

During the flight we had spoken to the crew, who confirmed that the old fivers were out of fashion, but that we would probably get a better rate for goods in dollars in Beirut, and when we landed and discovered that we would be stopping for a few hours we went off into the town for a quick look round. We purchased a few small souvenirs with some of our dollars, which were accepted without question (it felt as if I had been in Iran for too long, suspicious as I was), and we even changed some dollars into the small fivers, but not without some trepidation.

Back aboard again, and another spectacular take-off, another large meal, and landing in Rome (still under development), where we were invited ashore for tea in the airport building. On again, with a large dinner presented and consumed, and we landed in London, where we were quickly through the formalities and off in a taxi to a hotel in the West End, where a second dinner was waiting for us.

Absolutely gorged we staggered up to bed, and tried to get our sense of time back into order.

In the morning I went into the office, delivered some mail and a report of our efforts in Gach Saran and hired a small car, and we set off to Devon.

The car hire office were a bit put out when I presented my Iranian driving licence, which was all written in Arabic script, and when they asked me for the date of issue, I read out: "The fourth of the seventh month, 1327," which was what it said, or something like that. Anyway, the man believed me, and off we went, anticipating a six-week period of leave in a country with new small £5 notes. What was it all coming to?

We had a short spell in Devon, then travelled north, had a couple of weeks around Glasgow, a few days in the far north, where it had been snowing quite heavily, and returned to Devon to assist in the decoration of the house, and rest from our travels.

I had a note to ask me to call the office, and was advised that I would be asked to go to Sierra Leone on a dam construction project, and could expect to be in London for three months on the preparatory work before going out. We drove up to London and began flat hunting again, and found a pleasant house in Ealing available for a short stay, and moved in immediately.

PART FOUR –
SIERRA LEONE, NIGERIA AND LIBYA, 1962-1970
General map of Sierra Leone

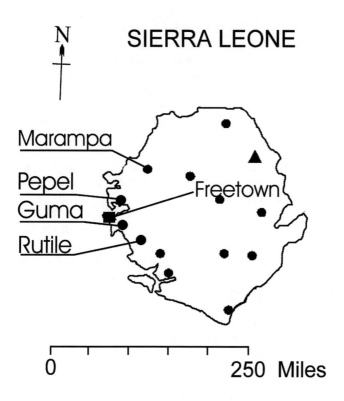

N

SIERRA LEONE

Marampa

Pepel

Guma

Rutile

Freetown

0   250 Miles

SECTION 1 – GUMA DAM

CHAPTER 82
Preparations

When I returned to the office in London I was advised that I could expect to depart for Freetown in about two weeks, and in the meantime I had to sort out our domestic arrangements, get myself revaccinated, collect my tropical kit and try to catch up with the head office work in connection with the ordering and selection of equipment, and the establishment of shipping routes and documentation through the Colonial Office, including work permits and all the other bureaucracy which seems deemed necessary to start any enterprise.

Once through all this I was off to Devon for a long weekend, and then set off again for London by train, and soon found myself at Heathrow preparing to depart overseas again.

The route this time was to Lisbon, Las Palmas, Bathurst (now Banjul) and Freetown, with an overnight stop in Las Palmas, and we flew in a Douglas DC-4.

First stop was Lisbon, which looked most attractive from the air as we came in. We stopped for refuelling, and on leaving were presented with a small bottle of port; very civilised.

On again over the sea for a long hop to Gran Canaria, where we landed in the midst of airport construction – dust everywhere, confusion over the baggage, and a frantic journey by minibus into the town of Las Palmas. The demon driver strikes again.

We were taken into a very elegant old hotel in one of the attractive town squares, and allocated rooms. My room was on the shady side of the building, and from the little balcony I could see out onto the square, where there were pleasant gardens, with seats, and the citizens were taking the evening air.

This was my first visit to Spanish territories, and I was intrigued by the lettering on the taps in the bathroom. I worked out, by trial and error, that 'C' meant cold and 'F' was for freezing. This apart, the room was very comfortable, with a cool tiled floor and dark, polished furniture.

When we assembled for dinner later, we were told that the bus would depart at 4:30 in the morning, and someone would call on each passenger at about 3:30. After a pleasant dinner, with a tame band, and specially imported Spanish dancers, I turned in, and was woken a very short time later by a great thundering on the door and shouting by the room service staff, announcing that the bus was already loading. This was, of course, a joke, but it certainly roused most of the customers, and breakfast was rather restrained.

Out again to the airfield, and off towards the south, heading for Bathurst. Before landing, the captain advised us that there would be a bit of a strange noise when we landed, as the runway was made up of steel planks and wire mesh, and it rattled a bit. Right enough, there was a great clattering racket as we touched down, which continued all the way down the runway, and right up to the airfield buildings.

This was a bit more like it. The buildings were about the size of a large bungalow, with not much effort wasted on separating passengers from spectators, and very little in the way of formalities concerning customs and immigration.

Outside, in the shade of some large trees, some locals were selling such things as snake skins, hand drums, small paintings, shrunken heads carved in ebony, and the usual tourist trap miscellany of wooden carved animals.

Back aboard again, and after the hellish noise, away again, headed for Freetown. As we approached we could see the brown stain of the main river extending far out to sea, and for several miles up the coast, and on the other, southern side of the estuary the town of Freetown against its background of wooded hills, with the deep blue of the sea running right up to the point of entry for the estuary. (NB. Only go swimming in the sea to the south of the town, where you might have a sporting chance of seeing a shark coming for you.)

The airfield was what was left of the old RAF station, and had a few large corrugated iron sheds, a few Nissen huts, and the regulation control tower. The customs were a bit more active here; either it was later in the day, and not so hot as Bathurst, or they were a little bit concerned about smuggling diamonds. Incoming passengers were not much troubled, but outgoing passengers were examined fairly closely.

Once through the formalities we were herded onto a ramshackle bus, and ground our way down to a small jetty, where we boarded the ferry to take us across the river to Freetown itself. In the town we were met by our local manager, and taken first to his office, and then to the accommodation that had been prepared for us.

For the first few days we were billeted with other members of the Freetown staff in their fine houses overlooking the ocean and the golf course.

After a couple of days we had organised our transport, and moved into the government rest house in the town. This was a relic from the days when Freetown was a transit camp for troops going round the cape to India, and the rest house was constructed on military lines. The rooms were all in a number of long buildings, with open verandas on one side, and the toilet and washing facilities at each end of each block. All the doors opened directly onto this veranda, and there was a constant flow of patrons passing your door. The mess and reading rooms, etc., were in a central block and kept spotlessly clean, and the food was pretty fair.

The grounds were kept in order by gangs of prisoners under supervision from the nearby jail, fondly referred to as 'King George's Hotel', and it appeared that the warders looked a lot more criminal than most of their customers.

There was a tradition of lunchtime meals on Saturdays at the City Hotel, and we were encouraged to attend this ceremony. The City Hotel was a large square building, near the town centre, which had the dining room and main bar some ten feet or so above the surrounding ground, and access was gained by mounting a flight of stone steps onto a long open balcony. As you climbed up, it was possible to catch a glimpse of the operations in progress in the kitchen, or basement, as it was.

No signs here of modern hygiene; no fancy electric cookers, nor stainless steel utensils. Indeed, it looked like a scene from Dante's *Inferno*, with large wood- or coal-fired furnaces belching black smoke, half-naked stokers with shovels, and cooks brandishing knives and choppers, while large black cooking pots, suitable for a small missionary or two, were bubbling away and white-coated waiters were standing in line at a counter, ready to collect their orders and carry them up to the hungry and thirsty customers upstairs.

Despite all this, the food was quite excellent, and the service really first-class. At this time, the new modern hotel was under construction and soon to be opened, and I suppose the old City was pulling out all the stops to maintain its customer base.

We were fairly busy for the first week or so in town, seeing the bank, the customs office, the public health office, various shipping agents and suppliers, and making out orders for the initial supply of locally manufactured equipment and supplies, including the buildings for our construction camp, which were being made at a place called Bo, some distance up-country.

The site for our camp had already been selected, about 12 miles out of town, on the coast, and our first job was to make a start on clearing the bush and setting out roads and services, and house footings, and engaging the services of a local contractor.

During this period we were allocated a group of four prefabricated houses nearby, which had been the camp of a drilling contractor some years previously.

CHAPTER 83
The Prefab Story

The four houses had been in use for about a year, and had been left properly closed up, but unattended for about two years, when we arrived. We obtained the keys from the Public Works Department engineer, who advised us that the headman in the nearby village had been asked to keep an eye on the camp, and said that it might be worth our while to speak to him before we started work.

We travelled out to the site, which was on a slight rise in the main coast road, and at the junction where the existing track led up to the site of the proposed dam. The local headman was pleased to see us, and we arranged for some of his men to come out the following day to assist in clearing the bush around the compound and make a check on the state of the premises and the services.

We forced our way into the compound, which was very badly overgrown with bushes and shrubs, and made a quick survey of the situation. The buildings, from the exterior, looked in reasonable shape, with most of the windows intact, and all the doors secure, although there were creepers growing up the walls and over the roof. Each house had an outside brick-built kitchen with a corrugated iron roof, and the vegetation had taken a good hold both inside and out, with interesting plants sprouting from the sinks. The doors and window frames were infested by termites, and in a rather delicate state.

The water supply came from a small dam in a stream behind the houses, and when we had cleared out some leaves and assorted vegetation it seemed to be in order. The taps ran clear water in the sinks, and the sinks drained away once they had been cleared out.

Next day we were met on the site by the headman and a number of his men, and after agreeing rates of pay we set them to work clearing the bush, taking the worst of the growth off the walls and roofs, and clearing out the external kitchens.

Each kitchen was fitted with a wood-burning range-type stove, with a firebox, ovens on either side, and hot plates on the top surface, and we directed that any useful wood cut from the bush-clearing should be cut and stacked ready for the stoves outside each kitchen.

The men were well aware that there was a chance of long-term employment on the construction work to follow, and worked away cheerfully and willingly, pausing every now and then to chase snakes or lizards which had taken up residence in the campsite, and at the end of the day the road access was clear, and most of the walls and roofs had been exposed.

Towards evening we ventured inside the houses. There were signs of damp in the areas where windows had been cracked, or forced open by the creepers, but generally things were not as bad as we might have expected, and we planned that each house would have a good scrub-out and be left with the doors and windows open during the day, while we investigated the electrics and the plumbing.

Next morning about 20 men appeared, and we conducted a quick interview to select those who had any previous experience of domestic work. Most of those present had testimonials, many written in the same hand, and with the only distinguishing feature the man's name and age, all declaring that they had been good and faithful servants, etc., etc. We selected eight men, and directed the remainder to start work clearing the bush at the side of the road up to the site, and turned our attention to the tasks in hand within the camp. Three men were detailed to each house to sweep out the floors, clear away the cobwebs and open the windows and doors. Another four were selected as more suitable for skilled operations, and set to cut firewood, light up a couple of the ranges, and boil up some water for the house cleaners.

I enquired if they understood the workings of the ranges, and they replied: "Yes, pa, we savvy them fine."

Believing, in my innocence, that there was not much to go wrong with lighting a fire in a range, we set off up the track to supervise the work of clearing the bush from the roadside.

This track had been in use about two years before, and had suffered from lack of maintenance. Most of the culverts had been blocked, and the water had overtopped the head walls and scoured out numerous sections. The trees and bushes had overgrown the verges and some had fallen across the track, causing further wash-outs and denying wheeled access any further up the hill.

We made an estimate of the work that would be required to allow normal traffic up to the site, about three miles away, and what equipment we would require to undertake this work.

When we returned to the campsite we could hear the chopping of wood and the general murmur of chatter normally associated with men at work, punctuated by severe coughing, and as we drew nearer we could see a cloud of smoke arising.

"My God, they've set the place on fire!" was the first thought. We broke into a run, and saw when we entered the compound that the smoke was coming from one of the kitchen buildings. Our two skilled operators were standing by the door, coughing and spluttering, and the whole building was full of thick smoke, but none of the other men seemed to be much concerned.

The second thought was: "At least they have only wrecked one kitchen. We might be able to explain that away to the termites." When we had caught our breath the interrogation began, more or less as follows.

Me: "What is your name?"

Reply: "My name is Kamara, pa."

Me: "Well, Kamara, why are you setting the house on fire?"

Kamara: "Pa, we are not setting the house on fire. This stove is humbugging us, he does not agree to draw."

This seemed quite a reasonable answer.

Me: "Perhaps the chimney is full of birds' nests. Climb you up onto the roof, and poke it down with a long stick."

Short pause while a stick was cut, and this stratagem applied.

Still no success.

By this time the smoke inside had cleared a bit, and we could see into the stove.

There were three doors along the front of the stove, the firebox in the centre, and two ovens, one on either side. What the poor chap had done was to open the larger oven, and tried to light the fire inside that. No wonder it failed to draw.

This discussion had halted all other work on the site, and we were now the centre of a small crowd of interested spectators.

"Kamara," I said, "I thought you said you savvied this stove? All morning you should have been boiling water here, and all you have done is smoked out a few ants and spiders."

Kamara: "Yes, pa."

"You had better go on up the road, and help with the bush-cutting."

One of the other men in the group said that he had worked in these houses when they had been occupied earlier, and he knew all about the stoves, so we gave him his chance, and soon the fire was burning brightly and the water heating up.

Inside the houses all the electrical cables had been run in small-bore pipes, and each pipe was the home of a different creature, which had spent the past two years devouring the insulation. We decided that the whole place would need to be rewired, and hired a local electrical contractor to come out and make a start. On his first visit the first cable he pulled out was the nest of a family of scorpions, which fell onto the floor, and caused immediate evacuation. Thereafter he was a bit more careful.

The work progressed well, and within three weeks we had two houses ready for occupation; another engineer and I moved into one, and the accountant and the senior plant fitter moved into the other. We were still without electricity, but managed fine on the wood stove and tilley lamps while the power was being installed and the generator set up.

CHAPTER 84
A fishy Story

At this time the territory was being delivered from colonial rule (?), and this had a number of direct effects on our progress on the construction – or, at least, on the preliminary works towards the construction.

The main construction camp was about three miles away from our prefabs, right on the beach. We had prepared the area by cutting down the bush, and made roads and paths to all the prospective buildings, laid the pipes for water and drainage, and erected most of the small concrete piers which would keep the floors off the ground. The houses were to be of timber construction, and were being prefabricated by a contractor in the town of Bo, some distance up-country.

Due to the celebrations about to take place, all the floors for our houses had been commandeered for use in the transfer ceremony, and our work on the camp construction was held up. In addition to that, everyone and his brother had come to Freetown for the celebrations, and all accommodation was fully booked up, and there were no house servants prepared to come out and work for us in the country when such good rates were available in town. Furthermore, a public holiday had been declared, but some confusion had arisen and it was not very clear which days were officially designated, so we reckoned that there would be no work done for about three days, plus a weekend.

My colleague, Eric, had bought a small fishing rod set in Freetown, and we thought it might be fun if we spent the holiday period fishing, so I went into town for a similar set of gear. Like practically everything else in Freetown, they were sold out, and I returned rodless, but managed to buy some line and hooks and a small reel.

In the morning I spoke to my friend Kamara and asked him if he savvied fishing.

"Yes, pa," he replied; "I savvy him fine."

Not much joy there.

"I need some bamboo canes." I tried again: "do you savvy bamboo?"

"Oh yes, pa. I can find plenty good bamboo."

"All right, let's see if you can find me some good bamboo. It is for a stick, for fishing."

This gave him to think. How the hell was the pa going to catch a fish with a bamboo stick?

"I will go off and find you some good bamboo, pa." And off he went.

In the evening, as we were sitting outside the bungalow, I could hear the tramp of feet, and then, through a gap in the bush, a large pole appeared, then a head, followed by a lot more pole, then another head, and still more pole – and this repeated itself several times.

With some difficulty, Kamara manoeuvred this bamboo into our compound, and laid it out for inspection, followed by another three large bamboo trees.

These bamboos were about 30 feet long, and seven or eight inches in diameter at the large end.

Kamara spoke. "Pa, these bamboos are the very best in the forest. We caught them especially for you, for the fishing." (A bit mystified here.)

"Thank you very much, Kamara; these are indeed splendid pieces of bamboo, and will do very well." (No further explanation given.)

Next day, being a holiday, I went out into the forest and found some bamboo of a more manageable size, and Eric and I sat fishing in the small stream nearby for most of the afternoon – without much luck, I must say.

What could we do with these trees?

We had a bit of a brainstorm, and decided that they would make an excellent template for the low-loader lorry that we would soon be using on our road to transport our machinery up to the site, and when work started again after the holiday we made the bamboos into a large rectangular frame, which we towed up the road behind the Land-Rover, and could then mark off any trees and narrow places that would restrict the passage of the machinery.

We saw Kamara working on the road from time to time, but he never made any comment as to the likelihood of catching fish by towing this framework up the track.

He probably put it down to the 'white man's madness'.

Meanwhile work on the main camp proceeded, and some houses were ready for occupation, so we moved out of our prefab and into our new timber house, right by the beach. We left our houseboy at the prefab for the new owner, and engaged a new man from the next village – much more convenient to our new residence.

CHAPTER 85
Offices on Site

The access track from the prefab campsite to the actual site of the dam was about three miles long, and climbed steeply up a narrow valley in the forest. The site of the dam had been cleared during the previous exploration contract, but secondary bush had regrown over the whole area, including the premises used as offices and works stores.

For our first accommodation on site we took over a timber shed of about 20 feet square, with a narrow veranda and shuttered window openings, and set up our office table and chairs with another set for the timekeepers and clerks. This hut was in the forest, and it was always very dark inside, so one of our first tasks was to fell a number of large trees, both to give us some light and also to provide room for the main office complex, and stores and workshop buildings.

As soon as the road was passable all manner of goods began to arrive on site, and soon we were able to move some of our equipment into another building, which we had cleared and repainted.

The trees in the office area were very large, over 150 feet tall, with huge buttresses at the roots going up about 20 feet, and very few branches on the trunk until over 100 feet up. Felling these trees was sometimes quite tricky, as either they were leaning towards one or other of the buildings, or were entwined with other trees at the top, and it was difficult to judge where they might fall.

At this stage of the construction we had no heavy machinery on site, and all our operations had to be carried out by hand. We found it best to set up a staging around the foot of the tree, where the buttresses were not too large, and then hack away with cross-cut saws and axes. During these cutting operations the resident population of the tree generally took fright, and there was a constant fall of caterpillars onto the woodsmen. These were readily collected, and kept for eating later.

We frequently disturbed flying foxes, and they would launch themselves out of the upper branches and fly gracefully down to the nearest available tree, and then scurry back up again into the canopy. This caused great excitement, as a flying fox was a good meal, and was known locally as 'tree beef'.

Sometimes there would not be a convenient tree, and the flying fox had to land on low scrub, or even on the roof of one of the buildings, and then there would be a mighty hunt, and all work stopped until either the poor thing had escaped or been killed.

When trees were likely to fall on our buildings we tried to get a rope around them, and guide the fall away from the buildings. Generally a promise of extra cash would be enough to provoke one of our crew to try to get up, but some trees were quite impossible to ascend, and novel methods were employed.

The first trick was by use of a fishing rod. Strong nylon line on a spinning reel and a heavy weight were cast up. This was reasonably successful, but sometimes the weight became caught up in the upper branches, and we lost a lot of line by this method. The next ploy was to use a bow and arrow, again with a fishing line, and, once more, the arrow frequently went up but would not come down (Isaac Newton, please note). Eventually it did work, and we pulled a heavier line, and then a heavy rope over the branch, and then sent a man up to make it fast. The use of a bow and arrow was new to the local population, and I think we started quite a fashion with them (and probably a few injuries, too).

One particular tree was giving us a lot of thought. It was leaning over towards our temporary office, and even with the rope on it there was a sporting chance that it might be deflected, or pull out the rope, and come down on the building. We took all reasonable precautions, and moved all the more delicate items out and away to a safe distance, and continued sawing.

With a great creaking noise the tree started to fall, breaking off the branches of adjacent trees, and showering the area with leaves and small insects, and then fell with a resounding crash. When it landed I saw a piece of grey metal flying through the air, and was sure it was the cover for our typewriter, and rushed over to try to salvage it. "How am I going to explain this away?" I thought; "bloody careless of me leaving it within range of the tree." Then I was greatly relieved to see that it was just a piece of corrugated iron sheeting off the roof of an old shed, which had become a bit squashed.

This shed contained the sample cores of rock from the earlier exploration, and they had been carefully arranged in long narrow boxes, all labelled and marked with location and depth, and stacked on racks in this shed.

After a smart piece of clearing of our fallen tree, we regretted to report to the client's engineer that there seemed to have been a sad case of the termites in his core shed, and the cores were now in some disarray, and asked if he would like us to try to put them back into some sort of order.

There was no hope of getting them back in the right order, but full details had been logged at the time of the drilling, and all the important decisions had already been taken on the logged and inspected results, so it was not of any real consequence any more.

The termites were really a great nuisance, and could eat away the supports of buildings at an incredible speed. We also discovered that one variety was very fond of red ink. In some of the record books of the earlier investigations notes and sketches had been drawn in red ink, and when what appeared to be a perfectly good book was opened some of the pages were destroyed. Everywhere a red line had been drawn the paper had been eaten away along that line, and, on inspection, it looked like a crazy form of origami gone wrong.

This was a warning to us, and from then on all our paper was kept in long, airtight cylinders until the main office furniture arrived, and we had proper airtight cupboards with heating elements installed.

CHAPTER 86
The Man with the Spider

When the trees in the immediate vicinity of the offices had been cleared, and the ground prepared for the foundations of the new offices, the work of tree-felling and bush-clearing continued further afield, making way for the haul roads and clearing sighting lines for the surveyors etc.

One day, when I was the only European left on site, I returned to the temporary office to have my lunch. When I arrived I found the chief clerk, a person rejoicing in the imposing name of Theopholous Bell, who had been seconded from our Freetown office, and had never worked in the bush before, standing arguing with one of the men from the wood-cutting squad. As it was the official lunch period I assumed that this was a private argument, and sat, drinking from my Thermos flask, and eating my sandwiches, in the cool of the office.

The noise continued outside, and seemed to be getting worse, and I called to Theopholous to take his discussion elsewhere and leave me to have my lunch in peace. The noise continued unabated, and eventually I asked him what all the fuss was about.

"This man, pa, he says he has been bitten by a large spider, and is in great pain, and in need of medical attention."

"Put some iodine on it, and tell him to get back to his work," I declared.

"Pa, he says that it was a very large and poisonous spider, and it bit him on his foot, and now he is unable to walk."

Curiosity got the better of me, and I went out to see what it was all about.

"What is your name?" I asked. "And what is all this palaver about?"

"Pa," he replied, "my name is Molo, and I have been bitten by a large spider, on my foot. See, it is all swollen up, and I am unable to walk properly."

I called him up into the office and examined his foot, which was indeed all swollen up.

"Look, pa," he said; "I have him here in my hand. The spider which done bite me."

In his hand was an enormous spider. Its body was about the size of a billiard ball, and its legs hung over the edge of his hand, and it was very dead (I hoped).

"Well Molo, we will wash your foot and put some medicine on it, and you will get a lift back to your village when the Land-Rover returns."

There followed some rapid discussion between Molo and Bell.

"What was all that about?" I asked.

"Pa," said Bell, "this man Molo says that you must cut his foot, and make the poison come out, otherwise he will die."

"Tell him that I will not allow anyone to die in my office, nor am I about to perform major surgery on his foot. Anyway, I have no equipment, no steriliser, and only what is in the first aid box in the way of bandages and plasters. Put some iodine on it and chase him away."

Much further discussion and argument.

"What is he on about now?"

"He is still saying he is going to die if you don't cut his foot and remove the poison."

"Tell him he is more likely to die if I do cut him."

By this time I was becoming seriously concerned. If I neglected to treat him and he did die, his relations could follow me to my grave with law suits for inattention, lack of feeling, etc. But if I did cut him and he still died, they would be after me for manslaughter, impersonating a medical practitioner, and failing to provide proper care and attention.

This was what has become known to be a 'no win' situation.

As I sat pondering these weighty philosophical matters the moaning increased, and Theopholous became more and more anxious. "I really think, pa, that you will have to cut his foot, as he now appears to be much more sick than he was earlier. Also, it was a very big spider."

There was no arguing with the last comment.

Decision time. "All right, bring him into the office, and sit him down by the door, where I can see what I am doing, and bring me the first aid box and the water."

The only implement in the first aid box was a small pair of scissors.

I had with me a small penknife, and tried to sterilise the blade by holding it in the flame from our Primus stove, and while it was cooking I washed down Molo's foot and tried to see where best I could make my incision. The spider had bitten him on the side of his big toe; there were two nasty marks on the surface of his foot, and the swelling had reached up almost to his ankle.

With the somewhat fatuous remark that this would hurt him more than it would hurt me, I stuck my knife into his foot. Or, at least, I tried to. There were no immediate signs of damage. I had not reckoned with the fact that the skin on the feet of persons who normally go without shoes is very tough, and a timid jab with a penknife was not good enough.

The thought of the host of microbes running up and down the blade of my knife after this first stab, as it were, caused me to heat it up again in the flame, and this time I adopted a sawing action on the foot.

It is really very difficult to cut into people, and even with a sharp, hot knife it takes a lot of pressure. The skin parted, however, and I made a second cut in the opposite direction, so that I had a cross marked out, with some small amount of blood coming out. I washed out the surface, and squeezed, and a lot of nasty yellow stuff came out, followed by a lot more blood.

All this time Molo had sat quietly, looking up into the roof, and when I turned round I had an audience of a dozen men standing in a semicircle behind me.

"That does it," I thought; "look at all those witnesses."

After a bit more squeezing and washing, it looked that all I was getting out was blood, although in fairly copious quantities, and I thought that the worst part of the operation was over.

The only liquids left in the first aid box were Aquaflavine and iodine, and there seemed to be a lot more iodine, so I prepared a swab, doused it in iodine, and stuck it over the wound. The reaction was immediate. For a man about to die from a bite by a poisonous spider he displayed remarkable agility, hopping all round the office, out onto the veranda, and down amongst the spectators, who fled in alarm.

"Get him back here, Theopholous, so that I can put a proper bandage on him," I called, but no one seemed keen to approach him, although there were encouraging noises being made from a safe distance. Eventually he was persuaded to come back, and I patched him up as best as I could, and when the Land-Rover returned with Eric we took him down to his village, where the headman said he would look after him.

In the morning Molo appeared at the head of the sick parade. His foot had reduced almost to its normal size, and otherwise he seemed quite well. The news of this miraculous cure went round, and for the next few days the sick parade, which normally dispensed aspirins as headache cures for hangovers, was crowded by persons suffering from a great variety of sores and complaints. A strong dab of iodine usually did no harm, and generally discouraged malingerers.

Things would be much improved when the construction got fully under way, with a proper first aid station, a full-time attendant, and the back-up of available transport back to the hospital in Freetown, but in the early stages we just had to do what we could.

CHAPTER 87
This Water not for drinking

During the early stages of the construction of the dam the expatriate staff comprised five persons. One accountant, one plant fitter, and one general foreman, who were normally located in the area of our new camp, and Eric and I, whose duties were normally at the construction site, where we were busy with the clearing of bush, assembling the offices and stores, and preparing access and services for the buildings.

We were usually both on site all day, leaving the camp at about 7:30 with our packed lunches, and returning in the evening about five o'clock. There were times when one or other of us had to go into Freetown, and this meant that the other was left on site, without any means of transport and no communication with the outside world, except by runner with a cleft stick and a scribbled note.

After our experience with the man with the spider bite we expanded the first aid facilities to ensure that there was always a good supply of clean water available in the office, and all the clerks were made aware that this water was not for drinking, but should be used only for medical purposes. One man was appointed to see that this water was refilled each morning, and a clean towel was set out ready for use.

During the previous exploration contract some experiments had been undertaken to try to seal cracks in the underground rock by injecting cement grout under pressure, and other attempts had been made to inject a calcining compound – a clear and odourless liquid. Eric had discovered some of this liquid in drums lying in the forest, and had brought a bucketful back to the office to see what it could do.

He had been called away to Freetown, and had left strict instructions to the effect that this material was not water, and on no account should it be taken. I had been at the far end of the site when these instructions were given, and was not aware of this liquid in the office.

When I came in at lunchtime I could hear some argument in progress outside the office, but this was quite normal, as there were always complaints about pay and overtime, staged to be in our hearing so that we might be provoked to take part over the authority of the local clerks or timekeepers – usually without success, I might add. From what I could gather, this particular argument was about water, and one individual, a compressor driver, was making a great fuss.

I called in Theopholous, our chief clerk, and told him to arrange this argument somewhere else away from my lunch break – or, better still, to go out and settle the matter. This was his lunch break too, but he went out, and I could hear more shouting, and the noise of a crowd gathering to see the fun.

He came back in and reported that the compressor man, Ali, had been drinking the first aid water, despite the other clerk's insistence that it was not for drinking but for first aid purposes only.

This was obviously a disciplinary matter, and I went out onto the veranda and summoned Ali to come forward. I then asked the clerk to report what had happened, and he said that Ali had deliberately taken water from the office in defiance of his instructions that this water was special for the treatment of the wounded.

When challenged, Ali agreed that he had indeed taken the water, but he was not prepared to recognise the clerk's authority to forbid him, although he did admit that it was a good idea to have water available for the wounded. I gave him a bit of a lecture, told him to go back to work and not be such a nuisance, and off he went.

Some time later I heard him again, outside the office, complaining of pains in his stomach and being told by the local foreman to get back to his machine.

His complaint now was that he had drunk some of the water from the office, and the clerk told him that Mr Eric had said that it was bad water, and not for drinking on any account. I called out that Mr Eric had said it was bad water to discourage idiots like him from drinking it, and to get back to work.

Later still the other clerk came in and picked up the bucket containing the petrifying liquid from a corner of the office. I asked him what was in the bucket, and he said that this was the stuff the other pa wanted to take down to the camp, and had given strict instructions that it was not to be drunk on any account.

I had not known of this liquid before, and put my finger in to see what sort of stuff it was. It looked just like rather thick water, but when I took my finger out a sort of film formed on exposure to air, which flaked off when it had dried.

It then dawned on me that perhaps Ali had taken some of this in mistake for the water, and I sent a runner off to find him, and bring him back to the office at once.

Meanwhile I quizzed the clerk about which water Ali had taken, and, of course, he did not know. The more I questioned him the more confused he became, and while this interrogation was in progress word came back that Ali was not by his machine, and had again complained of stomach pains.

By this time I had become quite concerned. If Ali had taken some of the petrifying liquid it could well do him some harm. Look what it had done to my finger! I sent out search parties for him, but with no success.

Eric returned with the Land-Rover, and we decided there was not much more we could do, but took a small sample of the liquid just in case we might need to produce it at the hospital and give them a clue of what he might be like inside.

As it was then getting late, and darkness was falling, we closed the site for the night, and I left word with the watchmen to bring him down to the camp if he returned, and had messages left in his village and others on his way home to collect him and bring him down to the camp, or to send a runner and we would come and get him. When we arrived back in the camp we had a further discussion, and I felt that we had done all that was possible in the circumstances.

I had visions of him lying by the track somewhere, being bitten by snakes or spiders, or falling into the river and being taken by a crocodile, and his relatives and dependants forming fours outside my office in the morning with claims for support and sustenance.

Just in case he was brought in by one of our searchers, we looked up all our reference books to see what the correct procedure was for dealing with petrification of the insides, but without any clues whatever.

The best advice was found in the *Nautical Almanac*, a copy of which I had happened to bring along, and in the section dealing with medical emergencies, and in particular reference to childbirth, it said that Mother Nature is a wonderful thing, and when all else fails leave things to her. With this as some sort of comfort, we retired for the night.

Next day we arrived on site early, and at the head of the sick parade stood Ali, looking very grey, with bloodshot eyes and drawn cheeks, and generally very sorry for himself. Secretly delighted to see that he was still with us, I asked him gruffly why he had deserted his machine last evening and failed to sign off before leaving the site, and what he was doing in that state, on the sick parade.

"Well, pa," he said, "yesterday, I get plenty fire in the stomach, so I go out early. I no go home to my house, I go stay with my brother over that way," and he pointed into the mountains. "Last night he make plenty of palm wine and we drink it all night. Now I have fire in the head. Please, pa, can I have some aspirins?"

I then delivered my standard speech concerning persons appearing with hangovers, and gave him a handful of aspirin tablets, with a stern admonishment never to leave his machine unattended ever again.

Later in the day we had a grand court of inquiry, and at the end of hours of questioning, and trying to straighten out the conflicting answers, we concluded that Ali had not taken of the petrifying liquid (he was still alive, wasn't he?), and, even if he had, the palm wine was an effective antidote. Perhaps we should have advised the medical authorities of this possible cure, just in case a similar case might arise.

We also learnt our lesson, and had a drinking water-butt installed outside the office the next day, and banished all petrifying fluid to the rear of the main stores, with skull and crossbones signs stuck up all round.

In conversation with one of the doctors from the hospital on a later social occasion we enquired if swallowing such liquid was harmful, and were reassured to learn that it might cause some temporary problems with the bowels, but that he did not think it would be fatal.

We expressed a little surprise, as, after all, it was designed to be used for sealing up cracks.

Anyway, my finger did not appear to be damaged.

CHAPTER 88
The Rains

Work continued on the construction of our new camp, which was set out round a small bay on the beach about 15 miles south of Freetown. The houses were all of a similar size and the basic layout was a 'T' shape, with a large bedroom each side of a central sitting and dining room, with the bathroom and inside kitchen to the rear. Outside, there was a small kitchen, with a covered walkway.

Where houses had been allocated for bachelor accommodation, the two bedrooms were divided up, and four men were allocated to a house. The club had been build at the head of a small inlet, and had a bar, cinema room, library, reading room and indoor sports area, for table tennis and billiards, etc.

As the buildings were completed and furnished more staff arrived, and the work on the construction at the dam moved into its next phase. My particular responsibility was the construction of the tunnel, which had to be completed before much permanent work could be done on the foundations of the dam. The water currently running in the river would be diverted into the tunnel, and this would allow the foundations of the dam to be completed in the dry.

The dam was what is called an earth-fill dam, and the actual wall of the dam was constructed of excavated earth, compacted to a certain density by a variety of rollers. The material for the filling was carefully selected, and had to be excavated when the moisture content was within fairly narrow limits, and these conditions meant that the main excavation and filling could be undertaken only during the dry season. This, in turn, governed the programme for the temporary works, which had to be reasonably complete before the onset of the dry weather, which also meant that most of this work had to be undertaken during the main rainy period.

The main rains were during October, and for about six weeks either side of the middle of the month, and there were lesser rains round July, when the deluge was equally heavy but not so protracted.

The intensity of these rains has to be seen to be believed. They began with frequent distant thunderstorms out to sea, with the sheet lightning flashing away all night, and the rumble of thunder almost continuous. There would be frequent heavy showers, lasting perhaps an hour or so, and then the sun would come out again, and everything looked fresh and green.

Gradually the showers became continuous, and the thunder and lightning was a bit nearer home, and then there would be about three weeks of continuous heavy rain, day and night, and then the intensity reduced to frequent heavy thundery showers until, at the end of the period, the sun came out and the place dried out.

At the start of the rains, when the place was still fairly warm, the most sensible dress was a bathing costume, with a light plastic mac on top, and a large brimmed hat. No matter what was worn, the rain came in, and the less available to get wet the better. With the continuing rain the temperature dropped, and it was really quite cold, but, again, no matter what you wore you got very wet. The only sensible trick was to carry spare dry clothes in a plastic bag and change into them when you had the chance, or were likely to be inside the office for some time. During these heavy rains the roads were in constant danger of being washed away, and a close watch was maintained on all the culverts and drainage ditches.

We had a rain gauge outside the office that had a trip mechanism whereby a clockwork motor drove a small drum, and as the cup became full a pen drew a trace on the paper on the drum, and when the cup filled up it tripped and emptied itself into a large container. Some days the rain was so heavy that the lines on the paper were too close together to count, and we had to measure the volume in the large container to find the day's total.

On more than one day we recorded over 15 inches of rain.

This caused many problems, as sudden floods would form as debris filled the watercourses and was then carried away, and the volume of water in the main river was most impressive. We had a large pump supplying water into the drilling machines in the tunnel, and one day we had to lift it 25 feet up the hillside to keep it out of the water. The noise of the water in the river was most alarming, as there were boulders the size of small cars being carried down, and bounced about from side to side.

Most of the heavy equipment had been ferried up to the site before the onset of the long rains, and work on assembling and preparing these machines for the dry weather was progressing inside our newly assembled workshops.

CHAPTER 89
A Wash-out in the Bungalow

Eric and I had settled down to bachelor living in our bungalow. We had a houseboy who did the laundry and kept the house clean, and, under supervision, he was capable of making tea or instant coffee, but most of the cooking we undertook ourselves, with occasional forays into Freetown, where we dined at the City Hotel or the rest house, or sometimes at one of the beach clubs.

The favourite beach club was the last building on the end of the promontory at the south end of the Freetown river estuary. It had a block of double-storied rooms with open verandas, and overlooked a concreted and paved area, surrounded with palm trees. There were a few dim lights scattered about on the trees, and each table had either a candle or a small kerosene lamp, which needed constant rekindling due to the wind. The place also boasted a loudspeaker system the equal of which I have yet to see. There were three or four three-sided columns, each about eight feet high, full of loudspeakers, which the staff kept in a state of constant animation, moving them about the floor, and turning them in different directions to comply with the shouted instructions of the clients. The favourite local dance at that time was the High Life, and this bouncy noise was wound up to full power, with everybody, including the waiters, hopping about in time.

We were there one night and there was a sudden flash of lightning, followed by a crack of thunder, and an immediate deluge of rain. The staff had been drilled for such events, and they rushed out and carried the columns, still blasting forth, into the cover of the veranda, which by then was equally crowded by the clients, with handfuls of food, plates, bottles and glasses salvaged from the rain.

Some wag, with reference to the loudspeakers, noted that they were lying down, but by no means dead.

We were there another evening when there was a special cabaret billed, featuring Ali bin Hassan, a noted sword-swallower, fire-eater, and general odd body, performing acts of self-destruction.

He came on dressed in a turban, with long pantaloons, and a wide embroidered jacket, and did his sword-swallowing and fire-eating quite successfully. He then set up a ladder with cutlasses mounted on a frame, and, in his bare feet, he ascended very carefully, and took a bow on the top rung.

At this there was some commotion in the audience, and a large Lebanese gentleman, obviously very drunk, rushed out, took his shoes off, and was about to have a go. Ali and his assistant managed to restrain him before he had done himself much harm, but I noticed that they both spoke fluent Norwegian. Some Arabs these!

His final trick was to collect a number of bottles, smash them onto a blanket on the concrete floor, and then take off his jacket and shirt and lie down on the broken glass on his back, while his assistant stood on his chest.

Not satisfied with this, they invited members of the audience to come and stand on him as well. He then stood up, displaying his back with the comment: "Look, the wounded," and marched round the audience.

This proved too much for the Lebanese gentleman, who rushed out and dived onto the glass, and was dismayed to find himself bleeding profusely when his friends collected him.

He was led away by the management and never seen again that night. Maybe this was all part of the act.

Ali then stuck a few knitting needles through his cheeks, and departed with a flourish.

We found out later that they were both stokers off some ancient tramp steamer in the harbour, and were out earning a crumb on the side.

We came down from the site one evening at the start of the rainy season, and went into the house. The houseboy produced cold beers, and we sat relaxing in the living room, which had some slight damp patches on the floor along the front. I went into my room, which was on the south side of the building, and noticed that the far corner was a bit wet, where the rain had blown in from the front window, but otherwise everything seemed all right.

Next thing I heard a cry of anguish from Eric, and went through to his end of the house to see what all the fuss was about. The whole end of his room was awash, the floor flooded, the bed soaked, the mosquito net dripping, the chest of drawers and his small bookcase with all their contents very damp, and Eric, standing in the middle of the room, shouting at the houseboy.

It seemed that he had left all the louvred windows open, and had not come back in time to close them before the storm, which had blown in across Eric's end of the house.

Eric was really annoyed; he had had a trying day on site, and to find all his gear soaked and his bed untenable was more than he could stand. He called the houseboy idle, useless, irresponsible, careless, and that he was no longer in our service. And he took out some cash and told him that that was his residual pay, and that he was never to darken our door again.

After that palaver we decided to eat out, and with Eric in borrowed clothes we got into the Land-Rover. Our houseboy was in the back already, begging for a lift into Freetown, and as part of his pay-off we agreed to take him into town with us. On the way he began to complain about unjust dismissal, and compensation, and such things, so Eric drove us straight into the local police station and explained the situation.

The desk sergeant was quite marvellous. He observed that these were civil matters, and outside his jurisdiction, but advised us to leave the man near his home on our way into town, and had a few words with him in dialect, obviously suggesting that he made himself scarce, and not to go on about unfair dismissal, as he should have been there to shut the windows.

We thanked the policeman, took our late houseboy into town and continued into the City Hotel for our supper, and when we returned fixed Eric up with a bed in one of the other dry houses.

CHAPTER 90
The Bee Man

As soon as the heavy equipment was assembled and ready for service, more staff arrived on site and the work round the dam began. The first job in the immediate area was the clearing of trees and bush, so that the foundations of the dam could be inspected and prepared, and first one and then two bulldozers were engaged on this work.

The driver of one of these machines was a Canadian, who came highly recommended from the Forestry Department in Sierra Leone, and he set to work scraping all the low bush out of the area into a large heap, which was subsequently burned. He then began knocking trees down, mainly by cutting the ground on one side with his bulldozer, and then going round the other side and pushing the tree over. This system worked very well, and soon sizeable areas were clear, and the second machine began to dig out boulders and remove the loose and unsuitable material.

There were many bees living in their hives in the tall trees, and as soon as any attempts were made to attack their particular tree they would swarm out, making a lot of angry noise. This was generally recognised as a good signal to leave that tree alone until later. Eventually the time came to remove these trees, and the drivers took some additional precautions, such as wearing long trousers rather than shorts, and keeping their arms covered, and for some time this seemed to be satisfactory. Our Canadian seemed more susceptible to bee attack, and he draped himself in mosquito netting and wore Wellington boots and long gauntlets.

Even with this protection he was always being stung, and we made up a sort of tent over the driver's seat out of bamboo canes and more mosquito netting, which helped for a short time. It was all right when he was inside, but as soon as he stepped outside the bees descended on him and stung him quite badly. Not only that, but if he came into the office, or the stores, the swarm of bees followed, and stung indiscriminately everyone in the building. It got so bad that we had to take him into Freetown to the hospital, and he was off duty for a week recovering from his stings.

On the day he was officially pronounced fit again there was some small party in progress in the bar of the club, and he came in to join the celebrations. As he was standing at the bar, in the midst of a crowd of participants, a bee flew in the open door and went straight to him, and stung him on the back of his neck.

It must have been waiting for him especially.

The poor chap retired hurt and sick, and on the Monday requested to be transferred to another site, where he would not be known to the local bees.

Fortunately for him, another contract had just started in the city, and we were able to transfer him the same week.

CHAPTER 91
The Freetown Housing

Some of the staff working in Freetown on a number of other building contracts were housed in what had been the old naval barracks, and, as the Freetown office carried out some work on our campsite, so we became acquainted with most of them.

The buildings in Freetown were of timber construction, and probably even then about 20 years old. They had been modernised at some time and divided into separate houses, with one or two bedrooms, a sitting room, bathroom and kitchen, and all had a small veranda that ran along the front of each block.

During the years the roof spaces had been taken over by the local livestock, with bats, wasps, spiders, scorpions, lizards and all known insects.

The residents were always complaining about these creatures, and I was asked to go along and conduct an independent inspection of the premises and make suggestions for smartening the place up.

In all honesty, apart from a lick of new paint, the accommodation was in what appeared to be good order, and set in gardens of massed bougainvillea and Canna lilies it looked a real picture. The roof spaces were a different matter, and I suggested a concentrated attack with smoke bombs before any full inspection could take place.

There had been some minor trouble on one of the sites, and one of the foremen had fallen out with the general foreman and had taken offence at some of his remarks. He proceeded to get very drunk, and went back to the barracks, where he found my inspection ladder set up, so with some difficulty he climbed into the roof space and shut the hatch behind him.

This put him into a state of instant darkness, and during his panic, stumbling about, he put his leg through the ceiling of the apartment next door.

This room was occupied by the wife of another foreman, who ran out in fright, and soon all the wives were outside, commiserating with the first one, and shouting abuse at the man in the roof.

Someone telephoned the office, and the general manager came down to see if he could restore some peace, but by this time the fellow in the roof had travelled along the entire building of about eight houses, and had managed to damage the ceilings in most of them. Despite coaxing he declined to come down, and no one else was prepared to go up inside to try to persuade him, on account of the scorpions and things, so the situation rested until the next morning, with the staff living in some fear of a descending body all night.

He came down later the next morning, demanding a quick passage home, and was taken down to the docks and put aboard a freighter that was making its way north along the coast, and stopping at such desirable places as Conakry and Dakar on its way. And here was he, hoping for a comfortable flight via Las Palmas.

I had to write up a report on the houses, and sent in some photographs. The comments back from London were that people paid money to be able to stay in such delightful situations, and I was told to get on with the painting."

CHAPTER 92
The Thunderstorm

The situation on site developed to the extent that there were sufficient houses in the camp available to allow wives and families to come out, and, as you can well imagine, their presence made some alterations in the lifestyles inevitable.

When the first delight of their coming had faded a little, the comparisons of the fittings in each house began. What had happened was that a job lot of lampshades had been purchased, some red, some blue and some green, and a similar situation had arisen with the mats for the floors. These small differences were magnified out of all proportion, and the poor camp manager had a lot of sorting out to do, to try to keep the peace

As far as the lampshades were concerned we were not bothered, and due to some delays in the shipping our Persian carpets were consigned directly from Iran to Freetown, and there was no customs duty to pay, so we had the best-carpeted house on the camp.

We were sitting having a sundowner on our veranda, and watching the distant lightning flashing away over the sea. The lights were shining brightly from the nearby houses, and there was the noise of some music coming from the club down by the beach; a perfectly normal evening. Gradually, it seemed that there was a bit of tension in the air, the hair on the back of my hand was tingling slightly, and everything became very clear and distinct. Next there were a few large drops of rain, which made a loud noise on the tin roof, and then suddenly all went completely silent.

Next moment there was a brilliant white light behind us, which came into our view as a ball of lightning. It seemed to be about the size of our small car, and it rolled and bounced, rather like a large balloon, across the front of our house, and vanished against the rear of the house directly in front of us. At the same time there was the most violent crack of thunder, followed by a longer roaring noise, and then the rain came, falling in sheets and drowning out all other sounds with the noise on the roof.

All the lights in the camp went out, and I rushed inside for our torch and a waterproof coat and made a rapid inspection of our premises to see that they were intact, and then went back to the veranda to see if any damage had been done to our neighbour's house.

They were in the midst of a similar inspection, and I could see light shining through the walls of their outside kitchen.

I dashed across in the pouring rain, and found that their kitchen walls were full of holes. There had been a small bush growing alongside the house, which had now gone completely; only a smouldering stump remained, and where any of its leaves had been in contact with the building the lightning had burned away the aluminium sheeting.

By this time several others had assembled, and once we had made sure that there had been no other damage, and no casualties, we all repaired to the club, just to make sure that the lightning had not damaged the bar stores.

The electrician went off the fix the generator, and light and power were soon restored. The mood in the camp was quite subdued, and despite a lot of drink being taken a relatively quiet night was had by one and all.

The rain cleared during the night, and in the morning sunshine we made a more comprehensive examination of the damage.

There were sort of scorch marks still evident on the ground between our two houses, the bush had been reduced to a stump, the walls of the outside kitchen had been perforated, and the wiring for the cooker blown off the wall. One of the wooden poles carrying the local electricity had been split down the middle, the wires were dangling close to the ground, and there were signs of overloading in the switch room of our little power station and generator house.

It was, I suppose, quite miraculous that no one had been hurt, and that no serious damage had been done to the overall installation, but it left a positive respect for the powers of nature throughout the community.

For some time thereafter prudent householders put their lights out when thunderstorms threatened, and, although we had many more pretty violent storms, we never experienced the ball lightning again.

CHAPTER 93
Camp Life and Leisure Activities

The work on site settled down to a pretty constant grind, working a five-and-a-half-day week, and from 8:00 in the morning until 5:00 in the evening.

My section of the work in the tunnel was working round the clock on two shifts, and I was obliged to visit each shift, sign out the record of explosives used, and order up the materials for the following day.

One evening we were sitting on the veranda before dinner when I noticed a number of small birds fluttering about in the ground at the side of the house, and on further investigation found the reason for this strange display.

Due to some change in the weather, the flying ants had decided it was time to move out of their burrows, and the birds – many of them terns – were lining up, one behind the other, over the exit holes then swooping down collecting several ants in their beaks before flying off, and then returning again.

This operation continued into the dark for some time, and happened again on the following three or four days, and as suddenly stopped.

The next avian invasion was by egrets, and these graceful creatures came in flocks and spent about a week pecking about around the exits to the flying ants' nests, before they too went off elsewhere. There were also families of large bats roosting in a large tree near the club, and in the evenings they went swooping around the lights of the houses for moths and other insects. They had never had it so good before.

We saw the occasional small snake around the camp, and, as expected, these caused a lot of shouting from the houseboys, as it gave them an excuse for neglecting their work and running about to show how brave they were. They were always very careful not to corner the snakes, though, which nearly always managed to escape into the long grass around the outside of the compound.

We went swimming in the small bay by the club, but it was really just a cove, with rocky sides and a small sandy beach, and the swimming was strictly limited. There were a number of small fish in the rocks around the bay, and I once saw a large turtle swimming past. It gave me a bit of a fright, too.

There were other longer beaches not far away, and at weekends picnic parties were arranged, and there was quite good surfing onto the long sandy shore, with miles of open beach and no others present.

In Freetown there was the local yacht club, which we joined, and we sailed fairly often on Sundays, but always in other people's boats.

This could be a dangerous practice, as on one race the rescue boat, a large launch provided by the local navy, sailed through one of the competitors, and noticed the trouble only when his propeller became caught in some trailing ropes from the wreck. They had all been dozing on deck. There was a bit of a fuss about this incident, with disclaimers all round, but it must have been sorted out as there was a new boat in the next series of races.

We had a visit from some of our directors, and I was asked to take part in a golf match – to make up the numbers, I was told. We arranged to meet at the golf club, which was on one side of the road along the beach to the south of Freetown, and the wives and children went down onto the beach. Someone had managed to borrow enough clubs for everyone, and when I arrived in the clubhouse everyone else was dressed and ready to go out.

"Go on and tee off," I said, "while I change my shoes, and I'll be out directly."

I had heard that two of the party were regular golfers, and hoped that they were not in the same party as me, but when I arrived on the tee they had all driven off and were waiting for me.

This was the first time I had held a golf club for about nine months, and I was a bit timid about my approach to the tee. After a couple of practice swings I realised that I could prolong the agony no longer, and took a mighty swipe.

Much to my delight the ball went soaring away, straight up the first fairway. A very satisfactory shot by any account.

We all set off in different directions, two to the right, me up the middle and the other to the left, and it turned out – much to my relief – that we were all duffers, and we had a most enjoyable day of it.

At that time there was a visit by the Commonwealth Secretary – a Mr Bottomley, I believe it was – and he was billeted in one of the houses overlooking the golf course, with one of the senior officers of the High Commission.

It seems that, while he slept one night, all his clothes were lifted by a thief using a long pole through the ornamental ironwork grating of his bedroom window.

He was rather a large man, and we were all asked to see if we had any spare garments that might fit him until such times as his kit could be replaced.

As most of us had been working hard outside, we were all the wrong shape for cabinet ministers, and he had no luck from us on that visit.

The thieves in Freetown were fairly sophisticated, and made use of all the modern gadgets available to supplement their personal skills.

One day the accountant was in town collecting the wages, and had put the cash into the boot of his car. He was called back into one of the offices to clarify some minor point in the paperwork, and when he returned outside the car had been taken, complete with the cash.

The car was found a few minutes later, but the cash had gone for good.

We had very little pilfering and thieving either on the site or in the camp, although I suppose the odd shirt went missing from the line from time to time.

We did suffer from the depredations of the customs and excise, in the pursuit of their business. The contract was for the City of Freetown Water Department, which was, in fact, part of the government.

Under the contract all our equipment and apparatus for use on the site was declared duty-free. There was a bit of a hazy area concerning personal belongings of expatriate members of staff, but very few of us had brought in anything that was either valuable or dutiable, and apart from the odd minor argument things were generally settled in our favour.

The customs office were always very stroppy if they thought anyone was trying to work a flanker on them, and they examined all the documentation very closely, and one time they thought they were onto a winner.

Part of the laboratory equipment used for testing the moisture content of the earth fill was held in the customs shed because it contained an oven. This was a perfectly ordinary piece of household equipment, and the customs decreed that it was being passed as special laboratory equipment, as who would have imagined that we would need to dry out soil in an oven in order to build a dam, and they seized the entire laboratory equipment on that consignment.

The minor officials were adamant that this was all part of a conspiracy to avoid duty, and we had to escalate our complaints to a very high level before someone saw some reason, and the equipment was eventually released.

We invited some of the customs staff to come out and see how the oven was being used, and managed to arrange for a sudden blast of air in the laboratory, which blew red dust all over their best white uniforms.

We then took them back to the club, and after a few beers all seemed to be forgiven.

CHAPTER 94
Pack up and go Home

My section of work in the tunnel was nearing completion, and it was agreed that we should depart on home leave as soon as the tunnel was through and the river diverted into it. The work continued round the clock, and as always towards the completion of tunnel drives the anxiety about coming out in the correct place on the other side of the hill was a bit of a worry.

Frantic checks were made in the surveying both inside and out, and it was with great relief that, when we drilled a long pilot hole from the inside, the end burst out almost exactly where we had hoped, trusted, and – really – expected it to. After that it was just a matter of opening it out, and eventually I could walk through, and later drive through in one of the dumper trucks. There was a minor celebration to mark the achievement of this stage of the work, and we packed up our belongings and set off on leave, with the work on the construction of the filling of the dam progressing well.

We spent one night in the newly completed Ambassador Hotel in Freetown – all chrome plate and glass furniture, and freezing cold with the air-conditioning, and went off to the airfield on the BOAC bus and the ferry across the estuary.

The airfield had not changed much since our last visit, and the customs staff were very keen looking for smuggled diamonds, but we had no great problems and were soon off again: a repeat trip stopping at Bathurst, Las Palmas overnight, complete with Spanish dancers, and then Lisbon and Heathrow.

We stopped overnight in London, and took the mail and some other items into the office in the morning, where I was told to have a good break and come back in six weeks, when another job would be waiting for me.

We went off to Devon, where Jean's father had just purchased the hardwood block floor of a ballroom from a nearby hotel, and most of my leave was spent cutting, fitting and sanding the new floor of their dining room.

We took some time off and visited Scotland for a week, and then it was time to go back to work in London.

The next job was on the construction of an iron ore loading pier and conveyor system, again in Sierra Leone, and it was thought that I would work in the London office during the preparation and planning of the work, and go out to site at a later stage in the construction, when most of the preliminary site works – such as the establishment of the camp, stores, workshops, and temporary works – had been completed. This fitted in quite well with our domestic arrangements, as we had hoped to buy a house in England within travelling distance of the office, as much as anything to find a home for our carpets, and a place to keep the boat.

We took a flat in Ealing, and during the week Jean collected details of suitable houses within our price range and within sensible travelling time of the office, and at weekends we travelled all round the outskirts of London, discarding unsuitable locations or properties that were too expensive. The basis of our selection was to find a suitable design of house, in a reasonable neighbourhood, and then calculate the cost of the purchase against the probable cost of travel in and out of Ealing.

We spent quite a while looking at route maps, and studying the train and underground connections, before going out on inspections, and eventually we settled on a company-built house in Maldon in Essex. This had the advantages of a bit of discount, by being a company employee; it was on a direct railway line to Liverpool Street station, which also had an underground service with a station right outside the office, and there was a sailing club just down the road, where we could park the boat and sail at weekends. The disadvantages were that it was about 40 miles out of Ealing, by a route round the North Circular road, and either by road or rail I would be travelling about three and a half hours each day.

We completed the formalities about mortgages and solicitors and arranged to have our new furniture delivered, cleared out our flat, and with the boat in tow again made for Maldon, where we had booked in to stay the night in a local hotel.

Next morning we rolled up to our brand new house, and after the ceremony of carrying Jean over the threshold unpacked our belongings, and stood about waiting for our furniture.

At about lunchtime I went out and telephoned the shop about our stuff, and was advised that there had been some delay and nothing would arrive until the following morning. I then had to arrange for another day out of the office, and we booked into the hotel for a second night.

Before leaving the hotel I rang the shop again, and was assured that our furniture would be along shortly and that it was already being loaded onto the vehicle.

We went back to the empty house, and moved such items as we had into some of the available cupboards, and then sat about waiting for the rest of the day. When at last the vehicle arrived, all it brought was a bed and the bedroom furniture; no chairs, no tables – nothing else. We had a quick session fitting up the bed, and carrying the other items upstairs, and then returned to the hotel for dinner.

Our first night in our own bed in our very own house was a little spoilt by the thought of a standing-up breakfast in the morning, but such are the problems of moving house and we made the best of it. I went off to work the next day and raised a bit of trouble for the furniture salesman, and about ten days later we had most of the stuff delivered.

Our house was set on a corner of a small estate, and the garden had been the site of the cement mixer and stores compound for the estate.

The first clear weekend after moving in I made a determined attempt to dig over some of the garden. What I found was layer after layer of cement, up to about a foot deep in total, where the cement mixer had been washed out at the end of each day's work, and this had settled and set and was very difficult to remove.

I also came across the sections of broken concrete fireplace surrounds which had been discarded and lost in the mud and cement washings, and some of which were several feet down in the ground. This was all too much for hand digging, so I went down to the site office and, after pulling a bit of rank, persuaded them to let me borrow the small digger on site, and with this machine I managed to clear most of the garden to a satisfactory surface.

Before leaving Ealing we had bought a small Standard Eight car, and it was mainly used about the town by Jean and on family visits at weekends.

After some minor repairs in the local garage we were out one evening, and the back wheel came off as we were rolling gently down a small hill. The car lurched to one side, and remained leaning over at a pretty hairy angle. When I got out I found the wheel had become jammed under the brake drum, and it took quite a bit of leverage to get it clear.

When I put it back on again I thought it might be a good idea to check all the other wheel nuts, and I was dismayed to find them all loose. I went back to the garage and raised a bit of a fuss, as it was just luck that we were on a quiet road when the wheel had come off. Soon after that we disposed of that car, and bought a large Jaguar in its place.

Shortly after moving into our new house we were delighted to learn that Jean was pregnant, and our spare time was spent between digging the garden, preparing a nursery and sailing the boat on the Blackwater river.

The sailing was a lot of fun, but being tidal at Maldon our times on the water were strictly limited to about three hours either side of high water. The boats were kept on their trailers on the quayside and lifted down by a small motorised crane, and if you were back in time before the tide went out you could sail right under the crane, and be lifted out without too much trouble. If you were a bit late, however, the water drained away very quickly and you had to drag the boat through the black mud before getting it under the crane, and then slide the slings under the boat, disturbing and distributing yet more mud. A very messy business.

During our first winter in Maldon there was a serious flood warning issued over the wireless, that water levels in the Blackwater area could rise as much as five feet over the predicted tide. Our boat was parked on the quayside, alongside an iron fence with spikes, and I had visions of the boat floating up and landing on top of the spikes, and when the tide receded having a series of inch holes at regular intervals all along the bottom.

The high tide was due at three o'clock in the morning, and in the pouring rain I got down to the harbour by midnight. There was quite a crowd of people there, and we planned to move all the boats out of the compound and up the road above the predicted tide level if that was necessary. We sat about in the clubhouse watching the water level, still in the pouring rain, and aware that if the tide came up another five feet the clubhouse itself would probably float away. There was a debate in progress that we might as well drink all the stock of the bar at once rather than see it all go off on the tide, but as we watched the rate of rise reduced, and although the water came over the quay it was not going to reach where the boats were parked, and at three in the morning we all went home.

Jean had some problems with her pregnancy, and was taken into hospital at Chelmsford before Christmas for observation and treatment, and this caused some rearrangements to our activities.

I left home by 7:00 in the morning and returned to Chelmsford at about 6:30 in the evenings, and went straight into the visitations at the hospital. I normally left about 9:00, and by the time I arrived in the house it was about 10:00 p.m.

The next task was to collect and sort out the post, do the washing and ironing, have a bite of supper, and turn in. At weekends I spent most of the time at Chelmsford, where the hospital staff were very good and allowed visitors without any set limits.

During January there was a heavy snowfall and severe frost for about a month, and with all the dashing back and forth I had no time to clear the snow from the drive into the house. Being a new estate, all the householders were very house-proud, and without exception they all cleared their drives of snow at the first snowfall. Our drive was fairly flat, and it was not difficult to get the car in and out of the garage, and I neglected to shovel the snow away, mainly due to lack of time but also through sheer idleness.

After about a week of very hard frost, practically all the houses except ours were out of water. The inlet pipes were laid under the drive, and they had all frozen up, not being protected by a snow blanket like ours.

We agreed that they could draw water from our tap by arrangement during the times I was out of the house, and they commented about how smart I was for not having cleared the snow from our drive. Like Brer Rabbit, I smiled and said nuffin', and never even suggested that it was because I was too idle.

The late Dr Beeching caused a major change in my travelling habits by closing the branch line to Maldon (damn his eyes), as this meant that I had to travel by car to Chelmsford each day, park the car at great expense, and go on by train from there.

The first day I undertook this revised journey I climbed into a compartment of the train and sat down in the corner seat. I was followed in by four city gentlemen in dark suits and Homburg hats, who were obviously displeased at something.

They huffed and puffed as they hung up their coats and hats, and suddenly they all sat down, two on each side. Someone produced a sort of apron with strings at each corner, which they

305

spread out in a well-practised fashion, someone else fanned out a pack of cards, and a third man brought out a scorebook, and before the train had started to move they were immersed in a game of poker.

By sitting in the corner seat I seemed to have upset their routine, but they continued to play until the train was into the platform at Liverpool Street, and had a quick reckoning of the score before they donned their hats and coats and got off.

In February our daughter, Susan, was born, and about ten days later they both came back from the hospital to the house, which I had specially heated up for the occasion. The very cold weather persisted for another few weeks, and then we had a splendid spell of spring weather. At this time the gas bill came in, and I was a bit shocked at the size of it, even though I appreciated that it was wintertime and it had been exceptionally cold, but I still felt it a little bit excessive. Our neighbour also mentioned his gas bill, and when we compared notes he was paying about four times as much as me. This was obviously because I had never been at home except for sleeping, and a few hours in the late evening, but our neighbour rushed back into his house, shut all the doors and windows, and turned off his central heating before giving a long lecture to his wife on the conservation of energy – and the price of gas.

We had the christening of Susan in the church in Maldon, and had a large party of friends and relations present. They all departed except my parents, who elected to stay on for a short holiday and to assist with the new baby, as grandparents do.

We went out one day in the car and drove as far as Norwich, where my father had been billeted during his service in the 1914-18 war. He knew the name of the road where he had stayed, and the district of the city, but we were unable to find anyone who could direct us. We came upon a man guarding a manhole in the road, and asked him if he knew where this street was, and he remarked that he had heard of it and it was not far away, but he was not sure in which direction it lay. He said, "Hang on a minute, I'll see if Jack knows where it is," and so saying he shouted down the manhole.

Having had some experience with pipes in the road, and manholes and such, I expected the reply to be something like: "Go along the 36-incher until you come to the third 24 on the right, then go on up that until you reach the third nine-incher on the left."

This was not what happened, but the chap came up and gave us precise directions to the street, which was only a short distance away. We found the house, and after a short wait to gather up his courage father went in and spoke to the lady of the house. She was very kind, and said that she had not been in the house very long, but an old lady across the road had been there a very long time, and she suggested he should go and ask her.

He went up to her door and they stood talking for a long time. She said she could remember the soldiers being there, but she had been very young at the time and could not recall very much, as it had been such a long time ago.

Much encouraged by this visit, father then remembered another address, and we asked numerous people, including a policeman, but nobody could tell us where it might be, and we started on our way home.

After a few miles father suddenly said, "That other address; it was not in Norwich at all – I think it was in Newcastle."

"Well," I countered, "we won't be going there today, that's for sure!"

The garden flourished, the summer came and went, and I was still working out of London, but the time came to be thinking of going out to Sierra Leone again.

The plan was that I should go out and make sure that things were in order for Jean and the baby to come out later, and we set about arranging with an agent to look after the house during our absence, hopeful to rent it out to decent God-fearing persons, at a great rent.

With all the arrangements made we towed the boat down to Devon again and sold the car, and I left by train for London while Jean was busy getting the baby vaccinated and immunised, in readiness to come out by sea in about a month's time.

SECTION 2 – PEPEL PORT, NIGERIA AND LIBYA
General maps of Nigeria and Libya

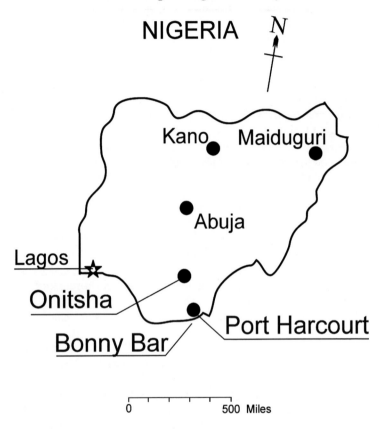

LIBYA

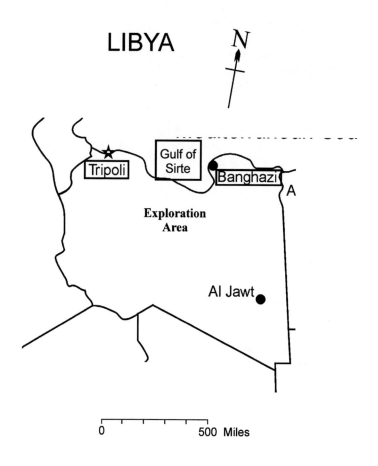

CHAPTER 95
London Work

At the start of any major construction work the normal sequence of events is as follows:

An invitation to tender is presented, usually with drawings, specifications and conditions of contract.

The contractor either decides to proceed to make a bid or he returns the tender. The contractor then prepares a programme for the submission of his tender, and allocates staff and resources to progress the presentation of the tender. This may include a site visit, and further visits to specialist manufacturers of equipment.

The tender team then prepares details of their proposals for the construction, including the main outline of the methods to be employed, equipment to be used, a detail programme and cost estimate, and then makes an addition for overheads and profit.

This proposal is then returned to the client, who considers it and either accepts or rejects it, or asks for explanations or variations in the conditions.

At the time I returned to the office, these stages had already been passed, and the contract for building an ore-loading jetty at Pepel Port had been signed.

The senior staff had been appointed, their tasks allocated, and the office and site organisations had been established, with programmes of meetings set up each month at fortnightly intervals so that the office minutes could relate to the site minutes two weeks earlier or later.

I had been appointed to be in charge of the civil works for the construction of a concrete jetty and a number of concrete piers, which would carry structural steel bridges for a conveyor system from the iron ore plant ashore to a pair of ship loaders on the new pier. These bridge sections and the ship loaders were to be built at St Helens, and, after a trial erection, dismantled, shipped out and reassembled on the site.

The outline selection of equipment had been made prior to my arrival, and I had to arrange for the inspection, selection and dispatch of the actual items we would require, which included tugs, barges, cranes, pumps, concrete mixers, pile hammers, etc., etc.

We engaged a marine specialist to make a summary of the suitable craft on the market, and after a study of his lists selected a number of vessels for inspection. These ranged from sugar barges, used for carrying sugar beet around East Anglia, to large flat-top barges with a capacity of 5,000 tons. We also spent hours chugging up and down the Thames testing tugs, and some time in Southampton looking at small passenger ferry boats for carrying our staff between Freetown and the site, some 12 miles upriver.

The selection of the barges was fairly simple, and was soon arranged, but the tugs were a bit more difficult, as we had to consider the availability of spares and the problems that breakdowns might cause in the course of the construction. We selected two tugs, and were referred to an ex-navy store on the south coast, where we were advised that some spares might be held for these two vessels, and made an expedition to see what was available.

This place was a real Aladdin's cave, and the storekeeper was extremely knowledgeable and forthcoming. We told him the names of our tugs, and he immediately said he knew of them both and had a selection of spares in his stores.

He told us the history of one engine, which had been designed for a submarine but had been used instead as a spare generator engine on an aircraft carrier, and had then been fitted to our tug, which had been used by the navy during the war in the Southampton area; he then led us to a section of his store containing enough bits to build a new engine for us, and spare engines, ex-trawlers, which would fit the same bedplates.

He had some spares for the other engine, which had been built in Norway and somehow come into the orbit of the Admiralty, and assured us that other spares were available from the makers in Norway and would be no problem. We enquired in a roundabout way about other naval stores where we might obtain such items as anchors, chain, and cable, buoys' rigging, old lifeboats and anything else nautical, and he directed us to a store near Harwich.

After another visit to another knowledgeable storeman, who knew where everything was, and took us round explaining where particular items had come from, we had ourselves equipped for most of the marine gear required, and then we left instructions for the loading and dispatch to site for the office logistics department.

We had a look at a number of small passenger vessels in Southampton and Bristol, but none were really what we were looking for, and eventually we settled on an ex-RAF fuelling launch hull, which we had reconditioned, fitted with two large diesel engines, a new wheelhouse, a passenger compartment with large bus-type windows, and London bus seating.

The construction of the steel bridges was progressing at St Helens, and I made regular visits there, and also to Retford, where the machinery for the conveyor system was being built.

In the works at St Helens they were also making domed ends for pressure vessels. These were made from thick steel plate, cut into a circle of about ten feet diameter, and shaped into a dome by the action of dropping a heavy weight onto them, while positioned over a sort of shaped sand-bed. This was thought to be the most skilled operation in the factory, and it was undertaken by two elderly gentlemen, one about 70 and his assistant about three years younger. The older man picked up the plate by a small crane, and positioned it over the correct part of the pit, and when he nodded his head his partner dropped the weight onto it (the plate, not his head). The standard of workmanship was really excellent, and when we had our trial assembly everything fitted perfectly.

The pier was to be reinforced by steel bars under tension, and we were invited to a demonstration of the process, first at Guildford Cathedral, and then at the Medway bridge, and set out for Guildford with the representative of the steel company one fine morning. As we travelled down he explained that their company had supplied some material and equipment for the cathedral, and we would pop in and see how they were getting on.

When we arrived we were told that the master mason and his crew were up on the roof, and we should just go on up. There was an outside scaffolding tower with an open lift cage in one corner of the building, and we ascended quite slowly. As we rose the horizons expanded, and the town came into view, then the hills and valleys beyond. This must have been like what Jacob experienced on his ladder.

When we arrived at the roof level and disembarked from the lift we were met by the master mason, an elderly gentleman dressed in dark trousers and a black overcoat, well buttoned up, and wearing a Homburg hat crammed down onto his head.

After the introductions were completed we stepped out of the shelter of the corner of the tower, and were almost lifted off our feet by the blast of wind. As we continued along the outside, we encountered another member of the construction staff, similarly garbed, and wheeling a small barrow. He paused, took off his hat very politely, and further introductions were made – all very formal.

We then entered the lower section of the tower, where the construction work was in progress. The intention was to build a spiral staircase up to the roof above, about 20 feet up, and they had already made the actual steps and had them set up in a scaffolding cage. Each step was rather in the form of a blade from an aircraft propeller, with the tread spreading out from a narrow, circular boss that had a hole through it, and an upstanding collar to provide the space between the treads above and below. The intention was to place a steel bar through the holes, and then tension it, to allow the stairway to be free-standing. This process is known as post tensioning, and a simple example for the non-technical can be demonstrated by assembling a row of books on a table, and then lifting them together by pressing in from both ends, so that what were individual narrow books, when pressed together, act like a thick beam.

The problem here was that, although the hole was open at the top and bottom, the bottom section was sitting on the floor, and there was a roof over the top, and it was not possible to insert the long bar, either without boring a hole in the floor and putting it in from below, or making a hole in the roof above and lowering in from the sky above. The only other obvious solution was to dismantle the whole erection, set up the bar, and thread all the treads in over the top. None of those solutions seemed to be very practical, and it was then suggested that a series of short threaded rods could be put in and connected up individually, rather like a chimney sweep's brush, and this suggestion was readily adopted.

We were then conducted on a grand tour of the other work in progress, and learnt that everything was very strictly controlled by financial restraints. Before any work could be undertaken the necessary funds had to be allocated, more or less on a weekly basis, and we retired to allow the financial and commercial discussions to be carried out between the interested parties. We descended in the open lift again, and met with some of the officials who guarded the bishop's purse, and after morning coffee set out for the large motorway bridge under construction near Rochester.

On the way we discussed the way some technical details were often overlooked by some architects, and the lack of planning which frequently caused long delays in construction and building projects. When we arrived at the bridge site we were met by the chief engineer, and taken round the work. This job was an excellent example of sensible construction and planning methods.

The large concrete beams were cast in an open yard and allowed to set. They were then tensioned by means of the steel bars, and then could be picked up and moved about on a system of rails and bogies. These completed beams, each weighing about 80 tons, were then marshalled into the correct order, and subsequently moved out onto the bridge and into their final position under the control of two men, working winches and controlling switches.

As I had seen in St Helens, this extremely skilled and responsible work was all done silently, with head nods and the occasional hand signal from one to the other.

After this demonstration we went back to the yard and saw the actual loading of the steel rods being carried out, and had some words with the engineers and foremen to try to see if there were any special tricks to be learned, then finally returned to London.

During the planning stages we had to decide which types and makes of piledriving hammers we should use. The piles we were about to drive were over 150 feet long, and the water depth was about 40 feet to the river bed. We decided that we should start the driving of each pile with a double-acting steam or air hammer, as this was easier to control during the early stages, especially if the ground was very soft.

This machine consisted of a cylinder with a heavy weight inside; high-pressure air was blasted into the bottom half of the chamber, making the weight rise up; the air was blown off, and high-pressure air was then introduced into the top chamber, driving the weight downwards. A skilful operator could work it so that relatively gentle taps were produced, and the pile could be guided in what was hoped to be the correct direction by wedges and ropes.

A German company had just introduced a bigger and better diesel-powered hammer, which had a different method of operation. This thing worked by having a small cup-shaped depression at the bottom of a firing cylinder. The actual weight was like a large piston, which could be lifted up, some fuel injected into the depression, and the weight released. The weight fell down, as weights normally do, and gave the bottom of the casing an almighty bash, and at the same time the fuel was ignited, and blasted the piston and weight back up again, and also gave the bottom of the casing another bash. I had never seen this large hammer in operation, and I got myself invited to Germany by the manufacturers to see it working.

I flew into Frankfurt and was met at the airfield by the general manager of the company, and we drove off at great speed in his large, expensive car to the site where the large hammer was supposed to be working.

After about an hour on the autobahn we could see some construction work on the river bank ahead of us, and turned down an access road to the site. The large hammer was there all right, but it had not yet been set up, and they were still working on the erection of cranes and other handling gear. There would be no piledriving there that day.

Following some fast talking with the site engineer we set off again, back onto the motorway, and drove another 100 miles. Again we could see cranes and piling works, but, alas, it was on the opposite side of the river, and we had to go another 20 miles up and back to find a bridge.

When we arrived the hammer had just completed its work, and was now dismantled for travelling.

Further fast talking. Perhaps we should return to Frankfurt for the night, where there just happened to be some sort of beer drinking performance in progress?

What a good idea. So off we went again, all the way back to Frankfurt, where I had been booked into a very posh hotel, and the general manager and his wife joined me for dinner.

In the morning we set out again, and, once more, the hammer we visited was not yet set up. Why don't we stop for some early lunch at a wayside inn?

Another good idea. While we were waiting, I heard the manager on the telephone to a series of his underlings, threatening dismissal, death and damnation to the next one who would fail to produce a hammer actually working, and if one was not working they had better get it going before we arrived.

During lunch he explained that this sort of thing was not uncommon, as the various sites had their own programmes to work to, and, after all, he was only a hammer supplier and not really in charge of the individual operations on each site.

Whatever he had said on the telephone, at the next site we visited we were welcomed like royalty and given a grandstand view of the large hammer in operation.

I seemed to detect some muttering in the background that it was just as well this one was working, and then we went over to the engineers and had a friendly discussion on the merits of the machine, and the problems associated with its working.

The main problem appeared to be in controlling the fuel. There was a pair of ropes connecting to the fuel control lever, and as long as these were accessible all was well, as the supply could be shut off; but there had been times when the strings had blown out of reach, or become snagged, and the hammer continued to drive until some hero managed to hook the lever with a boat-hook.

Satisfied with this demonstration we returned to Frankfurt in time to miss the last flight back to London, so I had to stay another night in great luxury, at the beer festival.

The time came when the preliminary work was well under way on site, all the major items of equipment had been ordered and were on their way out, the staff had been interviewed and engaged, and the London work had settled down to a regular routine of progressing the supply position and making certain that the drawings and details were coming forward at sufficient speed to allow other items to be resourced and purchased.

The last London meeting was a few days before my departure, and I was given the minutes of the meeting to take out with me. The procedure was that each department in the office gave its report on its progress and difficulties, and the London manager tried to sort out priorities and allocate blame as impartially as possible. Items that were not cleared from the previous month were allocated a star in the minutes, and this drew down pain and anguish on the department concerned. Two stars were very uncomfortable.

Before making my final arrangement for travel I had another session of vaccinations and all that, and started a course of anti-malarial pills. There were numerous different pills available: some you took once a day, others once a week; some made you quite yellow, and others had no apparent outward effect; and different pills affected different people in different ways. I settled for the once-a-day variety, working on the principle that, if for some reason I forgot to have one, there would probably be sufficient residue still inside me to provide the protection, and it was a sporting chance the mosquitoes would not know of my delinquency.

CHAPTER 96
The Scene on Site

I had the now familiar trip out via Lisbon, Las Palmas, and Bathurst to Freetown, but this time I took a bit more interest in the land to the north of the estuary at Freetown, where the site was located. The main river flowed brown in three main channels, with a number of islands scattered about in what was really a sort of delta. The land appeared to be low, and covered with dense bush with occasional tall trees, and looked quite uninhabited, with the exception of the buildings around the airfield and a few small fishing villages along the coast.

We landed and passed through the customs with the usual small niggles, and were soon on the bus, which had not improved since our previous visit, and over on the ferry to Freetown. At the ferry terminal we were met by people from the site, who had come across the river in the speedboat – a small fibreglass contraption with a very large engine, and which looked quite inadequate should it collide with the odd floating matchbox, never mind a tree trunk. I was assured that it had yet to come to harm, and soon we were off, speeding across the river to the site. Very exhilarating it was too.

On arrival I met the project manager, whom I had last seen in London about six months earlier, and we had the usual chat about the health of various acquaintances as we went up to the camp, which had been well set out in some large trees about a mile back from the works. My bungalow, of similar design to those in Iran, was quite comfortable, and a houseboy had been assigned to look after me and the house until either I engaged him full-time or I fired him and found someone else.

We went round the site, met all the people, and saw how things had progressed. As a result of my reading of the files in London, and my interest in the equipment sent out, things were much as I had expected and the main construction work was just beginning. A small landing platform had been built, with a jetty for foot passengers, and three of our five ships' lifeboats had been delivered and were in commission. The tugs and barges were all in Freetown harbour, being unloaded and made ready for their passage across to the site, and the initial surveying had been completed, so that we could make a start laying anchors for the moorings for our barges and the positioning of the other floating equipment.

One of our first tasks was to train the lifeboat drivers and engineers, and we had many enthusiastic applicants for these jobs, as this meant most of the time sitting down, shouting instructions to one another and their passengers. The eventual outcome, as a result of frequent near collisions, was to have one man as the coxswain, who steered the thing, and an other as the engineer, who worked the engine – which seemed capable only of 'full ahead' or 'full astern'.

Amidst all the clamour of engines roaring, passengers screaming and scrambling about, word-of-mouth commands were never very reliable, and a whistle system was developed; a long blast for go ahead, followed by a number of short toots for 'faster, faster', two long blasts and a heavy kick for ' full astern', etc., and three long blasts for put it in neutral, and hold onto everything, as we are about to ram something.

In addition to the lifeboats we had a very light aluminium boat with a large outboard engine, which was capable of about 25 knots, and we had one man specially trained to operate this vessel. He got to be pretty good, and normally could bring her alongside neatly without doing himself, my nerves, or the boat any damage. This boat was normally reserved for staff passengers, and some of them provided excellent entertainment.

When the boat was coming into the jetty, he came in fairly fast in a long curve, and as he got near he cut the engine and the stern settled down, with the bow almost touching the landing place. It was then necessary to secure a line to the staging before taking any other action, but unsuspecting passengers would frequently try to step or jump ashore, the result being that the bow swung away, and they failed to reach the staging and had to be fished out in an ignominious manner, and sent off to dry. Such people always tried to blame the driver, but it was rarely his fault. If he wanted to get you wet, he had other methods at his disposal, such as driving under low mooring ropes without giving much warning – always good for a laugh.

The camp at Pepel was built around in a circle, with the married quarters on little side roads, and the club and bachelor quarters around the main rim. There was a local rule that there should be no motor cars on the little island that was Pepel; there were no roads, anyway, but for the construction work, we had been allowed to have six motorcycles for the senior staff, and a tractor and trailer, known affectionately as 'the Pullman', for transport between the camp and the site – or as evening transport to the client's camp and club, which also boasted a fine swimming pool. The roads, as such, were mere tracks through the bush, and also seemed to be the main highway for snakes and large crocodiles in the dark, and were not recommended for motorcycling after dusk.

I had been allocated a motorcycle when I arrived, and after some practice became quite proficient, having learned how to fall off during my army service. My machine was pretty well worn out when I arrived, and when I drove down to the jetty I found that I could stop it by leaning it against a long, timber edge piece of the jetty, shutting the throttle, and stepping off.

One day the workshops took it in for service, and when I tried the same trick afterwards it just kept chuffing on towards the end, and a watery grave. I had to make a maximum-effort jump to catch up with it before it went over the edge, and was greeted by a round of applause by the assembled crowd, who had expected better, I suppose.

One of my young engineers asked if he could borrow the bike, and learn to drive on it. He said he could ride a push-bike, and I said it was much the same, only harder to lift up again when you fell off. I showed him how to start the engine, and explained about the gears and brakes, and suggested that he should confine his activities, at least for the first time, to going round the camp in first gear and practising getting on and off from a standing start, and keep well away from trees.

He took off during the lunch break, and came back shortly with a skinned elbow and a burn to his leg, where the exhaust pipe had leant on him after he had run into a tree. "Which tree was it?" I asked. "That one over there." He pointed to a small bush about 50 yards away. "It just seemed to keep coming at me, no matter what I did."

"Well," I advised, "you had better see the doc about your leg, and perhaps you should take more care when you try again tomorrow."

We had one rather nasty accident on motorcycles, when the client's engineer drove into a trench that someone had dug across the road. He came off pretty hard and broke his arm in the fall, and was off duty for a couple of weeks. By great good fortune, it was some of his own department who had dug the trench and there was no comeback on us, and by the time he had been to see the doc and come out again the trench had been filled in, and the evidence was not available. There was a furious row going on about this affair for some time, and for a while all motorcyclists proceeded with great caution.

The club was run by a committee elected every three months or so, and I was dragooned into serving shortly after my arrival. The duties of the committee were to make sure that the accounts were in order, that there was no fiddling with the petty cash, and that regular supplies of drinks etc. were maintained. At the end of each three-month period there was a complete accounting, and the rules stated that any profit had to be allocated to a party or other charitable cause. The following chapters describe some of these events.

CHAPTER 97
The new Beer

Part of my duties as club committee member was to ensure the proper ordering of drink from the suppliers in Freetown, and on my infrequent visits on other business I had to call at the brewery and discuss prices and deliveries, etc.

A new brewery had just been completed, and their salesman met up with me in the City Hotel one lunchtime and persuaded me to try out his new beer, at a reasonable discount, of course. This seemed not a bad deal, and I settled for a few dozen cases to see how it went, and, soon enough, these were delivered. It came in shiny new bottles with a modern label, and tasted absolutely foul. In fact, it almost caused a riot in the camp, and was withdrawn immediately.

I wrote a rude letter to the brewery, demanding our money back and threatening to sue for poisoning, causing a mutiny, and general disaffection in the camp. They replied, very contrite, asking me to call at their office on my next visit to town. About a week later I was in town, and went in to see them.

The full brass turned out, full of apology and contrition, and took me to lunch, where after some appeasement, and humble pie, they admitted that it had all been a terrible mistake, and we must have had some of the bottles that had been filled while they were cleaning out the pipes prior to proper production; of course, this had been a grave error, so they asked if we would accept replacement supplies, and as much again, and try to forget the matter.

As I was not coming to much harm drinking the beer in their boardroom I agreed, and returned to site, pleasantly hazed, and with a lot of free beer for the next party.

The organising of these parties took some time and effort, and on the next meeting of the committee we decided we would splash the profits, cut the prices in the bar, have a 'happy three hours' at the party and arrange a gambling night.

Two of the young engineers were detailed to construct a wheel of fortune, and to work out the odds so that no one could possibly lose; someone else was appointed blackjack croupier, and again instructed to pay out at 15; and various other games of chance and skill were arranged to ensure that the house could not win. This party was such a success that it managed to generate even more cash, despite being rigged as a give-away, so that we were compelled to arrange a similar event every month.

That particular committee was so good at making money and keeping the prices down that we were re-elected several times in succession, and had to introduce a new rule into the constitution that three times in a row were too many. The next lot made such a hash of things that we were welcomed back, and the club continued in profit.

With some of the loot we set up a bowling alley along one wall of the building, and had the bowls and pins made locally. For the floor we used some rubber that had been discarded from an old conveyor belt, and at the first session almost wrecked the place. Whether it was the mat or the bowl or the bowlers, great difficulty was experienced in keeping the bowls on line, and several innocent bystanders had been in danger of being maimed by low-flying misdirected bowls.

What we needed was a dwarf wall all along the side, and this was soon knocked up and fitted.

The other problem was now at the far end. It was not uncommon for fast deliveries to hit the back rubber and jump back into the floor, which made life interesting for the person detailed to reset the pins and return the bowls to the firing end – sometimes with a slight degree of malice. Soon these problems had been sorted out, and we established a team and issued challenges to the client, and other sites in the region.

We could now field teams in darts, tennis, bowling, snooker, table tennis and Scrabble as well as football and hockey (with each event raising still further our funds, despite all our efforts to just break even). In addition to all this sporting activity, we arranged trips to various points of interest within a day's travel.

There was a small island not far off the works, which had been a Portuguese settlement at one time, and we had a picnic there, where we found the ruins of an old fortress and some gravestones in a burial ground. This island was not considered a healthy place by the locals, and we had to go over without the local boatmen as crew on our launch. Whether their dislike was based on spirits or mosquitoes they would not say.

CHAPTER 98
The great Christmas Party

We had been budgeting for our Christmas party for some time, and the arrangements were to have a party for all the children in the camp, another for all the children in the local village, and a series of presents to be sent to all the children of the staff who were not present on site. The remainder were to go on a monumental beanfeast.

The resident wives all prepared special dishes, which were brought down to the club, and they supervised the general preparations of the dinner, decorated the room and tables, and generally smartened the whole place up.

The children's parties went off well, with Father Christmas sweating blood in a heavy red suit, complete with hood and whiskers, and handing out presents as if there was no tomorrow, and when this had settled down a bit, and the children departed, the serious eating and drinking got under way.

After the usual speeches and singing and dancing, most people trundled back to bed, but the hard core continued in the club for some time yet.

By common consent, it was deemed to be advisable to leave the party in groups according to houses, and it can be imagined that it is quite difficult to persuade three or four persons to move at any one time, all being involved in different conversations or activities. Nevertheless one group of four set off from the club to their house, which was on the other side of our circular compound, and they elected not to walk round the track but to go straight across the cleared area in the centre.

There was a very large tree in the centre of the area, where the nightwatchmen usually sat, and the group exchanged greetings with them as they passed. One of the men, George, brushed against what he took to be a piece of electric wire dangling from the tree, and jumped up in pain.

"What bloody fool left live wires hanging out of this tree? I've just had a hellish shock!"

The others in his group gathered round and commiserated with George, advised the watchmen to be careful of this live wire, and staggered off to their house.

One of them came to my house in the morning to say that George was not at all well (mind you, very few of us were at that time of that morning), and mentioned that he had been electrocuted during the night.

I dressed quickly and went over to the house, where George was still in bed and did indeed look dreadful.

"What is all this I hear about your being electrocuted last night; are you sure it was not just the demon drink?" I asked.

"Not likely," he replied. "Look at my elbow. It was that wire hanging from the tree that did it."

His elbow was all swollen up and red and puffy, and did not look like the results of electric shock. I thought it best to summon the doc, and in the meantime went out to see this alleged live wire hanging out of the big tree.

Of course, there were no wires anywhere near the tree, but on questioning the watchman he said that he had caught a large scorpion the previous day, and had hung it up on a string under the tree, for good luck. With this information, the doc was able to fix George up with some pretty strong medicine, but he took him off to the hospital, where he stayed until after the New Year party, well out of harms' way.

There were a lot of creepy-crawlies about the area, although the campsite itself was always kept well mown and sprayed regularly against mosquito infestation. From time to time we were plagued with flying ants, and there were always a lot of frogs about, croaking and blowing all night, so we presume there were many snakes as well, although we rarely saw them.

I was coming back from some late work one evening on my motorcycle when I saw a large black shadow on the track. It looked as if someone had just dug a trench about a foot wide across the road, but as I approached very slowly the shadow changed shape, and vanished into the bush at one side. It must have been a large python, or perhaps a crocodile with very short legs, but I realised that there was no room to turn the bike around, so I just opened up and roared past where it had been. In the morning, with a crew of local workmen, all making a lot of noise, they searched the area, but nothing was found.

Somehow, I don't think they were trying very hard.

CHAPTER 99
Famous Sporting Occasions

We had arranged a trip to Freetown that included a visit over the weekend to our colleagues at the Guma Dam contract. The objective was to get everyone off the island for a short change of scene, and to meet other people who were not directly concerned with the work, and also to show our superiority at all known games and pastimes.

We had invited some of the client's staff to come along, and thus were able to borrow their two launches, and all three vessels set out for the city early one Saturday morning, complete with wives and children and all our sporting equipment. A bus met us at the quay and we were back on the familiar road to the south of the city, through villages with names such as Wilberforce, Essex, and Sussex, and then to the Guma campsite.

Things had altered since we left, with gardens around each house blooming with Canna lilies, and bougainvillea round the doors. We were allocated beds in the various houses and went down to the club for a swim in the sea before lunch, and to size up the opposition before we settled on our teams.

The more gentle pursuits were undertaken first, such as billiards and snooker, and the card games were soon raging. Scrabble was also keenly contested, with violent arguments ensuing about words that might have been in the dictionary, had we but known that some of the pages had been removed. This was a dastardly ploy.

Later in the afternoon, when the heat had gone out of the sun, the football match started. The pitch was somewhat foreshortened by the sea on one side and a very large tree at one end. This tree was right on the edge of the beach, and the sand under some of the roots had been washed out, leaving them standing proud of the surface. The game flowed from end to end, with scores and brilliant saves, and a number of rest periods when exhausted players kicked the ball into the sea, and we had to wait until some brave swimmer went in to recover it.

During one of these imposed rest periods it was discovered that one unfortunate player appeared to be resting in amongst the roots of the large tree, and it was then discovered that he was not sleeping but had broken his leg and become unconscious. He was carted off to the first aid post, and later into the hospital in Freetown, and the game continued, but with a restriction of play in the area under the tree.

In the evening there was a grand party in the club, with singing and dancing until a late hour, and on the Sunday we had a lazy morning followed by a gentle sail back across the estuary from Freetown, all hands being much refreshed – except for the poor chap with the broken leg, but he was from the opposition.

Following this outing reports had been transmitted to the other branch of the client's operations, the iron ore mines, some 50 miles up-country by his private railway, and a challenge was sent down by them inviting us up to their compound some weekend later.

We had some pretty critical operations going on the site, and many of us were unable to escape, but we did send off a team, who departed late on a Friday evening and were due to return late on the Saturday. The train journey took about four hours, with frequent stops to allow trains to pass in the other direction on the mainly single-track line, and at each passing place there was a local store selling food and drink.

The crew departed with our good wishes, and arrived in pretty good shape and were billeted out around the camp. There was a good party in progress when they arrived, and it was very late before they turned in.

In the morning the games got under way, without any unusual incidents, followed by a large lunch, swimming races in the pool, general prize giving and speeches of goodwill and farewell, and they left on the evening train. During this trip everyone got off at each stop, and patronised the small shops at each station or lay-by, and when the train arrived in Pepel they were all very merry and bright. All except one poor man, who had tripped over the railway lines at the previous stop and broken his ankle.

He had been hoisted aboard and continued the journey, singing with all the others, and was assisted back to his house. When his wife noticed his swollen foot she came across to our bungalow, and after a quick inspection I called out the doc, who bandaged him up and took him round to his sickbay, and next morning he was off to Freetown for an X-ray, and further treatment. He came back a few days later with his foot encased in plaster, and spent the next three weeks hopping about, trying to keep out from under the feet of his wife.

CHAPTER 100
Marine Transportation

The ore port of Pepel was on a small island in the estuary of the main river system that flowed into the sea off Freetown, and the river was the main highway for trade in the area, although in more recent times the road network was being developed.

Most of the commerce was moved by 'trader launches', which were generally motor boats of about 50 feet in length, with a central wheelhouse, and covered with a canvas canopy over the decks. Some of them ran regular services and carried passengers and goods from the villages upriver down to the main market in Freetown, while others ran special runs for particular cargoes, or passengers for festivals, funerals, and suchlike social events.

In addition to these more modern vessels there were still many canoes operating, some with outboard motors and others using sails and paddles. These boatmen were very experienced, and made clever use of the tides and currents in their navigation, and often anchored or moored up to any convenient structure to await a favourable current or change in the tide.

As our construction work was out in the river, and we had many unattended small piers for the bridges, and also rafts for our cranes and the jetty itself, they frequently tied up against our work, then promptly fell asleep, and they were often difficult to move away when we found them to be interfering with our programme. They also suffered from mechanical breakdowns, and we often had to tow them clear of our works as the tide swept them up or down the river

One day, we were building one end of the jetty and had completed the piles, and were setting the wooden platform and the steel reinforcing for the concrete deck. This timber work had long beams projecting out over the water, and we were in the process of moving a barge carrying a crane into position when we saw a trader launch some distance away coming up the river towards us.

This was not an uncommon sight, and no particular attention was paid to it until it got pretty close. We then realised that it was not under control, and was likely to come into the underside of our timber deck. I called up one of our tugs on the radio and asked him to get round as soon as possible and haul this launch clear.

By this time the launch had been swept under our deck, and our long beams had gone under the canopy, the launch had come to rest against the piles of the jetty, and was heeling over under the force of the current. Some panic developed aboard the launch. The passengers all tried to climb onto our deck, but this was quite difficult, as there were no convenient ladders.

One man on the launch had a bicycle, and he threw it bodily a huge distance onto the deck of our floating crane. Another unfortunate man had a portable gramophone with him, and tried to throw it up onto the jetty, but the box opened in transit, and all the records cascaded out and vanished into the water.

The tide was still rising quite rapidly, and there was some danger of the launch becoming stuck under our deck, or even displacing the deck as it rose on the tide, but by then the tug had come round. I jumped into a lifeboat and went round and took a heaving line from him to the launch, and then the tug put a heavier rope down and began to pull him clear.

This took a considerable effort, as the launch was being pressed hard into the piles, but eventually it began to move, and the tug pulled it away from the jetty directly into the tidal stream. The two men of the crew of the launch realised that they were now reasonably safe, and came back and sat in the stern while the tug turned and began to pull the launch across the tide, with us following in the lifeboat.

By some mischance the tide carried the launch sufficiently far into our mooring pattern for it to run over one of our mooring wires, and as the wire came into contact with the rudder it forced it hard over, which in turn caused the tiller to sweep across the stern, bumping the two unsuspecting crewmen over the side and into the river. As chance would have it, they landed almost alongside our lifeboat, and we soon had them aboard and set off in pursuit of the now crewless launch. We put them back aboard, and directed the tug to take them into the small harbour in the village, and then had to arrange to ferry all the abandoned passengers back in our lifeboats.

When we started the main concreting operations on the jetty, we had a concrete mixing plant on a large barge, and took the mixed concrete by floating crane to its point of deposition, and once concreting operations on any one section had started we had to work on continuously until that part had been completed.

We worked a shift system for the meal breaks, and ferried the men ashore to the canteen and back again for the next shift.

Due to the tide our access points on the jetty were constantly changing, and sometimes the boats had to come in under the heavy beams that were supporting the temporary decking. The men crowded out onto these beams, and as the lifeboat came in dropped off and secured their seats. The coxswain was often bothered about overloading, and had to take extreme measures to prevent too many men from boarding. One trick was to come in slowly against the tide and, by reducing power, hold the boat steady while the men came on, and when the coxswain judged the boat full he would whistle for more power to the engineer, and the boat would steam away. Sometimes this process was not so successful. A number of men would drop aboard and then imitate the signals to the engineer, and the boat would steam off with most of the passengers standing up, and as it passed under the beams they all had to jump up and over them as the boat passed below.

Frequently, when this happened, the coxswain, standing in the stern, would jump up onto the beam and the boat would have passed before he could jump down on the other side, and he would be left standing, hopping up and down, and blowing on his whistle, while the boat careered about with no driver until someone else took control, or it drove into the structure and came to a halt. This caused delays in the men getting ashore for their lunch, and led to many acrimonious arguments, and sometimes disciplinary action, when the coxswain might be reduced to being a member of the concrete squad for a couple of weeks.

CHAPTER 101
"Pa, the Crane done go for Freetown"

We had laid a series of moorings for our floating plant, and were thus able to move the equipment about fairly freely, providing we arranged the moves to coincide with the direction of the tide.

One lunchtime, as I was relaxing after my lunch of scrambled eggs on toast, coffee and a short nap of half an hour, someone came running up to the door and announced: "Pa, the crane done go for Freetown!"

I knew that there were at least two of the marine staff on board, and that there were spare anchors ready rigged for dropping, so I was not too concerned.

Anyway, Freetown was about 15 miles away, and the tide had only about another two hours to run before it would bring the crane back again. Now fully awake, however, I set off down to the jetty, called out the tug, and went off on one of the lifeboats.

I could see the crane travelling sedately down the channel, and soon caught up with it. It appeared that two anchor wires had broken, and the barge had swung right round and tripped the other two anchors, which were still grinding along the bottom of the river, catching on rocks every now and again and causing the barge to lurch as the strain came on the wires, and they stretched visibly.

We were nearly in the middle of the channel and in no great danger of going aground, and other than drop the spare anchors there was not really much we could do until the tug arrived – and, besides, it was a good idea to keep the spare anchors ready just in case we might have trouble stopping on our return back to the jetty.

We continued in this manner with the tug in attendance until slack water, then lifted the anchors and were towed back towards the jetty, and were able to make all secure in the correct position before the full weight of the tide came on.

One evening I went out late in a lifeboat to make sure that some work had been completed on repairs to the concrete machine, which was also moored out in the river. It was a fine, calm night, with a strong ebb tide running, and I took care to keep well clear of the mooring buoys that marked our anchors. Suddenly there was a great rushing noise beside me, and one of the buoys came up to the surface, making a lot of waves, and appearing to be going through the water at a rate of knots. At first I thought it was a very large fish, and was about to have a go with the boat-hook until I noticed the shackle on the top. What had happened was that, due to some change in the force of the tide, the wire had become slacker and allowed the buoy to float up. It is all very well to rationalise these things later, but at the time it gave me a hell of a fright.

There were a lot of fish in the river, some very large indeed, and a number of the small villages existed on the fish trade alone. The fishermen would come out on one tide and allow themselves to be drifted up or down until the tide changed, and often boasted that they never had to use their paddles except when going out or coming in to their beaches. They were a hardy lot, spending ten or 12 hours afloat at a time, and generally had a small fire aboard for warmth and the cooking. The glow from the fire was also their idea of a navigation light.

CHAPTER 102
Sunday Lunches and a Burns Supper

Except when we were very pushed to maintain our works programme we tried to have Sunday as a free day. Most people rose late and went down to the club for lunch, went for a swim in the pool, played tennis or other indoor sports, and had a quiet evening and early to bed.

Sometimes we would go over to the client's club, which was about half a mile away, where they served good curry lunches every Sunday. The club was a large building, with a high roof and fine open windows, and lunch was served either outside on the terrace or inside in the large dining room. It was customary for the officers of ships loading ore at the existing terminal to visit on Sundays if they were not on duty, and it was very pleasant to meet people from the outside world from time to time.

The first Sunday Jean arrived on site with our baby daughter we decided to go over for lunch, and travelled across the island in the Pullman car. I had visited the club previously, but never for Sunday lunch, and was looking forward to it as a great change from my usual scrambled eggs on toast followed by a short nap.

Normally during the week we stopped for an hour at lunchtime, and this was carefully organised. A quick run ashore from the jetty, fast home on the motorcycle, and into the house. The houseboy would have heard me coming up the road, and by the time I had washed the scrambled eggs would be on the toast, the coffee made and consumed, and I would be on my back, where I rested for a further 20 minutes, and woke up feeling pretty dreadful. Then back on the bike, and out onto the water again until the evening.

Sundays were different, and this particular Sunday we relaxed by the pool, had a gentle swim, and went in for lunch. The curried meats were simmering away in a series of large pots at one end of the dining room, and on a long table down the centre was a magnificent display of small chop. There were nuts of all kinds, cooked, roasted, fried and raw, chopped and ground-up or whole, all sorts of fruit cut up into small segments, various cold vegetables, and little bowls of spices, sauces and chutneys, and at the end of the table an assortment of chapattis and flat bread. All this was set out in a large display of flowers, with attentive stewards on hand to assist and advise.

We selected our curry and then went down the long table, enquiring what was what and serving ourselves as we wanted, and then, after a full circuit of the table, moved over onto the terrace, where a cool wind was blowing and there was a bit of shade from the nearby trees; all very civilised. What I had not known was the material I had thought was fine chopped peanuts was in fact some sort of pepper – a sort of dehydrated Tabasco, it tasted like – and this rather ruined my taste buds for the rest of the day.

Some ice cream and a few beers later I felt slightly better, however, but – even yet – the thought of that lunch can still bring tears to my eyes. On following Sunday lunches I was most careful to sample all of these exotic substances – or, better still, have someone else taste them – before delving in.

As the client had been in occupation on that site for a considerable time, some traditions had become established, and it had been the custom of the Scottish contingent to have a quiet party on Burn's night each year. In our establishment there were about ten Scotsmen and their families, and it was decided that we should organise a proper Burns supper, with haggis and all that, and the correct sequence of speeches and recitations to follow.

Another engineer and I were directed to perform *Tam o' Shanter* (with actions), and we spent a few evenings rehearsing our parts. We had decided to speak alternate verses, and the non-speaker had to go through the motions, or hold up large display drawings of the scenery.

The carpenter's shop provided a hobby horse and a small bridge, and the props man was instructed to see what other appropriate items he could obtain.

Tom's wife was elected to play the part of Tam o' Shanter's wife, and the witch, and all other female parts, and the remainder of the Scots contingent who were still on their feet were drawn in as assorted warlocks, witches, and general hangers-on. We had one rehearsal, which was completely shambolic, but sufficient to give all the participants some idea of where and when they should come in, and in what guise, and we hoped it would be 'all right on the night,' as the saying goes.

Somehow, we managed to get all the equipment round to the club, and the evening began. The oldest inhabitant gave a welcome to the foreigners in a most eloquent little speech, and we settled down. The haggis was piped in, and someone else delivered the grace. The chief engineer announced the programme of events, and did a parody of *Holy Willie's Prayer*, bringing in a number of funny and topical comments, and proffering unsolicited advice to the management in the nicest possible way. Drink flowed freely. There was a bit of singing of *Bonny Mary* and *Afton Water*, the piper gave us his selection, and then it came to our turn.

Much to our surprise most of the cast appeared to be reasonably steady on their feet, and mustered on the correct sides of the bridge at the start. We had made a number of large drawings depicting the scenery and the principle events of the tale, but such had been the interest earlier that these had become a bit out of sequence, and some confusion was apparent in the audience.

As we were speaking in broad Scots, however, it was unlikely that anybody understood what we were saying anyway. The performance proceeded fast and furious, interspersed with additional gulps of Dutch courage, until the telling line: "Ah Tam, ah Tam, tha'll get thy farin', In Hell they'll roast thee like a herrin'".

This was the cue for the props man to come on with a herring, but unbeknown to us someone had caught an enormous fish in the river that day, and the man staggered in with this huge fish under his arm, about five feet long and its head as big as a bucket.

This apparition stopped the show for a few minutes, and we put it about that at *proper* Burns suppers it was frequently the custom for the props man to be offered a dram from time to time, to show appreciation for his efforts. This was hardly a good idea, as he had already had more than a few drams, and needed assistance to get himself and his large fish off the bridge.

The performance concluded to a round of applause, and total collapse of the performers. The rest of the evening I find hard to remember, but it was a great night out. We had managed to borrow the general manager's car to take us home, and we set out across the island.

The main purpose of the place was a collecting ground for iron ore from the mine, about 50 miles away, and the railway dominated the layout of all the buildings and roads. The incoming trains brought the ore down to one end of the island and the locomotive was disconnected. Thereafter, gravity did all the work.

The cars ran down to a tipping station, were stopped, overturned, righted and allowed to run further down the grade, where they were formed up into a train ready to return empty to the mine.

Because of this layout, the only road (?) had to go a long way out of the camp area to cross the line, and as we were about to cross someone came up and asked us to wait, as a train was just coming. We sat waiting for a few minutes but could not see or hear any signs of a train, so Tom opened his door and stepped out. Unknown to him, the car was on the edge of a monsoon ditch, and he stepped right into it. As I could not see him, I got out the other door of the car onto the road, and looked about. Still no sign of Tom (nor of the train, either).

Then I heard some muttering from below, and there he was, standing in the ditch, with his head below the road level, quite unhurt but very perplexed as to where the car had gone.

The next problem was where we could get him out; should we go uphill, where the ditch might change into a pipe, or down, where it would probably get deeper? This problem was resolved by the appearance of the train, and in the light from its headlamp we could see a place where some workmen had left a ladder set up in position, and we trudged along, calling encouragement to Tom below us in the dark, until we got him out and back to the car.

He was covered in iron ore dust – a very dense black colour – and his wife would not allow him back into the car, bringing all this black against her best dress, so we had to walk back to the club and commandeer the Pullman car to take us home.

When we went into the club to say we were borrowing the Pullman there were cries of delight, and calls for another number, this time from the Black and White Minstrels. As we were so late already, we did not believe that a few more bits of singing and drinking would get us any further into the mire, so we obliged, and arrived home very much later.

Next day we went over to recover our pictures, and found that someone had taken them off to be framed, and the intention was to hang them in the dining room as a memento of a memorable evening.

We next tried to prevail on the club committee to have a Welsh night (we had found a source of leeks) or a St George's night (we reckoned we could find the odd small crocodile), but there were no takers and life resumed its dreary routine, with Scrabble the main source of entertainment.

We also examined the railway approach where Tom had vanished, and saw how lucky he had been not to do himself a real mischief. He must have put his foot right on the top of the ditch and sort of glissaded down the slope, about six feet or so to the bottom. No wonder we could not see where he had got to.

There was a bit of a cool atmosphere about for some few days, but everyone came round to say how much they had enjoyed the performance of *Tam o' Shanter*, and soon all was forgiven and normal family harmony was restored.

CHAPTER 103
Fishing by Piledriver

When the construction work was fully under way our first job was to drive some temporary piles from a frame on a barge, and then set up a more substantial platform to drive the permanent piles. When the temporary piles were being driven the whole set up was quite fragile, and we used the light air-powered hammer, which gave a series of short blows in rapid succession.

When the platform was built, and we had a more stable base, we began to use the heavy hammers on the permanent piling.

We had some small problems making the hammers operate, probably because of the humidity, and also from their newness, and we had to try to get them going by injecting ether into the combustion chamber.

We started with small amounts, with no success, and gradually increased the dose, until suddenly there was a loud bang and the hammer fired.

It gave the pile such a knock that it drove it out of the frame, and the hammer was left dangling on the crane hook, much to everyone's alarm.

This caused us to pause and consider, and we decided to revert to using the smaller hammer, which we could keep under a certain amount of control, to start the driving, and change over when the piles appeared to offer some resistance into the ground.

The first time we tried this revised procedure the big hammer fired most successfully, and continued to bang away in a very satisfactory manner.

The next thing we knew was a great excitement from one of the lifeboats, which just happened to be passing.

It appeared that a large fish had surfaced alongside him, some way down the tide from the jetty, and it seemed to be quite inert, lying on its back and making no effort to swim away. At some risk to themselves the crew managed to pull it into the lifeboat, where it seemed to revive slightly, making feeble movements of its tail and fins and beginning to gasp a bit. The crew promptly knocked it on the head with some heavy weapon, and came alongside to display their catch. The fish was about six feet long, brown in colour, with a white to yellowish underside, and pronounced to be very good chop.

We later deduced that it must have been resting alongside our pile when the big hammer went off, and it had become concussed. The news of this giant fish soon travelled all round the nearby villages, and next day we found a ring of canoes all round the down-tide side of the jetty, waiting for the fishing to start.

The foreman came over and asked if he could station a lifeboat out there, to try to catch the fish – after all, he had done all the hard work by stunning it – and I agreed. When the hammer struck the foreman was perched on the highest part of the rig, on the lookout, and, right enough, a large fish came up some distance off.

There was a dash to try to catch it, and one man in a small canoe arrived there first, but the fish was much too large for his small boat and he had to abandon it to the lifeboat crew, who arrived very shortly later, precisely directed by the man on the rig. While this large fish was being collected, numerous smaller fish came up, equally stunned and pretty helpless, and violent arguments ensued as to which canoe got a hand to which fish first, with the lifeboat churning in and out and trying to frighten the canoes away. This fishing process continued until we had completed the piledriving, with some really large fish coming up almost every day, and the piling crew grew fat on the profits.

As far as progress on the site was concerned, this was much better than a bonus system, as all the crew were intent on getting the hammer going as soon as possible. The other work on the structure followed on behind the piling, and we were soon setting steel reinforcing bars for the concrete.

The steel-fixer was a Welshman – by the name of Jones, would you believe – and he had been damaged at some time and had lost the sight of one eye. The first time he came out onto the jetty with a new squad of local steel-fixers he arranged that they would continue to work during the lunch break, completing specific tasks, until he returned, and caused much alarm when he took out his glass eye, set it on a table by the window of the office, and told the foreman that he would be watching them. I suppose you can get away with that trick once or twice, but the easy solution was for some brave spirit to sneak into the office and put a hat over the eye, and there is some proverb about eyes seeing and hearts grieving which probably applies.

CHAPTER 104
Construction Problems

When the concrete for the intermediate piers had been completed, we then had to erect towers on each pier and then float out the bridge spans on a large barge.

The bridge sections, each about 180 feet long, were assembled in our yard ashore and rolled down to a small loading bay on railway lines, and then, at a suitable state of the tide, loaded out onto the barge and secured.

As the bridge sections were longer than the gaps between the concrete of the piers, we had to wait until the tide gave us clearance over the piers, move the barge in quickly, and land the bridge section, clear the fastenings, then move the empty barge out. While the barge was loading up the next section, the first bridge was being hoisted up by hand winches to its final position on top of the towers, and this process continued for the total number of 12 bridges.

The first one or two were the cause of some concern. What if the piers were too far apart? What if the tide was not high enough that day? But our fears were groundless, and the whole operation went smoothly.

On the construction of the main deck of the jetty we were tensioning up the special bars using a hydraulic jack, which pulled the bars to a set loading, and then a large nut was run down a threaded section to maintain the tension in the bar. The bars were led into ducts in the concrete, with the tensioning ends out over the front of the jetty, and the jacks were set in position by a small crane.

During the loading one of the bars broke quite early in the process, and the jack and the special gadget for running up the nuts were projected out into the river, along with about ten feet of thick bar.

Apart from being very dangerous to anyone who happened to be passing, we lost a vital part of the stressing equipment, which called the whole system into some doubt. The first task was to see if we could find out why the bar had broken, and fortunately we were able to pull out the other end, and could see that there was some physical damage on the surface of the steel that must have weakened it to a very large extent. An immediate system was applied for a rigorous inspection of all bars to be undertaken prior to installing.

We struggled on with the remaining set of jacks while waiting for a replacement set to be sent out, but as this would take some time we arranged for a diver to attend, and while we were waiting for him to come out made some attempts to recover the jack by use of a large magnet. We borrowed parts of a welding machine and fixed them into a frame, which we dangled over the side of a lifeboat. We then set up a welding machine on the jetty, connected to the magnet by welding cables, with a switch on the lifeboat.

The theory was to align the boat up with the trajectory of the jack, allow for the effects of the tide when the jack was sinking to the bottom, and then try to travel along this line, bumping the magnet along the bottom of the river.

This exercise was full of imponderables. How far had the jack gone before hitting the water? Which way and how strongly was the tide running at the time?

How far up-tide did we have to hold the lifeboat to allow for the current dragging the cable with the magnet to get it even into the right parish?

After a series of fine judgements, we were all set up for the first run. We lowered the magnet to the bottom and switched on. There was no apparent increase in weight, so we lifted it a little bit and lowered it again, having moved the boat a couple of feet out. At about the fourth try the magnet clanged onto something very heavy, nearly pulling some of the crew over the side. Here was another problem. This item was much heavier than the jack. Could we, or even should we, try to lift it up? If we brought it to the surface, could we handle it out of the water, or even get a sling onto it?

We decided against moving it at that time, and continued our search for the jack, but frequently kept on coming into contact with this heavy object. Eventually we pulled the magnet up to see what we had caught, and found it bristling like a hedgehog with the ends of welding rods which had been discarded, but no signs of the jack. This exercise was then abandoned and we waited impatiently for the diver. When he went down he reported almost nil visibility, and that the seabed was full of odd bits of steel sticking up – like a pincushion, he said. After several attempts, as he could only work for short periods about slack water, he found the large heavy object, which turned out to be a test pile that someone had lost on an earlier expedition, weighing about ten tons, so it was just as well we had not tried to salvage it. There was no sign of our jack, and we had to abandon the search, and press on in double shifts until a replacement unit was shipped out.

A feature of the design of the jetty was a special fendering system which used eight concrete fender units, each of about 80 tons weight, which were suspended below the deck by special high-tensioned bars, and which had four large rubber cylinders fitted into the front, connecting into a steel and timber frame. Due to some delays in our progress the fitting of these fender units became a major constraint on the programme.

We devised two alternative methods of construction, and by superimposing the best possible sequence of construction, by means of drawing out each operation on transparent paper, were able to decide what we should do next as each day's work was reviewed.

We had one set of equipment mounted on a large barge, whereby we could construct the complete fender on the barge and then sail it under the jetty, let the tide lift it into position, and secure it with the steel bars through the existing deck. The other method was to build the fender above the deck level, suspended from a bridge, then lower the fender into the final position and complete the deck on top of it, finally fixing it in position with the stressed bars.

Both of these methods were successful, but we did have the odd moment of excitement. When we floated the first unit under the deck, we had to engage 64 bars into 64 holes.

This matter had been recognised as potentially difficult. With all the bars projecting the same height above the fender, they would all enter their holes at the same time as the tide rose, and it was most unlikely that we would be able to align them all before entry. We fitted long screwed caps onto each bar, and by adjusting the caps were able to have them all engaged and started in their holes as the tide rose. We had a certain provision for minor delays, as we could lower the entire unit by winches on the barges and so keep pace with the tide, and if all failed we could still lower away and sail out to try again later, as they say on the telephone. The process of entering these bars was known as harping.

The other method was by use of towers and a bridge. We had some Bailey bridge equipment on site, and the towers were built on the deck of the jetty, with the bridge on top, some 40 feet up above the deck, and a temporary bridge to carry the construction at the deck level.

The units were concreted and allowed to set, and then lifted by winches on top of the bridge towers, and the lower bridge was removed. These were ten-ton hand winches, and by a system of blocks and tackles we were able to lift the 80 tons of fender. At the first attempt of lifting, with two expats and eight local men on the bridge, the load came on quite gently, with the odd creak as the towers and bridge settled to their work, but suddenly the wire rope on one of the winch drums forced adjacent turns on the drum apart, and effectively lowered itself by about an inch. This caused a violent jar throughout the structure, and gave us pause to consider.

After a careful inspection, with no evident damage, except to the nervous system, we continued. The same thing happened again, but it was not quite so alarming now we thought we knew what was causing the trouble. When we had completed the operation, I spoke to George, who was on the other winch.

"What were you going to do, George, when the whole contraption collapsed?"

"Well," he said, "I was going to hang on at the edge here, until we were nearly into the water, and then jump out that way."

"Fair enough," I replied; "I was going to do just the same, but jump out the other way, so we would not have bumped into one another."

During the construction we had to make provision for the electric cables for the navigation lights for the jetty, and we laid a small pipe through each section as it was concreted, leaving a length of thin rope inside for pulling the cable through when we were ready.

Sometime during the construction some clever fisherman had discovered this line and had removed most of it, leaving only the ends at each end of the jetty, where there was a small terminal box let into the deck, with a steel plate cover.

When the time came to install the cable this lack of line was discovered, and various attempts were made to poke a wire through, but without success. Some bright spark suggested that we might be able to blow a table tennis ball attached to a piece of fishing line, through the pipe, and this was immediately tried.

The ball went off up the pipe, pulling the line freely behind it, but then, despite the continued air pressure, no more line was being pulled in, and we were only about halfway to the other end. Much pondering about what was going on.

One of the mechanical foremen was working in one of the site huts, and was hearing a funny noise from under his feet, but only mentioned it some time later, when proposals were being considered for major digging-up operations.

"What kind of a noise was it, Jock?"

"Well, it was a funny sort of intermittent tapping, but it seemed to be coming from below the floor, and that is six feet of solid concrete."

Great relief all round. The hut had been landed over the central manhole for the electrical system, and the noise was of the ball bouncing about on the end of its line inside the chamber, and striking on the thin steel manhole cover.

Towards the end of the construction work we had to remove some of the long wooden baulks that had been used to support the timber deck, before the concrete was placed. These beams were about 30 feet long and 15 inches deep, and were fastened in position by a series of clamps, which had been secured around the piles. In many cases, it was possible to lower these beams down to the water level and tow them away by one of the lifeboats, but sometimes they had landed and jammed between some of the sloping piles, and were difficult to move.

The system employed in these cases was to fix an anchor point for a pulley system into the underside of the deck above, and by attaching the other end of the system to the beam it was possible to lift and slide it clear. Sometimes this worked better than others.

I was passing in a lifeboat when I saw this sequence of events.

The pulleys were set up correctly, and there were five men standing on the beam, which was just at water level, but not quite floating.

As they pulled in the fall of the pulley system, the beam began to move.

As the beam moved out, the load on their rope decreased, and they had to pull harder to keep their balance; for every foot the beam moved, they had to haul in four feet of rope.

Sooner, rather than later, the last man on the rope came to the end of the beam, and fell off into the water, where we scooped him up into the lifeboat, followed by all the others, one by one, while the beam went on out of the other side of the jetty like a torpedo, until it came to the end of the rope.

When the deck of the jetty was substantially completed, the assembly and erection of the ship loaders was put in hand. These were twin towers, which could each travel on a curved rail set into the deck, and which carried a sloping conveyor belt that could be raised or lowered, extended or contracted, to permit the flow of iron ore to be loaded into almost any part of a ship moored alongside.

Each structure was made up of six large sections, fastened together by large pins that fitted into machined holes. These pins were kept ashore until required.

During the assembly the first two sections had been set up, and one of the fitters was making sure that the holes were in exact alignment, filing off minute irregularities inside the bores. The pins were then called up from the shore, and were loaded and carefully wrapped up to prevent any damage to the surfaces.

The first two pins were the largest in the structure, and were about five feet long and 15 inches in diameter, but for some obscure reason the pins ordered out were for the next set of connections, and were about four feet long and 12 inches across, and when the man came to fit them he was very surprised to find that they were much too short and not nearly thick enough.

There was an immediate condemnation of practically everybody in the vicinity, but soon the correct pins were delivered and face was saved all round.

The arrangements of these two ship loaders were such that the operator of one machine was able to control the other machine from the master cabin, and a duplicate set of wiring had been fitted to the master machine.

When this was being fitted, what must have happened was that someone gave someone else two bundles of wires and cables under the floor of the cabin, saying, "Give me that bunch through this hole, and the other bunch goes through the other hole." That was the idea, anyway.

When the time came to test the controls and movements on the first loader, everybody was cleared away, just in case anything was going to go wrong, and the operator engaged the motors and saw the instruments indicating that his machine was working. He could not actually observe any movement, however, but there was great panic on the other loader, where men were working away, fitting bolts and doing what fitters are supposed to do. It was gently moving out over the water. Much shouting and apportioning of blame ensued.

Two days later the error had been corrected, and the first machine went through its testing programme without a hitch. Mind you, people were very reluctant to go up on the other structure during these tests.

The day approached when the first ship was due to load, and some overtime was worked to ensure that all the necessary work was completed before she docked. The first loading commenced on time, and we had a marvellous party to celebrate our completion of the main civil works.

During this loading one of the conveyors leading up to our section of the work broke down, and for some reason the conveyor feeding this failed to stop.

The feeding conveyor continued to run, carrying many tons of ore into the junction tower, and stopped only when the tower house was full, right up to the main outgoing belt. This caused some embarrassment, but as it was outside our scope of work we continued the party for another day.

CHAPTER 105
Freetown Activities

During our stay in Pepel a new contract for the upgrading of the airport was being prepared, and I made a couple of trips by our speedboat along to the ferry terminal and walked up to the airfield to confirm a number of details for the London manager in charge of estimating.

The place had not changed much since its inception as a base for the RAF during the war, except that the hangars had become more rusty and the drains had filled up a bit. I sent off the information back to London, and shortly thereafter we heard that the contract had been awarded to another UK contractor.

During one of my infrequent visits to Freetown I met the new site manager for the airport contract, congratulated him on the award of the contract, and offered any assistance or advice he might require. He had never worked overseas before, and did not yet believe that assistance and cooperation were freely given between overseas contractors, and thought that people were trying to put one over on him – perhaps out of envy at not themselves winning the contract.

I heard later that 50 tipper lorries were being shipped out, and that he had advertised for drivers to collect them from the docks and drive them round to the airport, and then to operate them during the construction of the field. At that time, a driving licence could be purchased in the market in Freetown for ten shillings, but this proposition raised the price to over a pound. It also was good business for the money-lenders.

The vehicles were unloaded from the ship and lined up on the dockside, where the usual palaver with customs was endured. The first 50 men to attend with driving licences were appointed and told to proceed to the nearest fuelling point, where arrangements had been made to fill them up, and they were directed to set off in groups of four or five to the site. This journey was about 80 miles around the estuary and over the first bridge in the river, and was expected to take a whole day, or perhaps two. Out of the 50 lorries that were dispatched, only about 30 had arrived within a week; some had come to grief on the journey, some had been stolen, and others had been doing a roaring trade in the outlying areas of the city.

By the end of a month most of these had been recovered, at great expense, and eventually only about 40 were in anything like good running order when they finally arrived.

The next time I met up with the manager he asked me if I knew of any good bulldozer drivers, and I was able to put him in touch with some good men who had been working on the dam a couple of years earlier.

It seems he must have got the message.

CHAPTER 106
Home again

After the first ship came alongside, and the loading had been completed without any serious snags, it was time for us to leave, and we packed up all our household goods and set off for home.

We had a fond farewell party, and left in state on our converted fuel launch, which had seen good service throughout our time on the island, and sailed across the harbour for the last time to Freetown.

Due to the uncertainties of the weather, and the normal state of organised confusion, we had elected to stop over in the new hotel, so that we would have somewhere fixed up if anything went wrong. The new hotel was quite splendid, full as it was of chromium plate, potted palms, and air-conditioning – a far cry from the government rest house standards of our introduction to the country.

The kitchens were kept out of bounds to customers, so for all we knew the denizens of the old City Hotel were ensconced below, and the place was still like Dante's *Inferno*.

The food, however, was quite good, and we passed the night in unaccustomed cool, thanks to the air-conditioning. In the morning our bus arrived and we departed on schedule for the airfield, and after a quick passage through the customs we were off on our way to Bathurst, Las Palmas, Lisbon and London.

In Bathurst we bought a few small crayon pictures on black paper, but otherwise had an uneventful stop, except for the noise on the runway, and were soon off again to Las Palmas.

At the airfield we disembarked, collected our overnight luggage, and were quickly through the formalities. Outside stood a line of Volkswagen minibuses, and with some care we selected the second one in the line, on the grounds that, if the first one hit something, we might have a sporting chance either to stop in time or leap out before the accident.

We loaded our bags and climbed aboard, with me sitting in the front alongside the driver. I seemed to have done it again; another demon driver. When all the passengers were aboard, he revved the engine, hooted the horn, and it sort of exploded at his touch.

A selection of discs and springs flew up; I managed to field one of the largest parts, but the remainder were all scattered around the vehicle – some on the floor, some in the rear passenger's compartment, and, for all we knew, some out of the open window. A pair of wires were still dangling out of the hub of the steering wheel, and when the driver attempted to join them together there were some sparks and he got a bit of an electric shock.

What a dilemma.

How could he possibly drive all the way into town without a hooter?

The van behind was tooting furiously, and we were obliged to move out.

Determined not to lose his place in the procession, our driver was off like a tiger, but at each corner, vainly trying to blow his hooter and getting mild electric shocks from the exposed wires, he was no match for the vehicles behind, and most of the other passengers had arrived at the hotel before us.

We were ushered into the same old pleasant hotel, and had to undergo the penance of the Spanish dancers during dinner once again. In the morning we had a different driver, who had a working hooter, and we were off at great speed through the quiet and deserted countryside back to the airfield, tooting all the way in a regular fashion.

Then off to Lisbon, where we were again presented with a small bottle of port (very hospitable people, the Portuguese), and then on to London, where we had been booked into a hotel in the city. In the morning we went back to Maldon by train, and were back into our house by noon.

The place was in good order, the tenants had made a good job of looking after it, and nothing significant seemed to be damaged or broken.

The garden was another matter. The grass was knee-high, and the so-called vegetable patch was not a success. I went into the office for a couple of days, and was then told to go away for a month and enjoy myself.

We did a spot of visiting the relatives, and then went home and I tackled the garden, and undertook other work about the house. The grass was no great problem, but very hard work, and when I came to the raspberry canes I had a quick look in the book, where I read something about cutting them down to nine inches. This seemed a bit drastic, but following the maxim 'when in doubt, consult the manual' I attacked them with gusto, and soon had the whole issue reduced to the regulation nine inches.

Then came the retribution. What I had not done was to look over onto the next page, where it stated quite specifically that only the new wood should be cut down to nine inches.

I couldn't stick them back, so we just had to hope for a better crop the following year.

We learned of a Bendix washing machine for sale, and went off to one of the American Air Force bases to conclude the purchase and bring the thing home, and with some difficulty managed to get it into the back of the car. This machine had foundation bolts that required to be set into a concrete floor, and I duly made holes in our new house floor and set the bolts in good concrete, carefully mixed and placed. After waiting about a week to allow the concrete to set and harden, the great moment for the first trial run was upon us.

The machine was filled with washing and soap powder, and the power turned on.

It ground away quite happily until the time came for the 'spin' phase. For the first few minutes all appeared to be well, but the pitch of the noise changed, and I dashed out to see what was happening. The wretched machine was dancing across the floor, and was denying access to the cable and plug in the wall.

Grabbing a broom, I was able to switch the power off, and the noise ceased.

Here was another problem; the machine was still full of wet washing, and the door was not programmed to open until the spinning was completed.

The engineering solution was to block the machine up so that the dangling bolts would do no further damage, and then run the power in short bursts until the cycle was completed and the door would open.

This completed I examined the bolts, and found that my new concrete around the bolts had pulled out, leaving a sort of sandy crater in the floor. They must have been a bit short of cement when they laid that floor. The repairs called for a much bigger patch round each bolt, and I am sure that it would take an earthquake to shift them now.

Back at work again in London, I was engaged in the estimating office, and my first estimate was for a job in Ghana, draining a lagoon.

I was investigating the capacities of various large pumps, and was invited down to the clay pits in Cornwall for a demonstration of similar pumps in action.

The clay there is mined by hosing down the faces of steep clay banks, and pumping the residual slurry back up hill into settling ponds. It was a very messy business, everything being covered with grey, slippery mud, and the pumps were all working under very trying conditions.

After visiting the working face we went back to the workshops and spoke to the maintenance engineers, and then adjourned to a local hostelry for lunch, and what is now called 'corporate hospitality'.

I had been so impressed with the performance of these pumps that I proposed them for use on the contract. When the tender was being prepared for submission I had to bear a certain amount of scrutiny concerning their cost and performance, but managed to convince people that the proposal was correct, and the tender was agreed and sent off.

A few weeks later there was great panic in the office. We had been awarded the contract, and were some considerable amount of money below the next bidder.

Crisis meetings were called, the bid was re-evaluated.

Why were our costs so much less than anyone else's?

Who had made these ridiculous recommendations?

How could they be justified?

Why had we not put a large contingency factor into the bid?

It looked as if I was the fall guy, and I sweated a bit, checking all my figures again and again. I called in the manufacturer's chief engineer to confirm that I had not forgotten a few noughts on my totals, and thus assured I went into the meeting, with the knowledge that the managing director of the pump manufacturer was only a phone call away, and he would be happy to come along and confirm my estimates. This turned out to be unnecessary, as my figures were believed, and the heat was turned on others in the department for not allowing a larger contingency.

This argument was squashed when the contingency factor was disclosed, and the threat was that I had better go out and look after the job in Accra.

From all accounts Accra was quite a pleasant place to work, with good swimming, and golf and tennis, and I agreed to go if required.

Because of other people who had already been in Accra becoming available at the end of their leave and wishing to return, it was decided that their local knowledge would be an advantage, however, and I returned to my duties in the estimating office.

For a few weeks I continued on the Accra job, selecting equipment and preparing loading programmes for the various items to be shipped out, and dealing with the London end of establishing the contract organisation and reporting procedures.

The next job that required a site visit was in Port Harcourt in Nigeria, where it was proposed to build a number of oil pipelines across the delta of the river Niger, connecting up with a new submerged pipeline from Port Harcourt to the Bonny Bar. Two other members of our staff were already in the area, and I was to join them on location as soon as the formalities had been completed.

CHAPTER 107
A Visit to Nigeria

After a week on London based-investigations, I was off to Lagos to visit the site and obtain the local information required to complete the tender.

I stopped overnight in Lagos with one of our resident staff, and in the morning made arrangements to go to Port Harcourt by air. Sometime during the day the flight was cancelled, and I was asked if I would go up by the coal boat.

The distance between Lagos and PH is about 300 miles, and it was a day and a half steaming by coal boat, but it might be a week before the aircraft would be flying again, so I agreed, and was taken down to the loading wharf, where the steamer was lying and preparing for sea.

There was a certain amount of coal dust about the ship, but the passenger accommodation was spotless. I was taken to my cabin just below the bridge, and then down to the dining saloon for afternoon tea.

Going on deck while working coal was not recommended.

Quite soon we were under way, and steaming east, out into the darkness of the Bight of Benin. As soon as we had slipped our moorings the ship was washed down, all the coal dust cleared away, and we could go out on deck to see the lights of Lagos grow dimmer as they dropped below the horizon. We sailed on, the ship rising and falling on the long ocean swell, with phosphorescence streaming back from the bows, and the wash like a track made of stardust trailing away behind us.

In the morning we could just see the line of the coast in the far distance. This was the mouths of the river Niger, which came out into the sea for about 100 miles along the coast. The coastal plain was pretty flat, but there were many tall trees standing on the low sandbars, and from our distance off they gave the impression of a solid line of shore.

Soon we were turning into the Bonny river, and began to sail between the trees along a twisting passage for about 20 miles. From the upper works of the ship it was possible to see over the majority of the trees, and it was a bit surprising to see the masts and funnel of another vessel coming across our bows, only to find later that he had been in another part of the channel and there was not much chance that we would meet.

When I arrived in the port some of our staff were there to meet me, and I was taken in by the plant manager, Alan, and lodged in his house for the duration of my stay. His wife had recently returned home, so we had the place to ourselves, and lived quite royally in his fine house.

The others in our estimating party had been on site a few days previous, and we met up to arrange the division of tasks for the tender. I had been booked on the oil company aircraft for a trip out to one of the legs of the proposed pipeline the next day, and set off in a jeep to the airstrip.

Our aircraft was the executive Dove, well fitted out with walnut panels and drinks cabinets (unfortunately empty), and we took off and flew smoothly for some distance to a little town called Ugehli, where we transferred to a more primitive aircraft, a de Havilland Canada DHC-3 Otter. We droned on for another 20 minutes, and landed on a large sandbank in the middle of a river. A helicopter was waiting for us there – very primitive, no doors, canvas seats, no springs – and we took off after a quick word from the captain about flight safety and flew about five minutes and landed on another, much smaller sandbank, where we could see a small launch anchored off. When we got out, a large black man gestured that we should get into a small canoe that was beached nearby, and he would take us out to the launch.

As we stepped gingerly aboard he handed us both a calabash, and made fast bailing motions with his hands, indicating that if we wanted to arrive at the launch we had better get at it, and keep at it. Perched on the gunwales of this frail craft, with all our papers and notebooks, we travelled out to the launch without shipping too much water, and climbed thankfully aboard. We then chugged across a broad river with the canoe in tow, and anchored not far offshore.

Once more we had to get into the canoe and be paddled ashore, where there had been a trace cut in the jungle, about ten feet wide in parts, running straight across the island – about a mile, I suppose – and our friendly canoe man indicated that we should get on with the walking, and that he would pick us up on the other side.

There being not much sign of any alternative, we began walking, taking note of any very large trees that might obstruct the passage of the proposed pipeline, and trying to estimate the soil conditions underfoot.

The Niger delta is all sand, except where it is mud, and the sand is generally about three inches below the surface of the ground, so all the trees have very wide, shallow roots, which are obvious places for snakes and crocodiles to lurk, just waiting for the unwary feet of passing estimators. When the trace had been cut all the debris had been thrown to one side, and formed a sort of wall of spiky leaves and branches, while on the other side the virgin forest waited, round every tree a possible python.

Overhead, the branches spread out to provide complete cover from the sun, and noisy birds and troops of monkeys advertised our passage as we made our fearful way across the island. We emerged at the river bank on the other side, slightly bitten by insects, very hot and sticky, but otherwise undamaged, and were met by our friend with the calabashes.

Once more we did the ferry trick to another island, went ashore and began to march along the trace, perhaps with a little more confidence as, apart from the birds and monkeys, we had not seen a living thing.

Out again onto the other side of the river, and a different person awaited our arrival, and took us out to another launch, and we steamed across to another trace in the forest, across yet another island.

As far as I could see, it was just exactly like the previous two traces, and I declined to walk across that one; I could see from the map that it was about a mile across, and there would be another two similar marches to be made after that.

My colleague, from another bidder, was very keen, and set off in the canoe again, while I sat at ease on the deck of the launch, and we sailed gently round the end of the island to meet him coming out the other side.

"What was that one like?" I asked.

"Much the same as all the others, but I was a bit concerned that I would fall and break my ankle in the roots, and you might not be able to find me," he replied.

"Well, what about the next ones?" I asked. "Are you going ashore again, or should we just sail round, and look at the starts and stops of each trace? After all, the difficult places will be where the pipe crosses the water, and comes out or goes in again. Pulling it through the traces on relatively dry ground will not be too difficult."

I could see that the logic of this argument had got through to him, as well as a bit of fatigue, and we instructed the launch captain to take us on to the place where the helicopter would be collecting us; but we also said that we would like to see the ends of each trace, from his fine launch, as we went past.

Thus agreed, we sat back and relaxed, enjoying the passing scenery, and making up our notes on the inspection so far. When we came round the last bend in the river, we could see the helicopter waiting for us, and were soon aboard, and lifted off for our large sandbank.

It was interesting to see that there was no sign of any disturbance in the canopy of trees to indicate where the traces had been cut, except where the line came out at the banks of the river.

Next stop the Otter, and then the Dove, where our muddy boots made a bit of a mess on the fine Axminster carpeting, and so back to PH in time for tea.

The next few days we spent driving around the various bush tracks, looking for sites to set up welding camps, and stores depots, and landing places for heavy equipment, and generally prospecting, and chatting to the few pipeline people who were still completing the original spread of lines.

One chap related the story about a young engineer, just out from home on his first overseas tour, who was set into the bush to recover all the plant and machinery before the river rose in the rainy season.

He had one machine left to extract, and was winching out from tree to tree, but as he neared the river bank the trees were smaller and had shallower roots, and as he connected up and pulled, instead of pulling his machine out, he pulled the trees across the surface. This continued until there were no further trees within reach, and in answer to his plea: "What do I do now?" his foreman replied, "You had better plant another tree, laddie."

During our stay in Port Harcourt we were invited to the wedding of a Dutch engineer to a young lady from the oil company. The night before the wedding we all went out to some dreadful nightclub somewhere near the airfield, and there was a lot of drink taken.

When we returned to the house and were preparing to go to bed, the bridegroom appeared at the door, seemingly having some doubts about the wisdom of getting married. As both of us had some experience of the state, we told him not to worry, and that it would all turn out right in the end, and put him to bed.

Some time later there was another banging on the door, and the best man was there, very distraught. He asked if we had seen Pieter. He was not at his own house, nor at anyone else's house; they had tried. He had been last seen some hours ago, muttering about not really wanting to get married; was he gone? Jumped in the river, possibly?"

All getting on for being a bit hysterical.

We set him down, gave him a drink, and said all was well and under control. We had Pieter safe in bed in the spare room, and if he liked he could stay and sleep on a sofa in the lounge. He agreed, and lay down and went out like a light.

About an hour later we were awoken again by further ringing of bells and banging on doors. They were asking if we had seen Jan, the best man. He had gone out looking for Pieter, and had failed to return, and now both were lost.

"It is all under control; they are both here, sleeping like logs; come on in and have a drink, and we will wind up the party again – after all, almost everyone is here now." Somewhat to our relief, they said, "Thanks but no thanks," and went away.

We considered putting a notice on the door saying 'Pieter and Jan are here; go away', but it was nearly daylight by then, and we did not think we would be further disturbed. Very shortly thereafter our two Dutchmen woke up, not knowing where they were, and some confusion reigned until Alan appeared, and sorted them out. They phoned around announcing their return from beyond, and making arrangements to get back to their houses and get dressed, and also to confirm that the bride was still speaking to them. They were collected, and we next saw them in church, both looking a bit weary but performing well. By the time of the reception they had both made a remarkable recovery; come to think of it, we hadn't done too badly either.

Great stuff this champagne and schnapps.

After the official reception we moved on to the nightclub, and arrived back in Alan's house very much later.

We continued with our site inspections around the area, travelling mainly by Land-Rover, but with odd excursions in canoes, hired after inspection, on the banks of the river. Travelling in these vessels required a great deal of faith, as the freeboard was minimal, to say the least, and the stability not much better. Their only obvious advantage was that it was normally possible to alight onto the bank without getting your feet wet, but this was not really relevant, as your feet and most of the rest of you were pretty damp already, either from the residual water in the bottom of the canoe, or from the splashes made by the paddles, or the spillages from the bailing calabashes.

Towards the end of our visit all the contractors were mustered for a trip down the river to see the line of the submerged pipeline from Port Harcourt to a new loading point inside the Bonny Bar.

We all appeared with our notebooks and drawings fairly early one morning, and boarded a launch provided by the oil company and chugged off downriver.

Our first stop was at a small mangrove swamp, teeming with flies and mosquitoes, and with mudskippers skating about on thick, black mud. This was the start of the new pipeline. Why this spot had been selected rather than somewhere else was not explained, and, in any case, one bit of muddy river bank looked much like any other bit, and even if we had been looking at the wrong place it would have made no difference to our estimates.

A number of photographs were taken and copious notes made by the more enthusiastic members of the party, and we sailed on to a sharp turn in the river.

Much reference to plans and drawings, more photographs and more notes, and off again.

In all, I suppose, we must have stopped at half a dozen completely featureless points along the route, and each one was meticulously photographed by the estimators. I wondered what these photographs were going to be used for in preparing their estimates, as each one was almost identical with all the others, and I also wondered what they were making notes about.

My own notes were restricted to such comments as 'change of direction point l, river about l00 yards wide, some mangroves, muddy banks'. 'CDP 2, much the same', etc., etc...

As we proceeded down the estuary the mangroves gave way to higher banks of grey sand, some with fairly tall trees growing on them, and the river channels became much wider. We observed a number of large crocodiles lying sunning themselves on the shore, which were disturbed by the wash of the launch and slithered off into the muddy water, and there was an abundance of small bird life evident. As we neared the sea, the tops of the sandbanks were a bit higher, and the beaches on the little islands we passed were of a whiter colour. There were also signs of some habitation, with small huts set up on tall stilts every two or three hundred yards, the homes of fishermen, and often small patches of cultivation along the tops of the islands.

Our journey ended at the pilot station at Bonny, where we went ashore, and had an enormous lunch of T-bone steak in the pilot's mess. This was one of the real 'outposts of empire'. The building was of white blocks, with a corrugated iron roof, and had a sort of bunkhouse for five or six people, a pleasant mess room and library, a radio room and lookout post, and a bit of a garden with a large flagpole and signalling mast. Probably it had been a very busy station at one time, but the traffic was now reduced to the regular daily coal boat service and a few tramp steamers each week, with the possibility of some tanker traffic upriver to Port Harcourt refinery and oil loading dock.

With the proposed new pipeline the oil tanker traffic would all end at the new loading facility at Bonny, and the commercial river traffic would revert to being the main purpose of the pilot station.

On the way back upriver to the port we visited one of the oil-drilling platforms that was operating in the river delta. This was a large raft that was able to jack itself up a few feet above the tidal range, and comprised a drill, the associated drilling mud tanks and pumps, and a flare line to a stack a short distance off into the bush.

The accommodation was in an old riverboat, and the majority of the rig crew lived aboard in some comfort, despite the mosquitoes.

The remaining few days of our visit were spent in confirming information from our resident staff on the harbour works, and we were off again back to London, to finalise our report and assist in preparing the tender.

No sooner had this excitement died down a bit, and I was asked to go on another tender visit, this time to Libya, to see about the development of a newly discovered oilfield to the south of the Gulf of Sidra.

CHAPTER 108
A Visit to Libya

At this time Libya was still a kingdom, and diplomatic relations were still fairly cordial, so it was not too difficult to obtain the necessary work permits and visas, etc., and as my health documents were in order (they did not know of our binge in Nigeria) I was readily certified as being a suitable person to visit the country.

Our party consisted of three: one mechanical engineer, one welding engineer, and me, and we flew off to Tripoli early one morning, courtesy of BOAC.

On arrival we had arranged to be met by the Shell representative, and he rushed us through the formalities and took us to a rather fine hotel, with a good view over the bay.

Our programme was to have a few days in Tripoli finding out the state of the market for subcontractors, local suppliers of stores and equipment, the local labour laws, useful contacts in the government's Oil Department, and a list of all the local spivs.

We all had an interesting few days in these endeavours, and then flew down to Benghazi, which was to be our main base for the up-country surveys.

In Benghazi we were booked into a huge hotel on the promenade, all marble and crystal, but somewhat run-down. It was rumoured that Mussolini had lived there during his stay in Libya, and there were also pieces of graffiti indicating the subsequent presence of German and British occupation during the war.

Benghazi was a pretty awful place, suffering from the inroads of the desert on three sides, and the erosion of the sea on the other. Each day a squad of men appeared in the streets, shovelling sand into dumper lorries, which went out into the desert and tipped their loads. Either the same sand, or other sand very similar, had blown back into town before they had returned, and it seemed as if they had a job for life.

The harbour was still recovering from some storm damage, where great waves had displaced large sections of the breakwater and dumped them into the harbour, where they were plainly visible through the clear water, lying at odd angles on the bottom.

We went along to the Greek contractor who was working on the harbour, but he was not very helpful. We made contact with some persons whose names had been suggested to us from Tripoli, and began to sort out the local score. There was a survey aircraft in the area, and we made some charter arrangements for us to be flown around the desert, and inspect the terrain at the various proposed sites and fly the road and proposed pipeline routes.

I marked up on the pilot's charts the places we wanted to see, and he indicated the nearest places he was prepared to land, and a compromise was reached.

We set out on our first reconnaissance early one morning from the field at Benghazi in his Otter aircraft, and flew inland for some time. He then prepared to land, seemingly in the middle of nowhere, and no sooner had we touched down than a pick-up truck appeared.

He had come to collect the newspapers, and asked if we could carry a passenger back to town. This chap climbed into the back of the aircraft, and we were off again. We landed on our first location, a reasonably flat stretch of stony desert, and I took a few handfuls of sand as a typical soil sample. Then off again.

This procedure was repeated several times, and in general the material was very similar. The surface was quite hard, with small stones sticking out, not much vegetation, and suitable for running vehicles across without much trouble.

The loose sand had all been blown into gullies and wadis, or further off into a series of dunes, and we took some care in plotting proposed routes in and out of the various sites to reduce the distances over this loose material.

We had three days of this flying and landing in the desert, and very often people appeared out of what had seemed to be an empty landscape, requesting a lift or asking us to carry mail back to Benghazi for them.

The pilot came back to the hotel with us and we sat chatting about his experiences flying on these prospecting trips. One time he had landed, picked someone up, and gone off again, when another passenger asked how he had found that particular place.

"Well," he said, "when we were coming in, do you remember those two tall palm trees towards the end of the strip?"

"Oh yes, but I cannot see them now."

"No, you won't see them from here; they are a mirage, but they are always there as we come in, at this time of the day."

So much for instrument landings.

He told us of another time, just after the war, when he was flying pilgrims to Mecca in an Avro York. Shortly after they had taken off from Tripoli the captain asked him to go back and see that the passengers were all right. When he got through to the passenger compartment he discovered that the passengers had brought along some camel dung, and were brewing up their tea on a small fire in the middle of the floor.

He put an end to this performance very rapidly by use of the fire extinguisher, and when he returned to the captain and reported this incident he was calmly told to remember to indent for a replacement extinguisher when they landed.

"These guys are on their way to Mecca; if we don't get them there on this flight they expect to go straight to paradise, so they have no worries at all."

"That might be all right for them," our captain had replied, "but lighting fires on the floor of our aeroplane is unlikely to assist our passage. I'll just go back and keep an eye on them for the rest of the journey."

On the last night of our stay in Benghazi we went along the promenade in a pair of horse-drawn carriages to a nightclub some distance out of town, and after some persuasion arranged for them to come back for us some time later. There was a good crowd in the place, mostly Arab businessmen from the town, with a few of the resident expatriates and a dozen or so drillers on leave from the desert.

It seems that their routine was to spent six weeks on site in the desert, and then have two weeks on leave, mostly spent in Beirut, but often with an overnight stop in Benghazi.

They were a pretty wild crew, and the likely source of disturbance as the evening progressed. The food served was passable, the service not bad, but the music and floor show were somewhat repetitive.

Half a dozen well-built young ladies, well concealed behind yashmaks and miles of trailing gauze, hopped and lumbered about, while the band played its five notes, and the Arabs in the audience conducted their business and the drillers shouted out advice and instructions to the dancers.

The pièce de résistance was a very plump young lady, advertised as the best belly dancer this side of Cairo, who came on with her own musicians, and troop of minders, and performed the most incredible wrigglings and shakings I had ever seen. This roused the drillers to a frenzy, and there were all the signs of an imminent confrontation between them and the minders, which was diverted in the most skilful manner by the manager offering drinks on the house before the next item on the programme.

We left soon after the free drinks were dispensed, and organised a small bet on which carriage would get back to our hotel first.

Both horses refused to go at more than a walk, despite sizeable rewards offered to the coachmen, but they probably realised that the carriages would most likely collapse if they went any faster, and, in any case, a race is to the first one to finish, irrespective of the speed.

We flew back to Tripoli the next day by the local airline, and were booked into a different large hotel, also full of air-conditioning, chromium furniture, and European cuisine.

Our head office in London had contacted a local businessman, and he came along to the hotel to offer his assistance in dealing with the local government formalities and setting up contracts with local suppliers, and he took us round to meet these various persons. The working routine in Tripoli was very complicated, governed as it was by Islamic law, but also greatly influenced by the Italian occupation in the late 1930s.

Thursday was the equivalent of the Christian Saturday, and most places closed at noon. Friday was the Islamic holy day, and all work stopped.

Saturday was the Christian half-day, and work stopped at noon, and on Sundays no work was contemplated. On top of this arrangement, numerous public holidays relevant to either faith were observed, and there were a few additional holidays for the celebration of matters of national importance, such as the King's birthday.

When all these considerations were placed into a proposed work programme, the time extended towards the millennium, and additional costs had to be entertained for double-time payments, and unforeseen additional delays.

We happened to be in Tripoli over an extended holiday period, and having completed as much of our work then available had to suffer a couple of days lazing about on the beach, and driving into the countryside.

The area directly to the south of the city was fertile, laid out in orange groves and date palms, with small villages and well-kept little townships, and as we climbed higher into the mountains the air became cool and the groves of orange and lemon trees gave way to less frequent large coniferous trees, with the depredations of sheep and goats becoming evident. We had an excellent lunch in a small café high up in the hills, and then descended to the coast again, where we had been invited to visit some colleagues of our Libyan friend who worked in the large American air base at Wheeler's Field, some distance out along the coast.

This was a very extensive piece of America set down on the shores of the Med. Apart from the operational facilities, the living quarters were well set out and appeared to be meticulously maintained, with swimming pools, tennis courts, bowling alleys and numerous restaurants, and a very large PX store.

The main distraction was the noise of aircraft landing and taking off, and the almost continuous test running of jet engines, which sometimes drowned out the music from the loudspeaker system – which also seemed to be running all the time.

We sat around in the early evening beside a swimming pool, swatting flies and making polite conversation, then had an enormous dinner and retired early.

We had a Tuesday and a Wednesday to complete our business, and made our escape on the Thursday morning before facing the prospects of another four-day dead period when no one was in their offices.

During our frequent landings in the desert I had collected a number of samples of sand and pebbles, and had some explaining to do when questioned by the customs officer. Were these diamonds, perhaps? Was this a blind to divert his attention from other parts of my luggage? He had me empty out all my baggage, and still was not certain, but he let me go with some reluctance.

Back in London we tried to make some sense of our observations, and started the planning stage of the estimate. After about three weeks on this estimate the company was awarded a contract in Sierra Leone for the construction of a rutile mining complex, to be built about 100 miles south of Freetown, and I was appointed to take charge of the civil engineering works on this project.

For the next few weeks I was busy with drawings and programmes for the contract, and also making arrangements to rent out our house in Maldon during our absence.

As the work was all out in the bush, and there were very few existing facilities, the first three or four months would be on a bachelor establishment basis, until we had constructed our camp and sorted out the transport situation in this remote location.

Our estate agent in Maldon found us a suitable tenant, and we arranged that Jean and Susan would go back to her parents' house in Teignmouth until such times as we were ready for them on site.

SECTION 3 – RUTILE

CHAPTER 109
Early Days

Having had no previous experience of rutile, I looked it up in various textbooks, and found that it is a red dusty dioxide of titanium, found in alluvial deposits. The material is used to increase the reflective properties of some paints, and also as a source for the metal titanium. The client had been prospecting in southern Sierra Leone for some time, and had located a number of possible deposits in the laterite in the area of a district known as Gbangbatok, about 100 miles along the coast from Freetown.

This was a very sparsely populated region, with poor roads and no reliable telephone connections to Freetown, but there was a bauxite mine in the area, about 20 miles distant from our site, operated by a Swiss company, excavating and running the ore in large lorries to a small jetty on the river system nearby.

The client's geologist was living in what had previously been the district commissioner's compound, and there was some accommodation for our staff in this place, for a limited period of time. The first two members of our staff had arrived two weeks before me, and had settled into one of the small aluminium huts at the compound and made a start at the site of our camp by running a duckboard walkway some 100 yards off the main road and erecting a large marquee, where a few items of stores and tools had already been mustered.

I arrived from Freetown in a minibus, fully laden with essential food stores and equipment, and was followed by a couple of five-ton lorries with more tools and stores.

The directions for finding the place were pretty vague. "Go up the main road towards Bo, turn right across the railway lines, and keep going towards the south-west. You will see a Lebanese shop on the left, and a few houses with trees full of weaver-birds in front of them, about 40 miles on, and we will meet you there. Have a good trip."

We trundled up the main road towards Bo, which was not really too terrible, but when we turned off across the railway lines the road became more of a jungle track than anything else, with deep ruts and very narrow bridges. The bridges were all protected by steep humps at each end. Whether by accident or design, one side of the hump was steeper than the other, and if you went even a little too fast the vehicle was thrown up in the air, and partly turned away from the planks across the bridge.

In most of the river beds the results of vehicles coming off and going into the rivers could be seen: deep scores in the banks, and, frequently, bent pieces of truck left sticking out. At that time the rivers were fairly low, as the rainy season had not yet started, but it was obvious that great care would be needed later in the season.

When I arrived in the late afternoon at the place with the weaver-birds and the Lebanese shop, a guide met us on the roadside, and I went off on foot to meet the others in their tent. After the exchange of greetings, and the handing over of the mail, they organised a few men to come and unload the lorries.

The first item off was a wheelbarrow, and three men solemnly took a part each, one on each handle and the other at the front wheel, and proceeded to carry it along the duckboard track. I stopped them, and had one man sit in the barrow, and trundled him down the remainder of the way. This was a great revelation, and for the next ten minutes all the men in the area were taking turns in being wheeled up and down. (Not by me, I might add.)

When the vehicles had been unloaded and parked for the night we set off in a Land-Rover for the client's camp, about 20 miles further down the track. The main building was a large bungalow with mud brick walls, wide verandas and a corrugated iron roof, and contained the normal accommodation for the DC and his guests: a large dining room, a spacious lounge, and kitchens and other outbuildings.

A feature of the dining room was an enormous hardwood table, cut from a large tree, about three inches thick, eight feet wide and 25 feet long, maintained in perfect condition by the resident staff. The accommodation for lesser mortals was all in little round aluminium huts, about 20 feet in diameter, with connecting corridors between the rooms, most houses having one living room, two bedrooms, bathroom and kitchen.

The camp was on the top of a ridge, among large trees that provided pleasant shade, and there were spectacular views out over the bush in three directions. There was a small village at the foot of the ridge, and most of the camp staff lived there. In the evenings the smoke from their cooking fires drifted up towards the compound, and the lights from the houses were the only signs of life to be seen.

Our days were fairly busy, as we had to drive the 20-odd miles back and forth to the site each day, where we were setting up the basic facilities for our own compound and preparing for the main stores and workshops for the mine and the permanent camp. As we had no generator on site we could only work during the daylight hours, so this called for an early start in the mornings and a late finish at night, and when we arrived back in the evenings we had to write up reports and requisitions, and prepare drawings, estimates and programmes for our temporary work.

On Sundays we had the day off, and travelled back up the road to the bauxite mine, where we were made honorary members of their club, and could swim in their pool and enjoy the changes in diet served up in their mess.

The camp at the DC's compound was run by the client, and, when available, American food was served, whereas at the mine the meals had a marked continental flavour (garlic?). It was refreshing to meet other people who had been operating in the area for some time, and we learned a few wrinkles from them on local affairs and the various tribal relationships.

Our actual contract was to complete all the construction and testing work for the mine, and included the construction of a housing estate, complete with club, shops, leisure centre with bowling alley and all the attendant services, a stores and service compound with power station and processing plant, and the actual dredger itself.

Some distance away we had also to build a small jetty where the rutile would be loaded out onto barges and taken out to sea, where it was to be loaded onto the ocean-going ships for despatch to the UK or America.

We were also required to build an earth dam, and to install powerful pumps to form a pond for the dredger, and all the services necessary to maintain these items in operation.

Our first priority was to build our temporary camp and set up water and sewage services for our own use, and also to establish a radio link with our head office in Freetown. We had already ordered our houses from a company in the forest to the north of Bo, similar buildings to those we had used at the Guma Dam, and as soon as the remainder of our transport arrived a convoy was dispatched to collect the first consignment.

We had expected them to return in about a week, but when ten days had passed we thought we had better send out a search and rescue party, and a couple of Land-Rovers and two large lorries with four-wheel drive and fitted with winches were prepared and set off.

The first of the rains had started, and we waited for a further few days before the loaded convoy returned, all covered in mud, but undamaged and a little bit triumphant.

We beavered away, preparing footings for the houses, laying drains and water pipes, and made a start on a small dam in a nearby stream for our temporary water supply.

One weekend the club at the bauxite mine were holding a dance and party for one of their visiting dignitaries, and we were all invited to stop over the Saturday night. Five of us piled into the Land-Rover, all spruced up, and set off in the early evening. The first part of the road was down the side of the ridge, and the road was ledged into the side of the hill, with a considerable drop on the outside. We were going fairly slowly when a bird flew in through the open window, and began to flap about in the back.

Complete confusion reigned.

I was driving, and vaguely saw something go past, but the people in the back were frantic.

"It's a snake," somebody shouted, and we were all out, standing on the road in the dark. "How the hell are we going to get it out?" was the first question.

We searched about by the roadside to see if we could find a long stick to encourage it out, but no one was keen even to stand outside and bang on the doors, just in case the snake was provoked into coming out at them. While this council of war was in progress, the bird appeared on the front seat. It was a hoopoe, about the size of a large pigeon, and how it had managed to come in without hitting any of the passengers or the frame of the window I will never know. It was a great relief to find that our incomer was not a snake, and we soon managed to shoo it away onto the road, where it sat quite calmly watching us make an inspection of the inside of the vehicle.

Just think; it might have laid an egg. Also, just think what might have been chasing it!

371

We continued a very subdued journey with the windows all closed up until we were well away from the side hill cutting.

We had a very pleasant evening with our neighbours, exercising great skill in our mastery of various continental languages, and generally being completely misunderstood most of the time.

In the morning we were awakened by a troop of howler monkeys in a big tree near the swimming pool. There were about 20 of them, scoffing the flowers as if they were going out of fashion, and hooting away at one another, before they all went off into the nearby bush, where we could hear them crashing about and calling to one another all the time we were having breakfast.

Later we were taken round the mine workings and saw all the huge machines digging away, loading the material into dumpers, which then carried the spoil to a sorting plant, where the ore was graded and loaded into silos ready for transporting down to the dock.

We set off fairly early, to make sure that we would be past the potential flying snake area before dark.

I have seen snakes rear up and jump a considerable distance, and could well imagine that it would be quite possible for one to jump into the back of the vehicle. The ones I have seen have always jumped away from the vehicle, however.

There were always tales of someone driving over a snake, which was no longer visible on the road behind, and which had somehow wrapped itself around the back axle, then found a convenient hole in the floor and come up and bit him, but I always regarded this as an old wives' tale. I was always fairly careful to inspect the floor of the vehicle, however, and I liked to see proper rubber sleeves around the openings for the various pedals.

Meanwhile our equipment continued to arrive in Freetown, we had started to recruit local labour, the layout of our camp was progressing well, we had a sort of water supply, and the septic tank was working. Some survey work was started to locate the dredger pond, and to mark out the limits of the required excavation.

Some explanation of the principles involved is called for.

The rutile was lying in the ground, all mixed up with the basic laterite material.

The intention was to build a large barge in the dry inside a hole in the ground, this hole being about 20 feet deep by 100 feet wide and 300 feet long. Inside the barge we would assemble a large dredger.

We would also build a dam, and form a sizeable reservoir at a stream some distance away, and then pump a lot of water into the hole, and so float the dredger.

The dredger would then start to dig; the material excavated would be screened and most of it dumped at his back, the selected material being piped ashore to the process plant.

In this manner the dredger was able to move along the hillside, digging in front and dumping behind, and keeping afloat all the time. If necessary, the level in his pond could be topped up from the main reservoir. The material piped ashore was to be dried and graded, and the rutile collected by an electrostatic system, which could distinguish between zircon and rutile, and directed each material into separate hoppers. From these hoppers it would be loaded into lorries, carried to the small dock and loaded out by barge and thence to the ocean-going ships.

CHAPTER 110
House-hunting

We had a message through to tell us that the next consignment of houses was ready for collection, and it was decided that I should go along with the convoy. We prepared the vehicles, loaded up the winch lorries with ropes and chains, sleepers and various jacks and hand tools, and set off very early in the morning.

We went through a small village while it was still dark, and caused a goat, tethered to a small tree, to break loose and charge into a nearby house. This disturbed the owners, and their chickens, all of which dashed out into the road, and a couple of the chooks were run over. This was what I had come for, and without much argument I persuaded the owner to accept some money in compensation. The argument continued as to who now owned the scrawny heap of feathers, but I left that to the local foreman driver to sort out.

The rains were now with us, and the roads were very muddy. Every few miles we came across lorries that had gone off the road, blocking the passage, and we had to set to and winch them out. News of our passage got out, and all manner of broken-down vehicles were lined up to try to join our convoy, knowing that we would pull them clear if they got stuck in front of us. I had impressed on all our drivers to keep closed up behind the vehicle in front, but this was especially difficult on the slopes down to the narrow bridges, and frequently another lorry would somehow get into our column.

This was a double problem, as they would not keep closed up, and yet more vehicles would join in, and when they became stuck the whole enterprise ground to a halt.

When this happened we had to halt, get our winch lorry back to the stuck vehicle, and then pull him out. We gave them all one chance only; the second time they got stuck we put our winch cable through a snatch block on a tree and pulled them off the road to one side, while the rest of our vehicles got past. This constant pulling vehicles through meant that what should have been one day's journey turned into a three-day sortie, and we had to rely on the provisions from local shops by the wayside to keep ourselves and our drivers fed, and had to sleep in the vehicles when we stopped overnight.

We arrived at the sawmills late in the afternoon and were welcomed by the owner and his wife, who took us into their splendid house, which was set not far from the sawmill in a grove of large trees. The house was built entirely of timber (as was most of the furniture) and it was a large square building with wide verandas all round, and the living accommodation in a central core. The floors were beautifully polished and decorated with oriental rugs, and the tables and chairs were of a variety of different woods, each item with a particular history in its selection and manufacture.

I dumped my gear and went out to see that the drivers were being allocated suitable lodgings, and returned for a very hot bath and an excellent dinner to follow, then an evening of most entertaining conversation with my hosts. They had been mining chromium until a few years earlier, and had found that working timber was a much more civilised way to make a living, and certainly more profitable.

They had been living in the bush for a number of years, and had established themselves very comfortably away from the towns, living more or less off the land, but keeping fairly busy and able to take holidays in Europe from time to time.

The following day, while the lorries were being loaded up, I was taken round the premises, and saw that he had a very well-established and modern sawmill, with all the necessary equipment for felling, collecting, and cutting up timber at one end of his yard, and a woodworking factory at the other end, where complete houses were being made up from a set of jigs and then broken down for transporting and loading out as required.

After another night in this lovely place I was off again early in the morning, but news of our departure had been leaked abroad, and a long queue of hopeful vehicles was waiting at the end of the access road to the sawmill.

The average vehicle in Sierra Leone, up-country, was an ex-army Bedford truck, usually with a high wooden framework fixed (?) to the body, always grossly overloaded, and nearly always decorated with religious or profane slogans painted across the roof of the cab. The drivers were generally a cheerful bunch, and as most were self-employed, and owners as well, their main intention was to get on with the journey in hand, and if they could cram in a few more passengers – so much the better.

375

We tried to keep closed up as long as possible, but sooner rather than later we came across a vehicle stuck on the road ahead, and as soon as we altered our order of march to send up the winch lorry another vehicle would slide into our column, and we would have to pull him through to enable our later vehicles to keep going.

Sometimes the lorry would be so heavily laden that we had to unload some of its cargo before we could pull it through. There was always the thought of massive compensation claims if we happened to pull the axle off, and a certain amount of care was essential. In some places, where the road approached a river bridge, the mud on the surface was about two feet deep, and the Land-Rover was having trouble in pushing the mud in front of it as we went along.

The worst problems occurred when vehicles coming the other way had become stuck, and a line of them had formed. In such cases we had to find a place to pull off the road to let them past as soon as we had pulled them out, while making sure that no other enterprising chap would try to overtake our column from the rear.

This all meant frequent trips of our winch lorry up and down the track, which did nothing to improve the surface, and when the rain really started to fall tempers became short, and it was difficult to be objective about people keeping their place.

This was another three-day passage, and I was very happy to be back into the relative civilisation of the little round aluminium hut, a hot bath, and a good dinner.

When I told the story of our escapades that night, another young and keen engineer volunteered to go along with the next convoy, and I made out to be reluctant before agreeing that he should go.

"It will be good experience for you," I said, not bothering to tell him about the insects, the discomfort of sleeping in the back of a Land-Rover in the pouring rain, on top of a heap of damp and muddy ropes, or floundering about in the mud in a monsoon rainstorm, trying to direct suicidal drivers away from the edge of a flooded river bank, probably full of crocodiles, snakes, and the odd rabid dog.

Living in the rondavels, as our circular houses were called, was not much fun in the rain, either. The concrete floors were only marginally above the ground level, and the slightest irregularity caused in the generally muddy surrounding ground often caused minor diversions in the surface water, so it was not uncommon to discover a muddy flood inside the house after a heavy rainstorm. The noise of the rain on the roof was also extreme, and what with the rain, intermittent thunder and lightning causing the camp electrical supply to shut down, and the consequent dampness in airing and drying cupboards, it was a bit of a hard life.

There were often amusing events that tended to lighten up the atmosphere from time to time. One of the children had a small tame monkey, which was often secured to a large tree in the middle of the camp area. The monkey wore a belt, and the string allowed him about 50 feet of range, so he was able to get onto and into a number of the rondavels.

He had been discouraged from coming into our house but still came onto the roof, usually by way of the electric cables stretched between the houses, and was very adept at balancing on these cables.

I was watching him through the window one evening, and he came across the wire and sat down on our roof to eat something he had stolen from another house. Having completed his meal, he turned round and saw his string stretching behind him, back to the tree. Unknown to him, he was actually sitting on the string, and it appeared quite tight back to the tree. Just to make sure, he gave the string a bit of a pull and then stepped onto it, and disappeared from view with a wild shriek.

One morning, when the sun was shining brightly, I had been into the mess for an early breakfast, and was collecting my gear for the day when I saw people emerging from the mess door and suddenly breaking into a run, flailing their arms about above their heads, and rushing back to their houses. A few moments later there was a general evacuation from the mess – diners, cooks, waiters, the lot, each one performing the same rituals.

What had happened was that a swarm of bees had been disturbed, and they were buzzing around, stinging anyone in sight. I managed to get back inside before they arrived at my vicinity, but it was about an hour before peace was restored and we could get away down to the site. When we returned in the evening the bees had settled somewhere up in the big tree, and there was a constant humming noise until well after dark. Next morning they went off into the forest, and troubled us no more.

CHAPTER 111
Camp Construction

The site we had chosen for our temporary camp was fairly open, on reasonably flat and well-drained ground, with a few large trees and a lot of scrub bush. Our first jobs were to clear the bush and set out the road system around the camp, dig the ditches, lay the sewer pipes, build a septic tank, lay the water mains, and set up the generator.

The houses, when they arrived, were all designed to be set up on small concrete piers, and the manufacture and erection of these piers was a priority, so that the houses could be assembled and prepared for occupation as quickly as possible. When all the concrete blocks for the piers were completed we started production for our club building, and made a start on the erection of the block walls.

The clubhouse was a good size, with a large bar, a reading room, a billiard room, and a section set aside as a cinema, and it had space provided for stores, a cool room and a general shop. The dimensions were agreed by local consent; there was even a trial bar height, where potential customers were invited to stand against the wall and mark the place where their drinking elbow felt most comfortable. This led to some acrimonious discussion, the possible solutions being:

a) an inclined bar, where customers could select their preferred height;

b) a sloping trench along the outside; or

c) the height I found most convenient.

Proposition 'a' was ruled out on the basis of spillage; 'b' was ruled out due to difficulties of drainage in the floor, and the potential problems associated with persons trying to get out of a hole; 'c' was carried, as I was responsible for the drawings of the building.

The construction proceeded, with the walls plastered and painted on the inside, with particular attention being paid to the wall that would act as the cinema screen.

As part of the artistic licence afforded to the designer, I decided that we would have one wall in the bar made of stone, taken from the small quarry we had established, and I selected a heap of suitable large rocks and had them delivered to the masons in the club.

With my own fair hands I built a couple of courses of this wall, and explained carefully that this was what was required. Large stones, with very little cement showing.

"Do you understand?"

"Yes, pa."

"All right; get on with it, then, and I will come back later to see how you are getting on."

I was away on another part of the site all day, and it was late in the afternoon when I returned to the clubhouse.

My two masons were standing on a trestle, just completing the plastering of my beautiful freestone wall. They were operating in a sort of a trance, and were rudely aroused when I came in shouting and roaring, and then returned with a small tractor and knocked the wall down.

"I told you. Big stones, no cement. What the hell is this? Go round to the time office – you are paid off."

I had had rather a trying day, and this was the straw that broke the camel's back.

News of my outburst quickly got around, and soon there was quite a crowd of spectators and commentators standing around the ruined wall. I went into the office to calm down, and was visited by a deputation led by the senior local foreman, begging my indulgence, and asking if these two foolish masons could please have their jobs back. They had been trying so hard to please, and were intent on making a very smooth surface indeed on the wall.

Further explanations were of no use. "I even built a section to demonstrate what was required: big stones, very little cement."

Further discussion in the background.

"Well, pa," the foreman entreated, "these are foolish men, but they will work all night and rebuild the wall exactly as you showed them, if they can have their jobs back."

This was an offer I found difficult to refuse, and I agreed that I would inspect their work in the morning, and if I found any cement on the faces of any of the stones I would be most displeased. I went down later and found them hard at work, scrubbing the partially set cement off the stones and generally trying to rectify the situation, and as a gesture of my appreciation I arranged for a driver to bring his vehicle along and shine the lights into the building until they were finished.

They were all there in the morning, the wall was finished, all the stones were sparkling, hardly any cement was showing, and the floor had been washed down.

"Very good," I pronounced. "You are not paid off, but you must listen to my instructions, and, if you are not sure, you must ask. Do you understand?"

"Yes, pa. Thank you, pa. We understand," they replied in unison.

Shortly after this event we were making holes for the electricity poles along the roadside, and had a small single-cylinder auger machine set upon a tripod for this work. It was trying manfully but was not quite up to the job, and someone noticed an advertisement for a similar machine with a larger engine in one of the trade magazines.

We requisitioned the larger machine, and it was delivered fairly quickly and sent on up the road. I had seen the machine being test run in the workshops, and had set out the positions of the holes along the road before going off to another part of the site.

When I came back some time later, the operator was standing, scratching his head, with no holes dug.

"What have you been doing all day? This is still the first hole, and you have hardly started."

"Well, pa," he replied; "this machine, he is turning, but no agree for dig."

"Fire it up and let's have a look," I said.

The engine fired first time, the auger went round, but – surprise, surprise – it was turning the wrong way. On closer inspection, there was no obvious way to change direction; it was either driving or stopped.

I took it back to the yard and had a long argument with a disbelieving fitter, who was very reluctant to take it to pieces and change over the valves so that it would run the other way. After some persuasion, though, he agreed, and the machine was back in service the next day, and digging well.

I wrote a rude letter to the manufacturer, who had the grace to explain that that particular machine had somehow been mounted on the wrong side of the frame, and they had not realised that this would mean that the drive was turning the wrong way. Very sorry, much humble pie, and all that.

Meanwhile the houses continued to be delivered, despite the rains, and our little camp developed. The intention was to have about eight married quarters, and four-bedroom houses for the remainder of the staff, plus the club. A tennis court had been laid out, but the ground was still too wet for play, and a small golf course had been marked out in the centre of the camp buildings, which were set out round an oval-shaped road.

The golf course was quite a success, despite the frequent hazards of balls bouncing off rocks in the field and landing inside the houses (very difficult to play a nine iron from under the kitchen sink).

Soon the great day came when the place was declared fit for families, and general occupation. More staff arrived, we moved out of the rondavels, and families arrived by steamer in Freetown and had to be collected and brought up to site.

We had established a cold store in the club building, supplies were ordered up from Freetown each week, and the shop was formally opened.

As soon as the families arrived there was a marked change in the contents of the shop, and life became much more civilised. The office manager, who had previously conducted all the arrangements for food and domestic supplies on a single-mess basis, was a bit distraught when confronted by ladies asking for particular brands of soap and shampoo, etc. Very sensibly, he appointed a ladies committee to assist him, and when the camp was fully occupied a store manager was appointed from London, and he came out and ran the shop.

The rains had ceased by then, and with a full staff on site the construction work got properly under way. As this was a contract for an American client, most of the equipment had been ordered from US manufacturers in America, and since this was the time of massive planning and programming a computer programme had been prepared that set out in meticulous detail the state of ordering, manufacture, delivery to US ports, dates of loading and expected dates of arrival by particular ships into the estuary not very far from the site of our new proposed dock. This programme was printed out and updated every two weeks, and a copy was pinned up all round the walls of my office. Our construction programme had to fit into the deliveries, and this was another programme, not nearly so complicated, which was also displayed on my office wall.

To be able to accept the shipments from the sea we borrowed one of the large barges and one of the tugs that had become available from the previous contract at Pepel, and made a small loading dock near the site of the permanent dock works, and set up our stores compound near the office to hold the equipment until it was needed for the permanent work.

The entire planning situation was suddenly thrown into utter confusion. A dock and shipping strike had been declared in America, and it had been decided to change the programme of shipments from US-registered ships to one making use of smaller vessels, owned by continental shipping companies.

The main consequence of this change was that many items that previously had been packed for individual shipment were repacked inside other larger cases that had some spare space, but the documentation of most of this repacking was rather carelessly completed, and where there was no room in any one particular box the smaller items were frequently sent by road to other ports for other ships loading our cargoes.

As a brief example, we had notice of five large generators, packed individually, with their deliveries spread over about three months. When the first machine was eventually delivered, and the crate opened before the customs officer, we discovered a whole range of house fittings – sinks, baths, cookers, fans, cutlery, crockery – all stowed alongside the main generator.

The other problem was that all these items were documented in American; thus a tap for a bath was not a tap at all, but a faucet, and we had to arrange for one of the client's engineers to come and translate for us before the customs would release anything. On top of all that, we had to rearrange our stores compound to accept loosely packed items out of sequence, and then amend our construction programme to suit.

One of the ships had been trading into Russia before collecting our equipment, and the captain told of the difference between various ports in Russia.

In Murmansk, he said, the place was all cold and miserable, the officials were most unfriendly and the port labour uncooperative. He had a sentry posted at the foot of his gangway, and he was escorted by police everywhere he went ashore.

Some time later he was sent to Siberia for another cargo, and the contrast was very marked. He was made welcome by the harbour master, and was free to go where he pleased about the town. The sentry at his gangway was full of advice on where to go for entertainment, and could arrange taxis etc., and later came aboard and ate in the seaman's mess. It seems that the native Siberians were a happy-go-lucky lot, and those persons who had been sent there probably felt that nothing worse could happen to them, so they joined in and tried to make the best of it.

He also described the bitter relationships then current in the American shipping and transport industries, and the trouble with petty officials he had endured when loading our cargo. I told him that we were having a bit of trouble with our local officials too, as a result of the re-shipment.

Not only that, the changes in the computer programme were always about two ships behind. They might have been better with pigeons.

CHAPTER 112
Diamonds in the Dam

One of our major works on this contract was the construction of an earth dam across a stream a short distance away from the camp. The site was surveyed, the bush cleared, and excavation for the foundations commenced.

The construction required a deep trench to be cut right across the valley, to form the base for the impervious clay core of the dam, and after a few days' digging the foreman came over to the office and said that we seemed to be into blue clay. From my limited knowledge of diamond mining, blue clay sounded very interesting, so at lunchtime Doug, my foreman, and I carried out a close inspection of the trench walls.

There were indeed sections of blue clay, some containing small pebbles, and we managed to dig out some more promising-looking stones before the men came back. In the records of the foundation trench we reported 'stiff grey/blue clay'. If this was indeed diamond-bearing, the rutile project could well be cancelled in favour of a diamond mine – probably much more profitable, too.

We took these stones back to the office for a closer inspection, and after careful washing and drying tried them on the glass of the louvred windows of the office. Success. Most of them were hard enough to scratch the glass. For the next few weeks we continued our selection of specimens, trying them out on the windows, and keeping all those that scratched the glass in coffee jars in the office cupboard.

I began to wonder whether, if we had a severe thunderstorm, all the panes of scratched glass would shatter, but the weather continued fair.

At the rear of the dam we were required to lay a thick blanket of sand, and I spent some time prospecting for a suitable source of this material. There was a sizeable river about ten miles away, which, in the dry season, had large sandbanks exposed, and samples were taken and analysed to see if the material would be suitable. The river was on the opposite side of the main track in from Freetown, and ran roughly parallel to the road, about seven miles away through the bush. There was a track of sorts, just passable by Land-Rover (if you kept your elbows in), which twisted through the trees and served as a connecting route between a few small villages, and also the main artery of communications with a larger village on the opposite side of the river.

We were advised that the natives on the opposite bank were not friendly. I spoke with the headmen of the villages we had passed by, and learned that there were coffee trees in the forest owned by various villagers, and that there were often arguments between villages and families as to who actually owned these trees, and other fruit-bearing bushes. Not wishing to become involved in inter-village strife, I arranged that we should meet with the local paramount chief and drive down the track, and he would adjudicate on the estimate of damage we might cause by opening up the road and disturbing a few coffee trees.

Some days later I went to the paramount chief's village and he came aboard the Land-Rover, followed by three attendants: his trumpeter, his clerk, and some other functionary whose duties were unclear to me. After the introductions we set off for the first village on our route, and when we arrived at the branch in the track that led off to the village we stopped. The chief spoke to his trumpeter, who climbed down and produced an elephant tusk from the nether regions of his garments, and blew a blast.

"Just giving them a toot," the chief said. "Must maintain some sort of authority, you know."

Immediately the bushes parted before us, and the village headman and his attendants were revealed as having been waiting there all the time.

The chief spoke at length, the headman replied, and the clerk announced: "The headman says that there are five coffee trees alongside the road belonging to people from his village."

"Ask him to come along with us and point them out," I said to the clerk.

Further dialogue between headman and clerk.

"What was all that about?" I enquired.

"He says maybe only four trees, pa."

Further violent argument between chief and headman.

"The man is a rascal, trying to cheat us out of the contents of a miserable tree," said the chief. "I offered him £1 each for the trees, and told him to be happy that he is getting anything at all."

The clerk produced a docket, I gave the man his £4, he put his thumbprint on the paper, I poured a shot of gin for the chief, and we moved on.

At the next branch in the track a similar procedure was adopted, but the news must have preceded us, as this time the headman claimed ten trees, and had men standing by each of them as evidence.

This additional expenditure seemed to cause a slight hesitation, and we got down and inspected each tree in turn, the chief making comments as we passed. Some further discussion with the headman ensued, and eventually the chief told me he had settled for a sum of £7. The cash was paid, the receipt signed, and off we went again, after another small gin.

This process continued at each small village on our route, with the price of trees falling as we progressed, until, at the last village, we settled for 20 trees at five shillings each.

We then climbed down the river bank, where the trumpeter gave another blast towards the forbidden village on the other bank. It seemed that they were within the jurisdiction of another paramount chief, and we were signalling our displeasure at this fact.

The hooting was not answered, so we had another drink and began to discuss the royalty for the sand we proposed to dig out and carry away. When we had reached an amicable price we had another drink, the trumpeter gave a final blast of contempt, saying "They really expect it, you know," and we loaded up and drove back to the chief's compound, tooting as we passed each branch junction in the road.

A few days later we had widened and strengthened the track, and were running our large scrapers through to the sandbank. These machines were able to travel at about 30 mph, and could carry about 20 tons of sand, so we had to arrange a lookout system at each major corner to allow them to pass one another, and to keep the local children off the roads. The presence of these children was a constant worry, as every time a vehicle stopped the children would clamber all over it, and it was quite likely that one of them would be carried away, or even stow away, to see what the rest of the world looked like.

The system seemed to work very well. The headmen were engaged to supply the lookouts, and this brought some cash into the local economy; also, the improved road meant that they could get their produce to the pick-up points for the nearby markets much more easily, we got our sand, and the paramount chief collected his royalty.

During the digging of this sand we found other very hard pebbles, and my windows were subjected to more and more scratching, and the stock of coffee jars full of potential diamonds overflowed onto the veranda of our bungalow. We had taken discreet advice on the quality of these stones, but as diamonds were very carefully controlled (in fact, it was said to be illegal to walk about with your head down, just in case you might discover one by accident) we found it difficult to obtain any sensible advice.

As luck would have it, we had arranged a sports weekend with our colleagues in the diamond mine contract, beyond Bo, and when they arrived I found that the person who had been allocated to stay with us over the weekend was the chief of security at the mine. He made some jocular comments about our coffee jar collection, and when sufficiently mellowed was able to tell us that they were all a bit premature, and we would need to wait a few million years before they would be of any value.

He did advise us how to tell a diamond from any old bit of pebble.

What you need is a black ashtray, some water, and bright sunlight.

You place the handful of potential millions into the ashtray, top up with water, and then hold it up in the sunlight, while swilling the water around.

If there are any diamonds there, they will shine out like a lighthouse.

We tried this trick with all our samples, and actually managed to find one microscopic speck of diamond, but certainly not in commercial quantities, and the stones were consigned to the footpath around the house.

Persons who might be reading this record a million years hence should note that our house was the third on the right, after the club building.

One of the young engineers from the diamond mine contract told us that, some time before, the concrete-cube-crushing machine on his site had broken down, and it had been agreed with the client that the samples should be taken to the university laboratory in Freetown for treatment.

The procedure for testing concrete is that it is made up into cubes, either four or six inches across, and then placed in a machine that imposes a crushing load, applied by hydraulic jacks, until failure occurs. The load is noted, and the strength of the concrete is then declared at so many pounds per square inch, or kilograms per square whatever, if you have gone metric.

Our friend had made the cubes out of gravel dug out in the mining process, as was the normal practice, and was on his way to Freetown when he was stopped by the security police.

"'Ullo, 'ullo, what have we here?"

"These are test cubes of concrete, our machine has broken down, and they are going to the university to be tested there."

"How do we know that they are not full of illicit diamonds?"

A bit of panic developing, because there might be a chance that a diamond had somehow got into the mix, and smuggling diamonds is a pretty serious offence.

Frantic telephone calls all round, until it was agreed that the cubes would be escorted to Freetown and all the broken pieces returned to site, also under escort – and sign here to confirm the weight before leaving.

A bit of a tricky one this, as the concrete could dry out during transit, and appear to lose some weight, but there was not much else to do but agree.

All was well in the end, however, and the broken bits did not disclose any diamonds.

Our dam was growing quickly, and it became necessary to cut a spillway channel to deal with the overflow likely to occur during the rainy season. We completed the survey and selected the site for the overflow channel, and started to clear away the large trees and bush in the area. An old man came along and asked us what we were doing, and when we told him that the water would come up to here, in a couple of months, and then start running down our channel he was convinced we were mad.

"The water never comes up here," he declared, "it all stays in the river down there. How can it get up here?"

We tried to explain to him that our dam would cause the water to pond up, but he could not imagine such a circumstance and went off muttering. Right enough, when the rains came the dam flooded the valley, and the water went coursing through our channel.

We sent out for the ancient inhabitant, and he came over and was suitably amazed.

Just before the onset of the rains our water supply to the camp began to dry up, and we had to undertake somewhat desperate measures to maintain health and safety. A decree was issued that all water for drinking and cooking should be boiled and filtered (this was normally done anyway).

Restrictions on the use of baths and showers were not formally declared, but everybody tied stockings around all the taps, to contain the larger leaves and more substantial forms of life, such as tadpoles and small lizards, from joining you at your ablutions.

The shortage persisted for a couple of weeks, with the colour of the water becoming darker and darker each day, although it seemed to clear quite well when filtered.

One of the ladies became very concerned about this state of affairs, and despite demonstrations of the purity of the boiled and filtered water by means of the Horrock's Box test results, she took a sample and despatched it to one of the official testing houses in Freeetown. (The Horrock's box consisted of a set of glass sample bottles, a small measuring spoon, and a set of coloured panels. By adding chlorine to the filtered sample it caused a change in colour to compare with the coloured panels. When the colours matched you could tell if sufficient chlorine had been added to kill all the bugs without killing the customers.)

Some time later, the results of this examination came back. The sample was formally acknowledged, an analyst's report set out the descriptions and percentages by weight of the various impurities, and the final opinion was:

"This horse should not be run."

What a marvellous sense of humour!!

By the time the report was received the rain had fallen again, and our water supply was back to normal, but there was still a bit of disaffection about in certain quarters, and we always kept a stocking over the taps in the bath.

CHAPTER 113
Marine Activities

Part of the contract was the construction of a small jetty and loading base on one of the rivers in the delta, the intention being that the rutile, which had been reduced to a fine powder, was to be stored in large cylindrical hoppers, which could then be loaded into a special barge that had a loading device fitted aboard to elevate the product, so that it could be discharged into ocean-going ships further down the estuary.

This special loading barge, and a number of other large barges, had been delivered to the site, and were moored in the river near our work in progress on the base.

One morning, while visiting the site, I noticed one of these large barges sailing past. The weather was pretty awful, with pouring rain and a bit of a wind, and as the barge went clear of our area I could see that it had a broken rope trailing from the mooring bitts. The only vessel in the area was one of our old lifeboats from Pepel, so one of the foreman and I got aboard and set off in pursuit of the barge. It had gone about 200 yards before we caught up with it, and was moving gently in the tide, down the creek, towards the open sea.

We managed to get a line onto it, and were able to keep it more or less in mid-channel, and generally under some small measure of control, so that it would not go aground, and we stuck with it, still in the pouring rain, throughout the tide. When the tide changed we were able to direct it back towards our mooring area, and, by a bit of luck, secured it to a spare mooring buoy and came ashore.

It was then about eight o'clock in the evening; we were both soaked to the skin, hungry and thirsty, but happy to have salvaged the barge, which we cheerfully estimated to be worth about £200,000, and set out back to camp in my small Volkswagen car. This vehicle had been running about all through the dry season, and the entire ventilation system had become choked with red laterite dust, but never used.

We got going with the heater and blower on at full blast to try to get warm again, but nothing much seemed to be happening. Suddenly we went over a big bump in the road, the interior of the car was filled with dust, and we could hardly see out. The bump had obviously shaken loose a blockage in the system, and all the dry season's red dust had blown in on us. At least it began to warm up.

When I arrived home there was a bit of a situation developing. We had been away since before 7:30 in the morning, and had been reported as having gone out in the lifeboat at about 10:00. Here it was, 9:00 in the evening, and no signs of us.

Search parties were being organised. I went down to the site manager's house to tell him that we had saved the barge, and were back, looking for the salvage, and to call off the search parties and all that.

His wife came out and fell about laughing. I had not realised that I was covered in red dust, where it had settled on my skin and stuck to my wet clothes, and must have looked a right sight. I walked across to our bungalow, and was greeted by a similar welcome.

Where had I been? Why had I not let someone know? How had I got into that state?

Followed by: "You can't come into the house like that, you will have to undress on the veranda and get showered by the garden hose before you can come in."

"All right, but run me a bath while I clean up, and get me a large strong dram, please."

Some time later, wallowing in hot, clean water, with another strong dram, I heard someone at the door, and when I stepped out, clean and refreshed, learned that the client's harbour master had been in and left a crate of beer, to share with my foreman. In the morning I wrote a report of our activities the previous day, mildly pointing out that we had saved the client a lot of money by keeping his barge off the rocks, apart from the delay it might have cost the operations, and mentioning Lloyd's Form for Salvage etc.

This caused a bit of a stir, but we had a polite letter of thanks back from the client, saying how smart we had been, but that was all part of our duties under the contract, and they had expected no less from us.

What a miserable lot...

We did notice that his men were to be seen doubling up on all the moorings for his barges the next time we visited that part of the river, however.

CHAPTER 114
The visiting Delegation

As a secondary contract, we had build a small airstrip not far from our loading base, and as it was not officially in use and properly manned we were frequently asked to have someone there to speak on the radio, report the weather and cloud conditions, and to keep the sheep and cattle off the runway. As we had the only radio rigged to speak to Freetown, we were often used as links for the government officials who wished to visit the area, or contact the paramount chief.

One day we were asked to man the airfield and transport a visiting delegation of parliamentarians from Ghana down to our loading base, where the Prime Minister's launch would pick them up and take them out to his holiday mansion on a small island in the estuary.

I went down to the dock to see if the launch was there, but there was no sign of her. I returned to the airstrip, where the visitors had landed, and met with the chief of police.

We tried to raise the launch on the radio, and eventually picked up a very faint signal to the effect that they were aground, had damaged their propellers, and needed assistance.

We obtained their estimated position, and I suggested to the chief that, as it looked as if it was going to be a long wait, he should fly his deputation back to Freetown and try again on the radio in the morning to see how things were. This he agreed to do, and they all piled into the aircraft and flew off.

I returned to the dock, saw that the tug was all fuelled up, gave him the position of the stranded launch, was certain he understood where that was, and sent him off to wait for the tide and then to pull the launch back to our dock.

"Do you understand?"

"Yes, pa."

"All right; off you go, but keep your wireless open, and let us know what is happening."

We heard that he had found the launch, but was unable to pull him off the reef until high water, about midnight, and he would do that and then anchor, and come back in the daylight. A very sensible thing to do. He arrived back in the early afternoon next day with the launch in tow, and we lifted her stern out to study the damage. Both propellers were buckled, and the shafts probably bent, so there was not a lot we could do to get her going again quickly.

The Prime Minister, meanwhile, was marooned in his mansion, out of touch with the government, and still waiting for his visitors from Ghana. A signal came through; could we please ferry the visitors out to the PM on his island on the client's fine new launch?

We replied that we could, but we would need to fit a few more seats aboard, and find some extra lifejackets, etc.

"Yes, please – go ahead; we will send the delegation to the airfield again tomorrow morning."

We worked on collecting spare seats from the mess, had them secured on the deck of the launch, had her refuelled, had the coxswain briefed on the position of the island and the house, checked that he had all the charts, tide tables, etc., and had him standing by at the dock ready to go the next day.

In the morning I went down to the airfield to conduct the visitors to the dock, and see them safely onto the launch, but when they arrived it seemed that one of the local ministers was missing as he was coming by car, and we were told to await his arrival before setting off.

As this was likely to be a long wait, one of the delegation suggested that it might be possible to arrange a spot of lunch, and I went off and found the village headman – nominally the airport manager, air traffic controller, and the man who had to chase the goats off the runway as the aircraft approached.

He gave me to understand that he was not the catering manager, but on a bit of persuasion from the chief of police he agreed to see what he could rustle up. Some short time later he appeared with a goat on the end of a piece of string, and suggested that it was available for lunch and that he could produce some rice and vegetables, but there might be a problem in butchering the goat.

This proposition did not go down very well with the policeman, who might have thought that he might be required to shoot the poor animal, and I suggested that we might be able to fix lunch in the club if they cared to travel round to the main site. The policeman accepted immediately, and I spoke on the wireless back to the camp, loaded the members of the delegation into various minibuses, and escorted them round to the main site.

After lunch, and a quick run around the works, they all drove down to the dock again, met the lost minister, and were duly loaded onto the launch and waved away.

Sometime in the early evening we had a wireless message from the launch to say that they had found the island, but there was no house there, no Prime Minister, and they had become trapped by the tide and would not be able to sail out until daylight. He further reported that, as he had been sailing about all day, he was running short of fuel, and might not make it back to base.

We had a quick conference and decided that we should send the tug down in the morning with fuel, and that the delegation did not really want to go swanning about all over the delta for another day, and we would bring them back in the morning. We ordered the launch to sail out to the place we normally unloaded the ocean-going ships, and said we would get the tug there as soon as possible. Then we had a swift move down to the dock to brief the tug crew, load up with some spare fuel, and be ready to set off at first light in the morning.

Next thing was to advise Freetown to organise the aircraft to come in and collect the visitors. We told them of our problems, our immediate intentions, and then suffered a breakdown in communications (caused by pulling the plug out of the radio) and went off to bed.

I suppose there was a good bit of fast talking going on in Freetown, it being then about 9:00 p.m., and all officials having knocked off for the day, at 3:30. Anyway, if there was any drastic change of plans we could always call the tug up on our radio, and a similar breakdown of communications could be arranged.

In the morning we were roused by irate signals from Freetown.

Where was the Prime Minister?

Where were the delegates?

Who had the authority to call up the aircraft?

Who had decided to abandon the visit?

We replied that we had no idea where the Prime Minister was. We thought the delegates would have spent an unpleasant night at anchor in the delta, no food, no blankets, not much water, and many insects.

As for the aircraft, we expected the visitors back alongside about noon, and would someone please arrange to get them out of our hair?

Unless otherwise instructed we would meet them at the dock, bring them back to the campsite, and feed and wash them. Please advise. The messenger looked as if he stood a good chance of getting shot, but we were reasonably fireproof, from our safe distance, and left them to sort it all out.

I returned to the dock and met the delegates, who were hungry, bitten, unwashed and pretty fed up, and quite happy to get back to Freetown as quickly as possible, and to hell with meeting the Prime Minister.

The mountains must have moved, as the aircraft was scheduled to arrive after lunch, and we gladly saw the last of them as they boarded.

There was a big inquiry into the affair in Freetown, and we were exonerated.

CHAPTER 115
The Acihbu

Prior to our arrival on the site the client had a geologist prospecting for minerals in the region for a couple of years and more. He had come out with his wife and some of their younger children, and had been living in the local villages, moving on as the surveys progressed. Although they had all mod cons, as far as they could reasonably manage, it must have been pretty rough at times. Where possible they had been lodged in the local district offices, set up for the convenience of district officers on their circuits, but sometimes they had rented a house in the village, and had to provide all the amenities, such as kerosene fridges, lights, water supply, sanitation, and food.

During this period Mrs Spenser had been in correspondence with her daughter, who was studying in America, and she had kept the letters and some of the replies, and had then compiled them into a book, called *African Creeks I Have Been Up*.

This book had just been published, and we had a great celebration to mark the occasion.

The client's launch was delivered just about the same time, and it was felt appropriate to name her *Acihbu* in recognition of Mrs Spenser's contributions to the success of the project.

It certainly made a good talking point when the boat's name was mentioned.

CHAPTER 116
Boat-building

The buildings for the permanent camp for the mine were all made up in aluminium panels – two thin sheets with a spacing of corrugated cardboard between them. The panels were framed in aluminium extrusions, with holes prepared for bolting together, and recesses for such items as switches, plugs, etc.

The original intention was to have them covered in a polythene film for shipment, but in a number of cases this film had been badly damaged, and water had penetrated into the cardboard spacing material. As a result, when we came to handle the panels the glue had lost its stick, and the aluminium sheets just peeled off.

This caused some consternation; experts were called out, blame apportioned, changes made to the glue, new panels ordered. The old panels were condemned, and our programme altered to suit the revised deliveries.

The old sheets of aluminium were dumped at the rear of the stores, where they constituted some hazard in high winds, but some were passed over to the local villagers, and some of the longer ones were selected to be used as potential canoes by members of the staff.

The first attempts were a signal failure. The theory was that you took a long sheet and, with a semicircular former in the middle, folded the ends together. This produced a long, narrow boat (?), which was impossible to keep upright, and folded along the edge as soon as the occupant had been tipped out.

Back to the drawing board.

Variations on this theme, based on the least possible fabrication and most efficient use of the long sheets, were tried, and some were almost a success. The main problems were that the sheets were so thin that any working of them caused cracks, which led to leaks and the appearance of the rising damp. In any case, we had no proper tools and very little time. Some of us persevered, and I eventually came up with a catamaran type of craft, which was very stable but leaked like the proverbial sieve, and, when sailing, was always collapsing due to the fastenings pulling out.

I had made a sail out of old cotton flour bags, which took up all the room on the veranda during manufacture, and at the first trial the boat went so fast that it was swamped by the spray coming over the bows. A deck was necessary, but then it was impossible to bail her out, and any crew movements while under way were very fraught.

The sailing idea was abandoned, and the boat used as a platform for fishing in the dam, operated under paddles only. Needless to say we never caught anything, but we did see a few snakes, and this rather put us off.

At the end of the contract all the houseboys fell heir to the various canoes, and I expect some of them are still about today.

CHAPTER 117
The Watchmen

We had been at a party at the bauxite mine one evening, and were asked to escort the harbour master and his wife and daughters back to their house, which was quite near the dock. We took them back and went in for a drink, and then decided that it would be a good idea to check on the watchmen on the site before returning to camp.

We drove into the site in the Land-Rover, and found the two watchmen fast asleep in their tent. Rather than disturb them we took their lantern and a spare blanket away with us, and in the morning they reported that there had been a serious theft.

The lamp and blanket were produced as evidence, and I told the men that they would be paid off.

They grumbled a bit, but admitted that it had been a fair cop and that they had been sleeping on duty – but would I please reconsider?

In the afternoon, as I was sitting in the site office, I heard a lot of shouting going on outside, and when I went out I found the trade union representative in a furious argument with one of the local foremen. He was all steamed up about the watchmen, and was threatening to have me arrested for the theft of the lamp and the blanket. I had them both called into my presence, and, from behind my desk, asked the union man what all the fuss was about.

He adopted what I can only call the confrontational stance, and declared, "These two watchmen allege that you stole a lamp and a blanket from them last night. They are going to take this matter to the local court. They also state that they have been dismissed without due cause, and this will be taken up by the Labour Officer in Freetown."

As an immediate counter to these charges I asked the foreman, "Who is this fellow, and what is he doing on the site?"

"This man, pa, is the local union man. He has come to make trouble over the watchmen."

Turning to the union man, I asked him for some proof of his identity, and his standing in the union.

When he was unable to produce any such documentation I suggested politely that he should go away, and, if he felt it necessary, make an appointment to come back and see me.

I also reminded him that the lamp and blanket were the property of the company, and that as the company representative I was well within my rights in taking these items into my care, as it was obvious that they were not being looked after at the time of the incident.

This roused him to some fury, and he began shouting that he would see about this. It was long past the time when white colonialist imperialistic capitalists were able to grind the faces of the workers into the dust. He became even more annoyed when I laughed at his tirade, and seemed to enjoy his implied description of me.

'White' I could partly agree to, but as I had served a long time in the tropics most of me was a delicate shade of brown.

'Colonialist'? – well, the territory was now an independent member of the Commonwealth.

'Imperialist'? – yes, I probably was, and still am.

'Capitalist'? – well, there's a joke for you. If I really was a capitalist, what the hell was I doing sweating it out in the middle of darkest Africa?

My local foreman obviously thought that I might respond to these charges, and quickly pulled the union man out of the office and escorted him off the site.

A few moments later the two ex-watchmen appeared at the door. They apologised for the presence of the union man, saying that they were not members of any union anyway. They were poor villagers who needed the work, and were very sorry to have been sleeping on the watch. Could I possibly see if they could be reinstated, as labourers? They would work willingly at any tasks allocated to them, and God would bless me.

Appreciating their concern for my soul, I decreed that they should suffer a week's suspension, and could rejoin the workforce in seven days' time.

In my weekly report I mentioned that we had had a visit from a union official, but I had failed to recognise him, and there might be some palaver from Freetown, and some days later the Labour Office asked if they could come and see how we were getting on.

This situation was a bit delicate, as we were only contractors on the site, and would not be concerned with the permanent employment policy of the client when the mine came into operation. I was satisfied that I had behaved correctly in this case.

We suggested that we would be happy to see the Labour Officer at any time, but that any policy decisions concerning the future operational manning levels and rates of pay and conditions would best be discussed with the client in Freetown at a much higher level than mine. I had little doubt that our affair with the union man had provoked this visit, and the client was well aware of the situation. The village headman, who was also well aware of the situation, confirmed that the union man was out of order, and that we might expect a visit from another union man, who had recently been appointed and just happened to come from his village.

We had a different encounter with another watchman in the camp. This man was employed to keep watch on the buildings in the camp overnight, and he had a small hut, where he prepared his food and sat out of the rain. He was a pleasant old man who came round each evening and announced his presence at each house, and exchanged small talk with the domestic staff.

One evening he failed to appear, but his son, who also worked on the site, came along to say that his father was sick and he would stand in overnight as the watchman.

I told him to have his father appear at the sick parade in the morning, and we would see what we could do for him. In the morning I learnt from the nurse that the old man had very bad toothache, and had been given some aspirins and some sort of mouthwash, but he really needed to see a dentist, which meant going to the government dentist in Freetown. I spoke to his son later in the day and arranged for the old man to go down to Freetown on our ration truck the following day, and gave him directions as to how to find the hospital. This was really quite difficult, as neither of them had ever been further away from their village than about ten miles, and the concept of streets was alien to them.

We packed the old man off, however, with a letter in his hand stating that he was one of our employees and could get a lift back on any of our transport which was available, and with instructions to the driver to see that he reached the hospital straightaway.

A few days later the old man returned to his night-time duties, and I asked him how he had managed. He had been thrilled at being a passenger in the truck – the first time he had ever been off his feet on a journey. Freetown was beautiful beyond all his wildest dreams, with all the people, and lights, and the harbour, with all the ships lit up in the bay.

The hospital had been very kind, and had dealt with him quickly, and he was now suffering no pain or discomfort. As for the journey home, he had not wanted to bother the office in town, and had decided to walk back through the bush.

This was a direct distance of about 100 miles, with three major river crossings, and he had made it in two and a half days, and was there for work the same day.

CHAPTER 118
The Menagerie

During most of the construction period our friends the Spensers lived in the old commissioner's boma. A boma was the residence of the district commissioner whilst on tour, and was a well-built set of buildings with accommodation for the DC, his senior staff, and other members of his escort, etc.

The remainder of the client's staff lived in the aluminium rondavels. The Spenser family had been in Sierra Leone for a few years and had acquired a number of pets and strays during their stay, and any visit to the boma was quite likely to be livened up by the antics of these creatures.

The naturalist Gerald Durrell was visiting the area, and we were all invited to a tea party to meet him and his party one Sunday afternoon. The animals had all been fed and watered and reduced to a relatively calm condition by the time we arrived, and the party got under way, first with tea and cakes, and later developing into some stronger brews.

The pièce de résistance of the Spenser collection was a young leopard. This creature was about the size of a greyhound, with thick, furry legs and an extremely long tail. It had been leaping about, displaying its agility, and suddenly retired into a small bush, where it thought it was concealed.

The bush was only about six feet high, and the leopard had chosen to lie along one of the branches with his legs and tail dangling down, although his head was partly hidden by the foliage. As the bush bent under his weight his tail made contact with the ground and he started up in some alarm, rather lost his balance, and came tumbling out of the bush. Everyone laughed, and he looked reproachfully at the audience, then turned his back on the party and sat staring out over the valley.

Attention was now centred on the civet cat kitten, which was dancing about, jumping up on people's laps and pouncing on unattended fingers, while the monkeys were keeping at a safe distance and chattering wildly amongst themselves.

The two local kittens joined in the fun, and the dog, obviously feeling neglected but still wary of all the cats, tried to sidle into the party, and succeeded in overturning a table full of small cakes and drink. This was too much for the monkeys, which swooped onto the floor and made off back to their perches with armfuls of goodies.

The party continued, with short pauses to patch up scratches and refill the drinks, and many tall tales were told of the wild beasts to be found in the nearby forests. For myself, the only creatures I had seen were the odd flying fox, a number of monkeys, and a very small deer. Apart from snakes and crocodiles, it was not the most active place for wildlife.

Mr Durrell and his friends departed for another part of the district, and I believe the leopard and the civet cat were eventually collected and taken back to the UK.

The only beast we kept in our bungalow was a large chameleon, which I had found crossing the road, and in some danger of being run over. It took to living on our lampshade, and was adept at catching flies and moths as they went past. It had a look of intense satisfaction as it crunched them up, while still looking for further prey.

CHAPTER 119
The Rains came

The rainy season was upon us again, and life became a bit more trying. The basic material of the surrounding country was laterite, and as soon as the rains came the surface became very porous, and slippery. As you walked about large sections of the ground became detached and stuck to your shoes, and soon the offices and public areas were covered in red mud. The thunderstorms were quite spectacular, with sheet lightning flashing away for hours before the storm passed over the area, and flashes and bangs were almost instantaneous. The noise of the rain on our tin roofs was often overpowering, and drowned out all other noises.

Our house had been built over a small depression in the ground, and we had laid a 12-inch pipe as drainage for the immediate area behind the building. When the rain stopped there was still a loud rushing noise from outside, and I went out to investigate. Our pipe was running full bore and had excavated a large hole in the garden. When I went round to the other side of the house it looked as if the tide had come in, with acres of water lying all around the camp.

In the morning I was called out by one of the foremen to see what was happening some distance down the road. We had built a causeway across a low-lying area, and had laid four large-diameter steel pipes under the road to accommodate the projected flow. These pipes were made up from short curved sections of plate bolted together.

We had allowed for the road to be widened later, so the pipes were projecting well out from the edge of our embankment into what had developed into a large pond, now about eight feet deep.

The water was still rising, and was almost up to the top of our causeway, and it was still raining. What had happened was that the water level had risen more quickly than the pipes could carry it away, and the pipes had floated up and were now running only about a foot deep.

This situation called for pretty drastic action, and we called down a large travelling crane and a small bulldozer.

We picked the bulldozer up and landed it on top of the first row of pipes, and by some clever balancing act managed to get the pipe partly submerged, and as the flow increased inside the pipe we were able to prevent the main pond from rising any further. Other heavy items of equipment were called down and the same trick performed, until all four pipes were running at full bore and the water level in the pond was visibly being reduced. It was just as well that we had discovered this early in the day, and still had daylight to complete our temporary repairs; otherwise we might well have lost the causeway, and probably all of our large-diameter pipes.

CHAPTER 120
Piling Problems

During our piledriving operations at the loading jetty we had been able to drive the steel piles rather further into the river bed than had been planned, and we were short of about 60 feet of pile. These piles were about 16 inches across a hexagonal section, and had been specially ordered up from the UK at the start of the work. When it became obvious that we would be short we scouted about for alternative materials, and I remembered that there were a number of cut-off lengths left at Pepel, surplus to our requirements there. We came to an arrangement with the Pepel owners that they would sell us these sections, and also make them up in their workshops into one long length, and we would arrange shipping as soon as possible.

On the next visit of one of the Lykes Lines ships from America with more of our materials we had him call at Freetown, and load our extra section of pile as deck cargo and ferry it round.

I went out in the tug with our large cargo barge to supervise the loading from the ship, and to clear the customs and all that jazz, and, still in the pouring rain, we moored our barge alongside the ship with a few old heavy tractor tyres as fenders.

They had slung the pile before I was up the boarding ladder, and it was passing overhead when one of the wire slings slipped. The end of the pile swung down, the other sling broke, and like a classical diver the whole thing fell in the space between the barge and the ship, leaving a ring of bubbles and hardly a splash.

At first I could not believe that no damage had been done, and when I considered the possibilities realised just how lucky we had been. The pile could have gone straight through the deck and the bottom of the barge, leaving a large hole, which might not have sunk her but would certainly have caused a severe list, and probably tipped all the crew into the water. It also might have fallen onto the ship, and again gone right through several decks, and possibly the bottom too.

It had certainly given us all a hell of a fright, and would also cause us some further problems in finding enough other parts to make up a complete pile from Pepel, and then risk shipping it again. The remainder of the cargo was very delicately handled, and frantic forms were being filled in as to who was at fault and where the responsibility lay. As the thing fell into the water, it was obviously clear of the ship. At least, that was their story.

As a result of some of the piles driving further than expected we were asked by the design office in London to try to drill through them, and then insert a thick bar and concrete it in. The piles had a cast-iron shoe at their pointed ends, and we had to call up a specialist contractor to carry out the drilling for us.

These people arrived and set to work on the first pile, which was about 50 feet down and lying at an angle of about 15 degrees off the vertical. They had managed to bore through the shoe, and were well into the rock below, when the drill stem broke, down inside the pile. The first problem was to empty the water out of the pile, and this was soon accomplished by using an air-lift pump – a sort of Hoover in reverse.

It was very dark inside the pile, and the only satisfactory means of illumination was by the use of a mirror, directing the sunlight down the hole. This was remarkably successful, and we could see the broken bit of drill rod sticking up about five feet from the pointed end of the shoe. The next task was to get a grip on it, and we offered large sums of money to any volunteers who would venture down.

No takers.

Next we obtained a mesh cable-pulling sock from our electricians. This was a wire-woven section of material that could be fitted over large-diameter cables, and it was fitted with a loop and an eye at the other end. A good idea – but how to attach it?

After some experimental work with welding rods and some sticky tape, we discovered that it was possible to hold the mesh open with the welding rods as pins, and then enter the mesh section over the stump. When the pins were pulled out, the mesh sleeve extended and gripped the stump. This worked reasonably well outside, but could we work it down the hole, bearing in mind that we would need to have the crane attached before we started? And we had still to make sure that the sleeve would be centred over the stump, and we had to be able to see what was going on.

Nowadays, this could all be done by remote control using a television camera, but at that time, in the middle of a river in darkest Africa, we just had to muddle through.

We made up a sort of carriage, which we hoped might keep the contraption more or less in the middle of the pile, and when the sun was at the right angle lowered away, and after some guddling about pulled out the pins, and the sock connected.

So far so good.

Next we connected the crane to a tackle, which would increase the pulling power by a factor of four, and started to heave up. The crane was on a pontoon, and was rated at seven tons, so we estimated that we had a load of nearly 30 tons on the rod, and as this was getting near the safe working load of the hook at the end of our tackle we thought we had better stop trying. Just as well, too, for when we looked round at the crane half the pontoons were under the water.

We went back and sent a signal off to London saying that we had tested the pull-out capacity to 30 tons, and that would have to do. There was a bit of huffing and puffing from them, but they got the message, and eventually agreed that this was satisfactory.

Another of the piles had struck a rock during driving, and had been deflected well out of alignment, and we decided that we would try to pull it out and reposition it.

We tried pulling with the crane, and working it from side to side from the shore, but it declined to move.

Things were getting pretty desperate. We went down in the early evening and moved our barge over the pile at low tide, then attached the pile to one of the mooring bitts on the barge, and went home for tea.

When we returned later we were a bit alarmed to see the barge rearing up out of the water at the far end, with the place where the ropes were attached underwater, and still about another three feet of tide to make.

As there was not much we could do about this situation, we sat in the Land-Rover and watched. If the ropes broke, the end of the barge would come down with a bit of a whack, but there was nothing in the way, so that was not a real worry. If the bitts pulled out we could probably expect the barge to sink where it was, and, apart from having to talk a bit fast, it was no great calamity.

As we watched the tide rose, and the end of the barge tipped higher and higher, until suddenly there was a groaning noise and it began to level out again, and soon it was floating happily, with the recovered pile hanging in its ropes, all as if nothing had happened.

We drove round to the camp and had a few drams of celebration in the club before going back to the bungalow for a well-earned rest.

CHAPTER 121
Fire in the Village

Our bungalow stood on the forward slope of a small escarpment, and at the end of the garden the ground fell away steeply. At the foot of this slope, about half a mile away, there was a small village of about 30 houses, and most of the men from the village were now working on the site. The houses were built of mud bricks and had thatched roofs, and were all very close to one another. Each house had a small garden, and the bush had been cleared for a distance of about 100 yards all round, and this area was under cultivation.

This may have been accidental, but it made a good firebreak.

I had been visiting another part of the site one day, and when I returned to the office I could see a cloud of smoke rising up from the bush in front of the camp.

I went over at once, and found that the entire village had been burnt to the ground, and all the inhabitants were standing about in stunned silence in a crowd.

By some minor miracle only one person had been injured during the fire – an old man who had run back into his house while the roof was ablaze, and it had fallen in on him. The children had all been at school, and the women were working in the fields nearby. We took the injured man up to the first aid post, but he was very badly burned, and apart from plastering him with cream and covering up the exposed flesh with some sheets there was not much we could do for him on site.

The nearest hospital was about 30 miles away, in a large mission station, but it was on the opposite side of a large river, and about five miles away from the ferry crossing point. We put a wireless message in to Freetown, asking them to alert the hospital if possible, and having made up a bed in the back of the Land-Rover, and made the old man as comfortable as we could, set off fairly gently towards the ferry.

The road was in a poor condition, with deep ruts on either side, and it had been damaged recently by flood water, so we made very slow progress. Our patient seemed to be no worse, but was obviously in a lot of pain, and each bump in the road was rocking him about on the bed and rubbing the charred skin off his injuries.

When we arrived at the ferry we were met by two of the sisters from the hospital, who had walked down and come across to assist, and we took them into the back of the Land-Rover, and while they offered comfort and additional dressings to our patient we managed to load the vehicle onto the ferry, and were pulled across to the other side. News of our coming must have been broadcast, because there was a large crowd waiting at the landing stage, and they all walked up alongside us as we made our way slowly to the hospital.

Once inside the compound the old man was lifted out, and the doctors took him into the main part of the building, where his needs were attended to. The sisters brought us some tea, and when they heard of the extent of the fire they asked if we would take them back with us, so that they could do what they could to help the remaining villagers.

When we arrived back in camp we found that the club had been made available to house the people overnight, and a whip-round had been organised to provide them with clothes and cooking utensils, food and drink; arrangements were in hand to billet them out in nearby villages, and to assist with the reconstruction of their village. The damaged aluminium sheets came in very handy here. The two sisters stayed with us for a couple of nights, and then we took them back to the mission, where the old man was still very seriously ill and not making much progress towards recovery, and some few days later we heard that he had died of his burns.

For some time after the fire it was quite common to recognise various items of clothing being worn about the site that had been the proud possession of members of the staff, and these people became known as 'the man with Alec's blue shirt', etc.

In a couple of months the village had been reoccupied, and a few weeks later all signs of the fire had gone.

I had expected that there might be some sort of formal inquiry, both into the death of the old man and also into the cause of the fire, but as far as I knew no such proceedings were initiated. As far as our records were concerned, we had to write off a lot of ointment and aspirins from the first aid box, and a few bed sheets from the commissary, plus a donation from company funds for general relief and rehabilitation.

CHAPTER 122
The Otter

The client's harbour master lived with his wife and daughters in a little house not far from the dock, and it was our practice to call in as we went past from time to time. The girls had a pet monkey, a little owl chick and a young otter, all with the run of the house.

The owl sat on its perch most of the time, taking careful note of all that was going on, and ready to pounce on any tasty morsels of food that it saw lying about. It had the remarkable capacity to turn its head right round to look directly behind it without moving the rest of its body, and in this posture it looked positively deformed.

The monkey and the otter got on well together, and were often to be seen playing in the salt pans along the roadside. These pans were a series of shallow ponds, about the size of a bath, cut into the ground, and water was allowed in and then dried off in the sun, leaving a crust of salt all round the surface of the pond. When full of water, these formed a splendid playground for the two animals, the monkey jumping about from side to side and keeping its feet dry, and the otter diving in at one end and surging out at the other.

As it grew up the otter became very smart, and found it could open the fridge – or any other door, for that matter – and nothing was safe in the house thereafter. Its favourite trick was to get inside the fridge, eat what was within reach, and then curl up in the cool interior until it was discovered.

We called in one day in passing, and learned that the otter had somehow eaten some rat poison, and was very sick. We had a look at it, and could see that there was nothing we could do but put it out of its misery, so we arranged to come back immediately with a gun and shoot it.

When we returned it was lying by the side of the house, very sick indeed, and beyond all hope of recovery, so we took it across the road, and put a bullet into its head. Altogether a very distressing afternoon, but we took care to depart before the girls returned and left the explanations to their mother.

We were not flavour of the month with the girls for a couple of weeks, but they came round to accept that we had only done what was best.

CHAPTER 123
Other Happenings

Towards the end of the construction Jean became rather unwell and was advised to go home, so she packed up and went off with Susan, and I took them down to the airfield in Freetown and saw them onto the plane – this time a direct flight to London.

Some alterations to the domestic arrangements on site were required, as the construction work developed into the more intensive electrical and instrumentation phases of the installations, and another engineer joined me in my house for the remaining four months.

We had a very simple routine: up for breakfast at 6:30, start work at 7:30, back for a light lunch and 20 minutes' snooze at lunchtime, off again until about 6:00, a quick shower, dinner before 8:00, and then perhaps a beer in the club, or early to bed with a book.

On Sundays we could relax a bit, and often went up to the bauxite mine for a swim in their pool, sometimes had lunch in the club, or sometimes made a special effort for Sunday dinner.

Jean had left a number of her cookery books behind, and we developed a random selection method for suitable menus. One particular book was entitled *European Cooking*, and had sections devoted to each country, so it was a safe bet that, no matter where the book opened, some exotic dish would be displayed. We then had to find the ingredients, or suitable substitutes, and undertake the preparation and cooking, and eventual eating.

Frequently we had made some error in the measuring of the quantities, and had heaps too much, and on such occasions we invited our neighbours round to assist in the demolition.

At other times something had gone wrong, and we had to revert to dinner at the club, or go on the scrounge to other establishments.

The main intellectual passion in the camp was Scrabble, and games would develop into all-night sessions, with long, acrimonious discussions about permitted words.

We had built a tennis court, and this was well used to work up a thirst, and conveniently located next to the bar. Our golf course never really took off; it was, perhaps, too much trouble to go ferreting about in the scrub bush looking for wayward balls, with a good chance of disturbing an anthill, a swarm of bees, or even the odd, small snake

Part of the equipment in the client's new camp was a bowling alley, and this caused us some problems during the construction, as the planks for the floor had been prepared in America, under different conditions of temperature and humidity, and would not lie flat on our new concrete floor. After some treatment with a blowlamp and a plane it was put into reasonable shape, and this formed another source of entertainment on the site.

We also had a film show once a month, courtesy of the High Commission, and these were also good for a laugh – all very moral and upright.

CHAPTER 124
Home again

As the construction work came to an end, and all the paperwork was being completed, it came time to prepare to go home again. I made some enquiries about further work in Africa, and was not much impressed by any of the locations that were likely to come up, and as Jean's health was in question decided that it might be better to try to remain in the UK for a spell. Also, there was the question of schooling for Susan coming into the picture. I left the site with a bit of regret, as it had been a most interesting contract and I had made some new friends, but that is the way of contracting. You will never see the world if you stay in one place.

I had a couple of days in Freetown and stayed in the new hotel, but it was all changed from our previous times: the furniture had gone a bit shabby, and the ornamental palm trees had got out of hand. The food was not too bad, but I still hankered after the old City Hotel days, when you could actually see into the kitchens down below (even if it did look like Dante's *Inferno*).

Out to the airfield and off on a non-stop flight to London, and then home to Maldon, where Jean and Susan were waiting for me.

We had ordered a new car from Freetown, thus avoiding some tax, and next day we collected it and set off for London, to deliver some papers and see about my next posting.

**PART FIVE –
ANGLESEY, INVERGORDON, INDONESIA AND THE OIL
BUSINESS
1970-1989**

SECTION 1 – ANGLESEY

CHAPTER 125
The new Car

We collected our new car the day after I returned home to Maldon and duly completed all the documentation, then set out for London, where I was to deliver some papers and try to find out about my next posting. We drove very carefully through the East End of London, at one time managing to join a taxi rank in the middle of the street, but eventually arrived unscathed at the office in Ealing.

I had a chat with some of my colleagues, sounding out possible future locations, and was told to go off and enjoy my leave, and that they would get in touch in about a month.

I did suggest that I was not very keen to return to west Africa, but was always open to suggestions.

We started off back to Maldon in the early afternoon, and as we were passing along the Bayswater Road Jean asked what I would do if someone ran into our shiny new car.

"Get out and punch his nose," was the immediate reply.

After the comparative quiet of Freetown traffic this was horrendous, with large cars and vans cutting us up, and barging at the traffic lights.

We stopped behind a large American car at some traffic lights, and he rolled back onto us. Before I could get out the lights had changed, and we were off in hot pursuit to the next set of lights. Once more he moved off before I could get round to him, and I was becoming a bit paranoid at the thought that he might escape.

At the next set of lights I was right behind him, leapt out, and dashed up to the driver's door. Great surprise. The man sitting there at ease, and smoking a large cigar, had no wheel in front of him. It was a left-hand drive machine, and a lady was driving from the other side. This rather took the steam out of my approach, but he agreed that he had felt a slight bump, but did not think any damage had been done.

The lights changed again, and off we went to the next stop. This time he climbed out, and together we examined the cars for damage. Not a mark to be seen. He obviously thought that I was making a lot of fuss about nothing, and perhaps I was, but this was the first new car we had ever owned, and it was only its second day out of the showroom. For the remainder of the journey I kept a long way back from the cars in front, and signalled, both with flashing lights and waving arms, my every intention about turning and slowing down, and arrived home undamaged but mentally exhausted.

For the next few days the car remained in the garage, and work continued about the house and garden before we went on another visitation of relatives safari, and a short holiday in Cornwall.

CHAPTER 126
Anglesey

I was called into the office towards the end of my leave and was asked if I would like to go and work on the nuclear power station in Anglesey, north Wales. It was a bit of Hobson's choice – either there or back to west Africa – so I agreed, and arranged to travel through the next weekend and start work on the Monday.

The distance from Maldon in Essex to the west of Anglesey is about 260 miles, and crosses the country diagonally. I had booked into a hotel in Holyhead for the night, and set off early in the morning one Sunday, estimating that it would take me about six hours' steady travelling. This was based on previous journeys, where I had averaged about 50 mph, non-stop.

All my previous trips had been on main north-south roads, and this cross-country business was a much slower operation. It was a hot day, and by the time I reached Holyhead it was about seven o'clock in the evening and I felt quite weary.

I certainly did not fancy making that journey every weekend.

Next day I turned up at the site and was introduced around the staff, and later allocated a place in the staff hostel, some five miles away. The power station was an enormous undertaking, with about 1,000 workmen on site. I was asked to take on some supervisory work in the turbine hall until I found my way around the site, and caught up with the local methods of working, procedures, etc.

The shells of the buildings had been erected by the time I arrived, and work was now proceeding on the foundations for the machinery and all the other services that passed through and under the building. Across the road was the reactor building, and, again, the shell of the building was almost completed, but the steel pressure vessels that fitted inside were hardly started. At the rear of the site was the electrical switchyard, housed in another giant building, the roof area of which was claimed to be over seven acres.

The staff hostel was in the small village of Amlwch, about five miles from the site, and provided single rooms with beds, a recreation area, skittle alley, dining and television room. Most new arrivals spent some time in the hostel, which was very convenient, as there was a bus service to the site, but the general impression was that it was a good place to keep away from, and it was much better to stay in lodgings or find a flat or house nearby.

I made enquiries of the local estate agents in Bangor and Holyhead and was deluged with potential houses to buy or rent throughout the island, and down the mainland too.

Each weekend I returned to Maldon on the Friday evening, with bundles of house particulars, and we decided that (a) this living apart except for weekends was no good and (b) the driving back and forth was even worse, and we should try to buy a place within a reasonable distance of the site and sell the house in Maldon.

Each evening during the next few weeks I spent going round houses, and had whittled the selection down to three, so we arranged that Jean would come over for a week and we would live with one of my colleagues while we settled on a suitable place.

This arrangement worked well, and we agreed that we would take a house in the village of Llanfechell, which was handy for the site and had an infant school and a couple of shops. We completed all the formalities for this house, and had no trouble selling the Maldon house, and prepared to move in.

The weekend before our move I collected the boat from Maldon and towed it across the country to Holyhead; or, at least, that was the intention, but we had a small mishap with one of the wheels, and were forced to leave the boat and trailer in a garage forecourt overnight. After work on the Monday I scoured Anglesey for a Ford Seven wheel, and had some good luck at about the fourth place I tried, so the following evening we set out again to replace the wheel and recover the boat back to Holyhead.

This time all went well, the boat was parked up in the Holyhead Yacht Club yard, and we returned very late at night to the hostel.

There were a number of foreign gentlemen residing in the hostel, one of whom was an Indian surveyor. His task in life was to go around taking measurements of the reactor building, which meant that he spent most of the day in a dark basement, taking sights with a theodolite through four-inch holes in the floors above to special targets that were attached to the structure.

In the normal course of events, anything that was spilt on any of the floors above always came down through the hole he was working at. He was a bit of a philosopher, and reckoned that this job was a penance, and he must have done something pretty terrible in an earlier existence to justify his present status. In addition to his surveying duties, he cast horoscopes, and was by way of being a miracle curer for all manner of minor ailments, predictor of the sex of unborn babies, etc. – but would not give advice on the winner of the Grand National.

Another foreign member of the staff was an Iranian engineer, who happened to own a large red car, and who always drove at great speed and without due care and attention. I think he adopted the Middle Eastern practice of saying a few words of prayer before taking off, so that he was covered.

One morning I overheard these two proposing to travel to London for the weekend, and saw them set off from the hostel to the site in the large red car. Some time later in the day I spoke to our Indian colleague, saying that I thought he was going to be travelling with Ali, and would have been gone by now.

"Oh no," he replied. "I had another look at my destiny for today, and discovered that it was not a good day for me to be travelling in a red car."

I was considerate enough not to remind him that the trains from Holyhead to London all had red cars. I was sure that he had been frightened to death on the short ride from the hostel to the site, and had chickened out, but found a good excuse that was unlikely to offend his friend Ali.

Living in the hostel was an eye-opener in many respects. I overheard at breakfast one morning a conversation behind me, where three men were sitting round staring at a small tape recorder on the table. It seemed that they were playing back a recording of the old night express from London to Holyhead climbing over some summit or other on the line, and were all making knowledgeable comments on the number of puffs per minute, and the comparative rate of clickety-clicks against clickety-clacks.

Just think; these guys were building a nuclear power station.

There was another chap there who was building a concrete yacht, and the steel reinforcing was all set up in the car park outside. He was waiting for the better weather before he did the plastering, and his room was full of bits of furniture for the boat, which he had been making all winter.

He told me he had a small sailing dinghy in the bay nearby, and invited me for a sail early one evening. I came straight from the site and met him on the beach. I must confess that he had said 'a small dinghy', but this thing was minute.

By the time I arrived he had her all rigged and ready to go, and I was pushed out off the beach into the sea. Unfortunately my boots were just too long for the space between the side of the boat and the centreboard casing, and had become stuck, so that when I came to tack the boat, and had to sit out on the other side, I had to take my boot off and sail back with only one foot covered. This was not too bad until I came in to the beach, which was composed of large round cobblestones, and I had to hop out smartly into about a foot of water and stumble about for a bit.

CHAPTER 127
Our new Home

The day came when we were ready to move, and I returned to Maldon to assist in the packing up and loading out of all our belongings. By this time we had taken in a small grey kitten, and it was very active, rushing about after anything that moved, and following people about. When the men were loading our furniture one of them stepped back, and his foot came down on the kitten, breaking its back and killing it instantly.

Everyone was very upset, and while this was going on the wind got up, the garage door slammed shut, and all the glass in the window went flying into the kitchen.

These incidents rather clouded our departure from Maldon, and the replacement window had to come off the agreed price for the house.

We saw the pantechnicon off the premises and set out independently for Anglesey some time later, I having arranged to stop at my sister in Droitwich overnight, then go on to Llanfechell in the early afternoon.

We had the house opened and took formal possession, unloaded the car, and sat down to wait on the removal van. Of course, it failed to arrive, and we had to book into a local hotel for the night, leaving notices pinned to the door as to where we might be found if the van came along later.

We were back at the house again in the morning, and about midday the van arrived, and we started the unloading. We put the main items of furniture into their final location, more or less, hung up all the curtains that fitted, and dumped all the other boxes in the lounge. Quite quickly we had settled in, and the place looked reasonably civilised.

The main plan of operations was to fit central heating radiators at some time in the near future, and to replace a dreadful brick fireplace in the lounge.

The parts were ordered up for the radiators and pipes, I set about making holes in the walls and a trench in the concrete floor to carry the pipes into the lounge, and soon had the system operational, running off the coal-fired hot water boiler in the corner of the kitchen.

A design was prepared for the new fireplace, and suitable slate material taken from the old quarries at some place with an unpronounceable name on the mainland, and the demolition commenced.

The hole in the wall got larger and larger, and I even had to put a few props in, just in case it all came down. The soot from the chimney was everywhere, a new carpet was about to be delivered, and visitors were due any day now. By almost superhuman efforts the construction was completed, the floor scrubbed and dried out, the new carpet laid, and the fire lit the morning our first visitors arrived.

Great relief; the smoke went up the chimney as required, and we had a splendid house-warming party.

CHAPTER 128
Union Palaver

Very shortly after I started work at the power station I caused what could have been an expensive stoppage of work. I had been conducting some surveying on one of the outbuildings in the turbine hall, trying to fix the location of a piece of pipework where it came through an internal wall, and had set up my theodolite on the other side of the road.

Having surveyed in the actual location of my instrument, I had to turn off a specific angle and mark the wall, but when I turned the angle I found that my view was obscured by a piece of scaffolding pipe, forming a bit of a handrail across the face of the building. Rather than set up the instrument again in another position, with a sporting chance of the scaffold pipe still being in the way, I took a spanner from the car toolbox and moved the offending pipe to a position about two inches lower down, returned to the instrument, and made the necessary mark on the wall.

As I was packing the instrument back in its box, I was accosted by a very uncouth and rude individual, who first asked to see my union card, and when I suggested that he should go back to work became quite abusive.

When I arrived back in the office I was called in to see the agent, and was confronted by the senior shop steward. He alleged that I, not being a member of the union, had taken the unauthorised and outrageous action of moving a scaffold pipe, and that this action had compromised the safety of the works.

In my innocence, I said yes, of course I had moved the pipe, and later secured it so that it did not interfere with my sighting line.

The steward was then asked to leave the meeting.

The agent then asked me if I realised the effects of this action, and again, in all innocence, I replied that I was only trying to get on with my work, and had a squad of men waiting to start breaking out the wall, and I had taken particular care to make sure that the pipe was secure before I left.

"My God!" he exclaimed, "don't you know anything about unions and shop stewards? The whole place is about to come out on strike because of this." (It did occur to me that things were better organised in Africa, where engineers were expected to get things done as quickly and safely as possible, but I thought better of mentioning these revolutionary ideas.)

I replied that I was not aware of having done anything out of order, and had been quite polite to the man who had first spoken to me. (If I had not been, I would probably have told him to bugger off.)

The agent probably realised that this was a case of ignorance being bliss, and having read me a long session of the riot act told me to seek out the general foreman, who would put me wise to some of these restrictive practices. A meeting had been called by the site steward to discuss this action, but somehow it was circumvented, and peace was restored.

CHAPTER 129
Entertainment at Holyhead

We had joined the yacht club in Holyhead, which was one of the few places drink was legally available on a Sunday (although, of course, that was not why we joined), and tried to get out on the boat at weekends. The sailing was very keen, we were quite competitive, and we had a lot of fun, with the average run of minor accidents.

One day when we were leading the fleet, and had right of way over everything on the water, two catamarans collided a few yards away from us, in another race, and going in another direction. One of them had broken the connecting piece between the rudders, and he was completely out of control. He came very quickly down onto us, and bashed a triangle-shaped hole in the side of our boat. We fended him off, and were hanging over the side to inspect the damage when he came at us again, and put another hole about a foot further along the side. By good luck, we were on the right side of the wind to make it back to the club, and got back without shipping too much water, but it kept us out of the racing until the next weekend.

We had bought a new car in Anglesey during the summer, and Jean had to go out visiting in Holyhead one evening while I was left to babysit at home. Shortly after dark there was a telephone call from Jean, in some alarm, saying that a green light had lit up on the dashboard, and asking what she should do next.

"Don't worry, it is only to tell you that the lights are shining. If in doubt, read the handbook."

At the end of the summer there was a great wine-making bonanza, and most of the staff had a number of brews going. Our best effort was with raisins and oranges, and it turned out not unlike sherry (well, it fooled a few people). During the manufacture, the raisins were allowed to settle in the bottom of a bucket while the fermentation was going on, and when this process was complete I tipped the residue out on the lawn. Within a few minutes the birds had discovered this material, and a short time later they were all staggering about in the garden, unable to take off. The demon drink strikes again.

We had bought an Afghan hound puppy in Holyhead, called Flinders, and she was a bit of a handful and of a nervous disposition. Her favourite trick was to lie down in a field with horses and roll on her back, waving her feet in the air. When the horses came up to see what all this was about, she would jump past them, going in the opposite direction, catch hold of their tail and swing round, finishing backwards and then rushing off in delight.

One day I was working in the garden, which had an eight-foot high stone wall all round, and Flinders came over in a great leap. When I went round to see what had frightened her I found the farmer, greatly amused, and astonished at her agility.

On another day Jean had walked up the road to visit a friend, and had left Flinders tied up to a dustbin in her garden. Next thing she knew there was a great barking and howling. The coalman had dumped a bag of coal into the bunker, and Flinders had fled, taking the dustbin with her, halfway through a hedge.

She was a great worry, this dog, as when she saw something her attention seemed to concentrate only on that, and she would pay no heed to any shouts, whistles, or other remonstrations, and we were always very conscious that she might start chasing the sheep.

In the end, it was chickens she caught up with. A farmer from some distance away came in with a very dead chicken, saying that Flinders had somehow got into his hen house, and this mangled carcass was the result. He said he knew it was not a fox, because only one bird had been killed. Poor Flinders must have been panic-stricken when she found herself in amongst all those squawking birds, and probably trod on the casualty by accident.

Why is it that all dead chickens are prize Leghorns?

She was a most affectionate animal, and an excellent watchdog, barking furiously at anyone coming into the garden and always keeping a safe distance away, ready for instant flight but sounding very fierce.

One afternoon Jean noticed a small pool of water on the floor of the dining room, and blamed Flinders, of course. The dog was banished to the garden, the puddle was wiped up, and normal service resumed. The next time Jean was passing the dining room she observed another pool in the same place. Not the dog after all.

What she next discovered was a fine jet of water coming from one of our new radiators, passing over the table, and landing on the floor.

She stuck some plasticine over the hole, called the office, and I came home at once. When I arrived, all that was to be seen was this small piece of plasticine stuck on the radiator, but when I pulled it away it brought off a section of paint about the size of a penny, and I was deluged with red-hot water.

In dealing with plumbing emergencies it is always wise to turn the water off before conducting investigations, but if you have a fire in the boiler the first task is to draw the fire. I had to replace the plasticine, hold it in position (wedged in by a long stick against a chair) before I could isolate the radiator, draw the fire, and then examine the damage.

Most wives take exception to persons carrying red-hot coals over their kitchen floor, with the attendant dust and fumes and lack of warmth in the house, and Jean was no exception. When the dust had settled, so to speak, the radiator was inspected and found to be seriously thin in places, however; in fact, it was only the paint that was keeping the thing together.

The supplier was called, and appeared the next day, full of apologies and with a replacement radiator. "Just as well," I told him; "we had no carpet down, or you might have had to pay for that too."

CHAPTER 130
The Power Station

There was a large workman's camp on site, and each day one of the staff had to go round the canteen at mealtimes to see that the food was satisfactory, and to take up any complaints. One evening a week there was a camp committee meeting, where such complaints were aired, and at each meeting a young engineer was present, to broaden his experience. The first such meeting I attended there had been a complaint about the cost and quantity of the helpings of potatoes during the week, and I was dispatched to the village with a few shillings and a pair of scales to find the comparable cost of potatoes outside.

I obtained a sample from one of the hotels of a scoopful of mash, from the local chip shop some chips, and also from the pub a bag of crisps, and before going back into the meeting had prepared a little table of weight versus cost for the local supplies. This I handed to the chairman, and quick as a flash he announced that there would be an immediate rise in the cost of potatoes served in the canteen, justifying this increase by reference to the prices for similar goods available in the village. After a lot of argument it was agreed that, to keep the peace, there would be no immediate increase; but, really, the meals were pretty good value, and that complaint was dismissed.

As the work progressed so different trades were engaged and others were paid off, and at such times there were some disaffected persons who tried to make trouble by telephoning bomb threats. These were always taken very seriously, and notices were displayed at the gate, advising persons entering that there was a bomb warning in effect and that access to certain areas was restricted while a search was being made.

The senior staff were mustered to make the search, they being most likely to notice anything unusual, and also because of their knowledge of the routes into and out of the various buildings. We were all given whistles, and some of us hand-held radio sets, and we were allocated discrete sections of the works to search.

My first inspection was in the basement of the reactor halls – a dark area full of pipes and electric cables, with so many corners and places of concealment for small devices that it was almost impossible to be sure that nothing had been missed. What does a bomb look like, anyway? Would we find a ball-shaped device with a smoking wick, or would it be in a plastic washing-up container?

On this first search a number of suspicious articles were located, but all were found to be discarded pieces of equipment, or the normal residue left by untidy workmen.

Finding any suspicious article posed a bit of a dilemma.

How carefully should you examine it?

Should you listen for ticking noises?

Should you call for assistance, or perhaps a second opinion, without rousing the whole place?

Should you call in the so-called experts?

Should you blow the whistle and get the hell out immediately?

Obviously different people had different ideas as to what 'suspicious' meant, and no one wanted to get blown up, but neither did they want to be seen as panicmongers, and most people called quietly for a second opinion. Although a number of items were found, none of them proved to be a bomb, and most were removed and disposed of before the work was allowed to continue. If nothing else, this provoked a much tidier attitude throughout the site, and also heightened the attention paid to the removal of rubbish.

When potential bombs were discovered, whistles blew, and all searchers were pulled out of their areas while a more careful inspection of the suspicious object was undertaken.

Because of the nature of the work, a large area of the site was what was declared to be 'clean conditions'.

These areas were screened off by hardboard partitions, and all the seams taped over to prevent the contamination by dust from the outside. At the entrances to these areas were changing rooms, and everyone had to strip down completely and don overalls, caps, special shoes, and sometimes masks before entering the restricted zones, and on leaving the process had to be reversed, with the overalls being consigned to the laundry. The thought of having to go through the change rooms was a consideration in the decision to summon help, but there were many places outside the clean conditions where suspicious objects were found, and the process of evacuating different sections became somewhat selective.

At times of public holidays various members of the staff who were not travelling away were detailed to act as special inspectors for any incidents, and I happened to be on this list, as our house was not far from the site.

Over Christmas one year I had a telephone call to come into the site on an urgent alert. When I arrived, full of the Christmas spirit and a bit choked off at being turned out, I found that the crisis was that the clean conditions canteen cat, very pregnant, had vanished, and was probably having her kittens in some obscure corner of the reactor, and the damage this might do to the future performance of the reactor was incalculable.

The next four hours were spent crawling about inside the works calling: "Pussy, pussy; come out, you little black bugger" (or words to that effect), until we were recalled by someone announcing that she had been found on top of one of the lockers in the changing room, still very pregnant. Such was the relief that she had not managed to get inside, however, and no harm had been done to the reactor, that we were all invited to a celebration and thanksgiving in the local hotel.

One Saturday we had been out sailing, and on our way home we noticed a lot of smoke coming out of the top of the reactor building. We went straight to the site and into the building, and found a good fire going in one of the external rooms of the structure, where some gas bottles were stored. It seemed that the fire had started in some rubbish, which had been swept out onto a terrace but not yet collected, and the fire had somehow burnt through a gas hose and ignited the contents of the bottle.

When we arrived on the scene the fire brigade were just arriving, and we were ordered to act as guides to lead the firemen into the building, by then full of black smoke, and to point out the location of our temporary fire water connections.

The fire was burning some distance away from the reactor, and we were instructed to keep the firemen away from the internals, which were all in clean conditions, and on no account to let them put any water inside without specific authority. The construction had reached the stage of fitting specially shaped graphite bricks into the floor, and it was imperative that they were kept dry.

Meanwhile the fire had increased in intensity, and the gas bottles were exploding in the heat, adding more fuel and neat oxygen to the blaze. The main intention was to localise the fire within the gas bottle store, and as there was not much that could burn there, except the gas, this was accomplished after two or three hours. We were all stood down at about midnight, and went down to the clean conditions for a shower, as we were all black from the smoke, and returned home in our overalls very much later.

In the morning, when the place had cooled down, we could see that it must have been very hot in the room. There were grooves in the floor where the concrete had been burnt away by the jets of gas coming out of the bottles, where all the brass valves and fittings had melted, and there were pockmarks all round the walls where bits of the casings had dug in when the bottles had exploded.

No one had been hurt, however, and the damage had been limited to the one small room and its terrace; no water had entered the reactor, and work was going ahead as usual. After that, there was another blitz on general tidiness, and all rubbish had to be cleared before work closed for the day.

We had a visit from one of the television programmes that were showing us to be at the sharp edge of technology, and they were filming in the basement of the turbine hall. Some painters were working inside the roof overhead, and we had suggested that the film crew kept under the side galleries as a sensible measure of protection, both for them and their equipment.

Just as I was crossing the main bridge across the hall, a can of paint went past, followed by a shout from above. I looked over the bridge, and could see a large white area on the floor, and a number of television operatives hopping about.

None of them had been damaged, but a couple of them had been spattered by paint, and their boxes and recording equipment were all covered in white.

The paint can itself had collapsed in a concertina pattern, and one of the men said that he saw it land and, for an instant, the can collapsed and the paint stood there in a column, before splashing out all round.

Once more there was a safety inquiry, and recriminations made all round, the main conclusion being that it is not a good idea to walk about underneath where men are working on the roof.

The roof of the electrical substation was about seven acres in extent, and the architect, in his wisdom, had arranged all the drainpipes from the roof to be brought down to ground level inside the building.

One day it was observed that a lot of water was ponding on the roof, and I was asked to go up and have a look. I climbed up a long ladder and opened the hatch onto the roof. What I saw had me very worried. One whole valley, right across the building, was about seven feet deep in water, and I could see that a large polythene sheet had been pulled onto the grating around the downpipe and very little water was getting away.

I climbed down and reported to the office, and was instructed to take a few men and go and clear the obstruction, and some time later I was back on the roof with my four men and a selection of assorted garden implements and a very long boat-hook from the offshore works department.

After a bit of fishing about we managed to tear pieces of the offending sheet clear, and there was a most satisfactory gurgling noise as the water disappeared down the pipe.

When we descended to the floor level we found that we were not the popular heroes we had imagined we should have been. The floor of the building had been designed to accommodate the odd spill of water, and perhaps some rain blowing in through opened doors, but the deluge we had released was a lot more than that. All the trenches in the floor were full of water, floating up plastic cups and other rubbish, and the working floor itself was inches deep, with a sort of tidemark forming of broken pieces of polystyrene foam round the more elevated items of equipment.

The electrical department were going mad at this unannounced flood, but when I told them that they were lucky all the water was coming down a pipe, and that the roof itself had not collapsed, they simmered down a bit.

Next day there was another inquiry, with blame again being apportioned.

Why had I released all this water so quickly?

I answered that I thought that the roof was in some danger of collapse with this great weight upon it, and the sooner the weight was relieved the happier I would be.

Did I not realise that the water would come down the downpipe and into the building?

That is what downpipes are for, but the whole intention of having a building is to keep the water out, and only an idiot would

435

consider collecting all the water off the roof and then conducting it down inside.

There was not much opposition to that proposition, no architects being present.

The design of the reactor container called for steel wires to be set all round the structure, and then when the concreting had been completed these wires were stretched by means of large hydraulic jacks. The details of our particular system were based on developments by a French gentleman called M. Fressinet, and he requested that, as this was the largest structure to be treated in this manner, he be called to site to see how the work was progressing, and if we were having any particular problems.

The meeting was arranged, and M. Fressinet and his party duly arrived on site.

M. Fressinet was an elderly gentleman, officially recognised in France as a genius, and his party included three senior engineers and a personal assistant. They arrived about lunchtime, and we adjourned to a local hotel for a meal before returning to site for the technical meetings. When the party assembled we produced some drawings and listed out some minor problems we had been having with the equipment, and during the discussion which followed the personal assistant announced that it was time for M. Fressinet to have a short nap, so we all sat about very quietly for about 20 minutes, and when the old gentleman woke up the meeting continued.

We went round the site, and he made numerous comments on our methods of operation, and we then took him into our 'black museum' where some of the problems we had experienced could be seen, and he offered sensible clues on the causes of these problems, and possible solutions.

At the end of the visit I was asked if I could drive the party to Chester in time for the afternoon train, and we set out in some haste. On the way through Wales we crossed the Waterloo Bridge in Betws-y-coed, where the name was plainly visible from the approach road. "Look at that," the old gentleman said; "Waterloo. If only Marshal Ney had started his charges a little bit earlier, before the Prussians arrived, the whole of history would have been different." He went on to explain exactly what had happened, and from the depth of his arguments it appeared that he knew the names of most of the combatants, and where individually they had erred.

My knowledge of the history of that period was pretty scant, being limited to the names of Quatre-Bras, where action had been joined the previous day, and some of the Scottish regiments that took part in the campaign, so it was rather a one-sided argument, but interesting to contemplate.

We arrived in the station at Chester just as the train was pulling in, and I saw them aboard and waved them goodbye. That was my only claim to fame, having driven a recognised genius from Anglesey to Chester, and discussed the Napoleonic Wars en route.

We had been having some problems with the steel wire rope for the reactor, and had to make a pilgrimage to the manufacturer's works in south Wales. The factory was spread out over a considerable area, with rail tracks running between sheds and across open areas, and at the end of the meeting, after some light refreshment, we all piled into the car and started off out of the works.

Disaster struck almost at once. Somehow we had come off a so-called level crossing, and were now bumping along the track from sleeper to sleeper. As soon as this was realised, everyone but the driver baled out, expecting that the next light we saw would be from an oncoming locomotive. From a safe position well away from the rails we directed the driver until he could find another level crossing to climb out, but eventually we were off the rails, and soon onto a road leading out of the works.

There had been some serious noises coming from the underside of the car during its transit of the railway, and we thought it prudent to find a garage with a car lift and have a good look round to see if there was any obvious damage.

Such a garage was found straightaway, and our inspection indicated that one or two of the bolts under the springs seemed to have been bashed over slightly, but as this might lessen the chances of the nuts coming off we decided that all was well, and continued our journey, stopping for dinner in Hereford.

After an excellent meal we were back on the road, and I fell fast asleep in the back seat, taking little interest in the scenery, but observing from time to time that we were going north. Some considerable time later I was awoken, and noticed that the car had stopped – and, further, it had stopped in the middle of a field.

"You are the navigator; get us back to Anglesey from here," I was instructed.

"Where the hell are we anyway?" I asked, still half asleep.

"We think we are in Wales, but, wherever we are, we are in a field."

"Go carefully round the field until you come to a gate, and we will see what way that leads," was my next instruction.

We found a gate onto a lane, which came out on a metalled road and led us down to a junction on the outskirts of a village.

"Which way now?" the driver asked.

"Go straight down that road and turn right at the next crossroads," I ordered.

"Can't do that," said the driver; "it's a one-way street, going the other way."

"Bloody hell, it's two o'clock in the morning; who is going to see us? Just get on."

No sooner had we started down this road than a little man appeared out of the shadows (lurking, he must have been), waving his arms about and shouting about this being a one-way street.

"Put the lights out and get a move on," I told the driver, and, indeed, he did just that.

A few miles further on I was relieved to see a signpost for somewhere I recognised, and also could pronounce; we were going in the right direction, and only had another 80 miles to go.

Towards the end of the construction period we were required to conduct a pressure test on the structure. When you realise that the inside of the pressure vessel was a sphere of about 98 feet across, and we had to pressurise it to about 400 pounds per square inch, you will see that it was going to take a lot of air, and a lot of machinery. Compressor stations were set up, all known holes in the shell blocked up, initial measurements taken, and the job started. As it was to be a continuous operation, sleeping and eating quarters were established, and the staff installed.

The plan was to take the pressure up in stages, with a hold at each stage while measurements were taken, and the results compared with the predictions before going on to the next hold. The first two inspections revealed minor leaks, but the odd bolt was tightened, and work continued. Across the top of the structure was a series of standpipes, with bolted flanges on top, and two men were fitted out with directional microphones that could detect the merest whisper of a leak. In fact, it was possible to point the gadget across about 150 feet and hear the other man's heartbeat, or the ticking of his watch.

A few more small leaks were located and dealt with by this method, and the next stage was started. Sometime during this procedure a bolt came loose in one of the standpipes, and it vanished into the roof space somewhere. When we went to conduct our measurements there was the most frightful noise of air escaping under high pressure. Even with ear protectors the noise was very frightening, and steps were taken to fit another bolt into the hole.

The problem was that it was impossible to place the bolt into the hole, as there was a pressure of nearly 400 pounds escaping. Some very clever engineering was rattled up, and by making a trolley that could be rolled into position, and a movable weight of about half a ton, it was possible to get the bolt into its hole and eventually secured.

What a relief.

At the end of the test, the structure behaved as predicted, the pressure was gradually released, and we all went home.

At an earlier stage in the construction the steel shells of the reactors had been made up in the yard nearby. The lower halves of each vessel were erected in sequence, lifted out of the yard, and placed in position by a very large gantry crane onto prepared foundations within the reactor building.

Each of these sections was made up of inch-thick steel plate, and there were numerous cooling water pipes welded to the outside of the shells, so that all round the rim there were two-inch square pipes every nine inches, with open ends, waiting to be connected to similar pipes in the upper hemisphere.

Such a large, thin-walled structure was very flexible, and great care was taken to ensure that it maintained its shape, and, as concrete was placed around it, check measurements were taken and minor adjustments made to keep it in shape.

While the concreting was in progress the upper hemispheres were being made up in the yard, and, again, check surveys were being made to ensure that the pipes around the rims would match up with those already in position. The fabrication of these shells was carried out within a large shed with a removable roof, and out of direct sunlight, so all the measurements were taken and corrected for temperature and the necessary adjustments made.

As the time grew near for lifting the first upper hemisphere, the surveying became more intense, and it was discovered that the lower hemisphere, which was within the open building, seemed to have gone a bit out of shape. Further surveys were ordered, and these gave different degrees of ovality, and a lot of anxiety to the surveyors. Still more surveys were conducted, and again the results were inconsistent.

An inquiry was convened.

Why could the various surveyors not agree with one another, and sometimes with themselves?

Go and do it again.

Then someone rumbled that the sun shining on one side of the shell in the morning was pulling the vessel out of shape, and when measured again in the afternoon it had been pulled into a different shape.

The surveyors were forgiven.

A follow-up survey was undertaken after dark, when the temperature stresses had evened out, and the shape had returned to an acceptable degree of ovality, and it was then possible to fit the upper shell with a good chance that the pair would match up, more or less.

When the panic had subsided, and cool reason returned, a simple calculation revealed that with a diameter of 98 feet the differences we had been measuring were consistent with the effects of the sun on one side and the shadow on the other, and that all would be well on the day.

When the upper sections were eventually fitted, all the pipe joints could be accommodated, and that was one problem behind us.

Having spent most of my time surveying based on rectangular coordinates, the positioning of points within the sphere was a new diversion, and old textbooks on spherical trigonometry were dug out and appropriate tables made for ready reference.

As the internals of the reactors were being completed, with vertical holes through the core for the fuel rods, a precise location over each hole was required for the standpipes. Fixing the centres of these holes was relatively simple, but as the standpipes required to be set into the spherical surface of the shell each one was a different shape, and the surveying for cutting out the holes in the shell became very complicated, with no second chance for errors.

A very simple arrangement was set up on rails across the top of the standpipe levels, and a cutting machine was built that ran in a circle, but the actual nozzle was free to travel vertically as the surface of the hemisphere varied, thus making elliptical holes in the surface, which were true circles in plan.

During the construction we were the subject of numerous visits by foreign engineers and designers, and part of my job was to usher these people around the site and explain or otherwise comment on the procedures and operations in hand.

When we were visited by Japanese parties, one of the first things they did was to send a man with a camera round all the corridors of the offices, photographing all the nameplates on the doors. This probably led to considerable overmanning of Japanese construction projects, as, in many cases, the labels referred to previous occupants who had moved on as the construction progressed, and no one had bothered to alter the inscriptions on the doors.

By the end of the main civil engineering works we had built up considerable practical experience of the system of prestressing the steel wire cables, and one day we were invited to visit the works of a company making an alternative system of loading and anchoring the cables.

We arrived at the works and, after coffee, were given a presentation by the chief engineer, and then taken down to the test bay to observe a practical demonstration.

As in most such performances, everything went very well until the end, when they were trying to demonstrate how the system could be unloaded and, of course, it would not work.

"Why don't we go out for lunch?" said the managing director, with a savage aside to his staff: "Get that bloody thing working before we come back."

In fact they had not managed to get it clear when we returned, and someone suggested that we should go off and see some large job they had just completed on the motorway. This seemed a good idea, as it gave them a chance to get their system sorted out without spectators, and we set off, having been given precise instructions on how to find the place.

"Get onto the motorway under construction, and go along about five miles; you can't fail to see the bridge, and we still have some people there."

We set off and found the motorway under construction, and raced along, looking for a large bridge. When we had gone about ten miles we realised we must have missed it, and went back a bit more slowly. Right enough, there was a bit of a bridge, with people working on it, and when we went down they were very happy to show us over the works.

We had to explain that we had gone on a bit to see if we could see where one of our party used to live, rather than mention that we had driven over their large bridge without noticing it.

We returned to their office, where they had managed to sort out the unloading, and we had a bit of a brainstorming session, telling them of some of our problems, and asking how their system would overcome them. At the end of the meeting we had agreed that their system had certain advantages in some applications, and we would come back again to see if the modifications we had suggested had been of any value.

The concept of traceability of components was just coming into its own at this time, and we were obliged to maintain extensive records of the individual location of numerous items of equipment now incorporated in the building. Towards the end of the construction we had rooms full of filing cabinets stuffed with records, and a comprehensive system of cross-referencing, so that, no matter what question was posed, we had an answer for it.

From time to time we had visits from technical auditors, who would come in and ask obscure questions about the location of particular items, and then trace back where they had been made, their relevant test certificates, when they had been fitted into the structure, and a whole host of other particulars – almost down to the colour of the installation engineer's eyes.

CHAPTER 131
Moving out

As the construction work was almost complete the time came to prepare to move out to another contract.

At this time, in the heat of the technological revolution, aluminium smelters were being designed for Holyhead, Hartlepool and Invergordon, and the company was tendering for all of them.

You might have thought it would have made sense for us to be awarded the contract at Holyhead, as many of the proposed staff were already resident in the area and knew the difficulties of operating there, but perhaps this knowledge had some influence on our bid, as our tender was not accepted.

Our tender for Invergordon was accepted, however, and about the same time we were awarded the contract for another nuclear power station near Hartlepool. I was given the choice of going either to Invergordon or Hartlepool, and took a few days' leave to have a reconnaissance in both places.

We set out on a fine sunny morning, driving gently across England in a north-easterly direction, taking the tourist route and avoiding the major towns and cities.

As we approached the conurbation of Teesside we could see a great cloud of industrial haze and smoke hanging over the area, and decided against spending the next few years living in such a smog, so we turned north again and continued into Scotland, passing through Perth and Inverness, and on to Invergordon, on the Cromarty Firth.

It was another beautiful day, and we did not take much persuading to agree that this would be a much healthier place to live.

Back again to Anglesey to see about selling the house, clearing out the office, and making ready to move again.

We advertised the house in the local paper, and had interested potential customers round in a very short time. The first people who came were very pleased with the property, and when the lady saw the greenhouse, full of tomatoes, and the small orchard loaded with fruit she turned to her husband and said, "We'll have it; go and settle up with the man, before anyone else comes along and makes him a better offer."

This suited us fine, and we agreed suitable dates for our moving out.

Meanwhile I was transferred to the London office to assist on the design and programming of the work at the smelter in Invergordon, while Jean and Susan remained in Anglesey.

Together with another engineer I travelled up and back from London each weekend by the train from Holyhead, until we were ready to leave for the site in Invergordon.

The local authority had started the development of a number of housing estates specially for the influx of employees during the construction and operation of the smelter, and we had been allocated a house in Alness, a small village about four miles from the site, and on the chosen day we loaded all our goods into a large furniture van, and set off.

By this time I had reached the dizzy heights of having a company car, so we were a two-car-and-one-boat family travelling in convoy to the north.

The removal company had indicated that they would take two days for the journey, so we travelled quite sedately, stopping in Glasgow with my parents overnight, and arriving in Invergordon about noon the following day.

SECTION 2 – INVERGORDON

CHAPTER 132
Moving in again

Our first night in the north was spent in the County Hotel in Dingwall, an old building in the centre of the town, not yet modernised to the city standards but very comfortable and homely. We retired after dinner, slept well, but were woken by the dog barking at the maid bringing in the tea. This was an intrusion beyond her ken; it also gave the maid a bit of a fright.

We set off to see our new house, in an estate of houses and bungalows still very much under construction, and were assured that it would be ready for us on the following morning. This was just as well, as our furniture would be arriving then, we hoped. We stayed over in the Royal Hotel in Invergordon for the next night, and camped out on the doorstep of our new house the following morning.

Looking at it in the cold light of morning, I wondered where we were going to put all our furniture, having just left a rather large four-bedroomed house in Anglesey to come to this two-bedroomed house in Alness. The furniture arrived, and we set out all there was room for where we expected to be able to use it, but were still left with a considerable number of items, and no obvious home for them. One of our neighbours, a colleague on the site, came to our rescue, and we managed to stow all our excess belongings into the spare room in his house.

Almost all the occupants of these houses were working on the site, and there was a good spirit in the neighbourhood.

In the springtime all the young children were learning to ride their bicycles, and the roads were full of children wobbling along and anxious and breathless parents trying to keep up with them.

We had decided to try to find a larger house, with a garden and a garage, and spent some time looking for a suitable place. The house we eventually settled on was in Invergordon itself, not too far from the site, right by the golf course, and with a splendid view, looking south down the Cromarty Firth.

During the sales spiel from the previous owner they announced that it really was a very warm house; sometimes they could shut down the central heating in June and parts of July.

The move from Alness was conducted in parts, courtesy of the site van and some assistance from friends, and we were soon happily installed, with our furniture around us, and a garden to dig. Our neighbours were very kind and made us most welcome, the only possible snag being that they were very keen gardeners, and rather put our efforts in the shade. Soon we had joined the sailing club and the golf club, and were taking an interest in the community.

CHAPTER 133
Site Work

The site of the smelter had been selected on the only reasonably level piece of ground in the area. This turned out to have been the bed of a large, swampy pond, which had been drained by some Dutch engineers in the late 1800s, and was now cultivated for potatoes and grain.

The ground conditions were not of the best for heavy construction, comprising a layer of turf on top of some grey, soft clay with the consistency of toothpaste. The design called for a lot of piling, and the subcontractor had been appointed and was making his preparations to come to the site. Some large earth-moving machines had been brought in to make the access roads for the piling gear, and this worked quite well until the surface was broken. For the next few days practically no digging was done; all the time was spent pulling one machine after another out of the mire.

A special brainstorming meeting was called, and all the London experts had their say with their possible solutions. While these discussions were being costed, and possible programme alterations considered, there came a very hard frost, and we were able to run at will all over the site, getting roads filled with stone, ditches dug and preparing for the piling equipment. For about ten days of hard frost we worked round the clock, and at about this time the experts came up with their opinion.

"Don't worry," we signalled back, "the problem has been solved on site."

The smelter itself was divided into three main sections: a reduction shop, where aluminium oxide was placed into tanks lined with carbon, and electricity passed through, which resulted in the production of molten aluminium; a casting shop, where the molten aluminium was cast into moulds, forming ingots; and a carbon plant, where the various components of the reduction line were made up, repaired, and refurbished as necessary.

In support of these sections there was a large conveyor system, designed to carry the aluminium oxide from a jetty in the firth into storage tanks on site, a services section, with laboratories, office buildings, canteen and welfare facilities, and a site building for workshops and stores.

During the construction period we had about 700 men on site, mostly recruited locally and with little experience of heavy engineering. The work was on a tight time schedule, and the planning and programming were of a high order, enlivened by a number of unexpected complications.

On the day when Mr Wilson conceded defeat in the general election I noticed all the steel erectors trooping back to their bothies. When I enquired of their agent why they had stopped work, I was advised that they were having a day's strike, in sympathy for Harold. As I had not yet heard the news, I imagined that Harold was one of the erection crew who had suffered some mishap, and asked how he was now. Such ignorance seldom goes undiscovered.

Our men were not all that bothered about Harold, and our work continued unabated.

The management of the construction activities was conducted by the various section agents, working to a joint programme, which was being altered continuously in the light of current progress. Each section had a weekly meeting that considered the work done to date and the details for the following seven days, and, in lesser detail, for the next three weeks. There was a weekly meeting of section agents, where the achievements of the previous week were discussed, and proposals for the following periods coordinated.

Any activities that had fallen behind were allocated a star in the minutes, and the object of these weekly meetings was to avoid the collection of stars. It soon became obvious to the section agents, however, that if any one item collected a number of stars the duration of the examination of this item took much longer, and if that particular item was not seriously significant in the overall scheme of things it was often convenient to have a long time on diversionary discussion of that particular problem while other shortcomings tended to be overlooked.

Each month there was a combined site meeting where the financial aspects of the contract were discussed, and following that, a general meeting on site, where the client's staff and the head office design staff attended. Because the job was going very smoothly, these meetings tended to be a fairly pleasant diversion from the daily grind, and included a slap-up special lunch in the works canteen after the troops had been fed.

CHAPTER 134
Sailing Club Activities

The club owned a small wooden shed adjacent to the harbour, and most of the members kept their boats at home, or moored them out in the corner of the harbour. This arrangement was fairly satisfactory; the boats were in a secure berth, floating through most of the tide, and access was available to members who had arranged to have a key to the Admiralty gate.

There was a bad storm one day, and our boat was swamped at her mooring. When I arrived in the evening and pulled her ashore I found that the inside was full of a dreadful mixture of chaff from the nearby distillery and fine sand. This mixture had clogged up the self-bailing outlets, and had penetrated into all the openings into the buoyancy compartments.

After some drastic action with a fire hose we had her cleaned out again, but there remained a faint alcoholic fragrance inside for a long time thereafter.

The club had some fine cups and trophies, and it was agreed that we would start a regular racing programme, and also hold special events, for the general entertainment. One such event was the seamanship race.

The boats were tied up at a long slipway, and the crews lined up on the shore.

On the given signal they had to run down to their boats, push off, hoist the sails, and then sail round the course. After the first mark, they had to take the sails down, drop anchor, re-hoist, lift the anchor, and sail to the next mark. From there, they had to go alongside another slipway, send a man ashore to collect a bottle of beer (while the other person sailed round a mark some short way off), recover the man from the slipway, sail out, drink the beer, drop the bottle overboard, go back and recover the bottle, sail to the next mark, take the sails down and row about 100 yards to the next point, re-hoist and then sail to the finish.

The logistics of this exercise took some arranging, it being considered a dangerous undertaking to allow some fortunate member to be landed at the slipway with a dozen bottles of beer, and have to wait half an hour before any customers arrived.

The event was proclaimed a great success, additional entertainment being provided by one pair of intrepid mariners who had borrowed a very fast boat. They came into the slipway to drop the crew, and when they thought they were near enough, the skipper shouted, "Jump!"

His willing crew leapt out, only to find himself in about ten feet of water, and he had to swim for the shore. He ran up the beach, collected his bottle, and came back down as his boat came rushing back. It executed a smart turn alongside the slipway, and the crew was instructed to: "Jump!" again. First he threw the bottle aboard, and then jumped onto the stern of the boat, but by this time the boat had gathered way, and our man was left holding onto the brass mainsheet horse across the stern, with his body streaming out behind. After a few yards his trousers came off; then the speed was reduced, and he climbed back aboard and completed the course in a state of undress.

When he came ashore at the end, he was all wrapped up in a sail, to cover his embarrassment – or whatever.

The club had owned an old shop in the harbour area, which had been used as a changing room, but it was falling down through neglect and a bit of planning blight. We negotiated an exchange of this building for a strip of land along the shore, straddling the old ferry slip, and erected a fence and a timber hut on this new site.

This was the operating headquarters of the club for a number of years, and then 'progress' arrived. Massive developments were forecast right across our site, and we had been completely ignored during the preparations of these plans. Battle was joined, we had a title, and the slipway was both an ancient monument and a public right of way.

After a great deal of haggling we agreed to move along the shore to a less convenient location, on condition that the developers built us a new slipway and bought us out of our present site at a price that would enable us to build a new clubhouse.

The new clubhouse stands there today as a monument to sheer persistence, and a bit of smart dealing, when we sold up the old shop. The Port Authority, newly formed, regarded us as a bit of a nuisance, and relationships were strained, to say the least.

On the opening of the new clubhouse we had borrowed numerous signal flags, and dressed the mast with these. When asked what the message was, we replied, "It says 'To Hell With The Port Authority'," but this was difficult to decipher, as we were missing a few letters.

From the new clubhouse we were able to carry out cadet training and arrange inter-club events and a bit of an instructional and social programme, and, with increasing revenue, we bought a generator.

CHAPTER 135
Horse-trading

While we were living in Invergordon we became involved with the local Pony Club, and were obliged to purchase a pony for our daughter. We took advice, looked at a number of large animals, and eventually settled on a brown horse called Fury. (Just as well his name did not describe his temperament.)

As a piece of advice to the non-horsy contemplating the purchase of a pony for their offspring, be warned that the horse is only the start. The animal needs feeding and watering, exercising, brushing and combing, and a warm place to live in the wintertime. It also needs shoes, a waterproof coat (which it always tears on the nearest piece of wire fencing), and a sailmaker's kit for the repairs to the coat.

Then there is the saddle, and this year's model of the bridle, together with all manner of other gadgets, bits, martingales and sweat rugs, and on top of all that there is the correct outfit for the offspring, which includes a number of different hats, breeches, jackets, stocks, the Pony Club tie, short and long boots, gloves, a riding crop, and various training aids. By the time all these items are taken care of, they have you pretty well trained already.

The next subject to be broached is riding lessons. These always take place several miles from where you live, and always in the opposite direction from where the horse lives, and, of course, you need a horsebox. This also means a tow bar for the car, and all the proper electrical connections to the trailer.

The first visit to one of these training establishments can be both painful and expensive. The person in charge examines the horse and the rider, points out several things that will not do, and produces a series of suggestions to put the affair to rights, and then takes the pupil away for indoctrination, suggesting that you return in a couple of hours.

Somewhat apprehensively you depart, leaving your daughter in the hands of some demented woman, cracking her whip and shouting incomprehensible commands at the circus of riders.

When you return you are advised of the progress and when the next meeting will be held, together with numerous comments about the animal, the equipment, and the items that should really be available before coming back. Then, after the tricky business of loading the horse into its box, and driving back, you say, "Well, did you enjoy all that?"

"Oh yes, but Mrs Black says that I need a bit more practice, and Sally has lovely long boots; can I have a pair like that, too?"

Before the next lesson word has gone around that there is an additional horsebox in the fleet, and arrangements have been made amongst the children to organise the ferrying of horses to and from their fields and the riding school, with somewhat imprecise details of the locations of the fields, and also the dimensions and colours of the individual horses to be collected (it would never do if you collected the wrong horse).

After a few weekends of this, it was time for the first event.

Special arrangements were made for the collection of horses, evenings were spent on polishing equipment, pressing clothes, brushing hats, shining boots, chanting out the sequences for dressage, checking the programme to make sure we would be there in good time, with all the other loose items, such as horse bandages, hoof protectors, brushes, combs, sweat blanket, hay, oats, water bucket, etc. all loaded in advance.

Pony Club events nearly always clashed with special sailing club races, and the logistics of getting the horse or horses to the site, parking the box, driving like a fiend back to the club, taking two cars out to where we hoped the sailing race would end, and then getting back to the club in the other car to get afloat were enough to tax the best planning engineer in the business. At the end of the performance, we had to end up with the horses back in their fields, the boat on her mooring, and both cars back at the house.

We became quite skilled at loading and unloading horses into the box, and generally, being a bit pressed for time, did the odd bit of pushing and shoving, but horses are very stubborn creatures, and sometimes underhand methods were employed.

I had been given a demonstration once on how to load a particularly excited beast into its box. You stretched two long pieces of bailer twine from the upper sides of the horsebox and laid the outer ends on the ground. Then some brave citizen took the horse's head, and began to lead it towards the box. As soon as it was within reach the second person picked up the strings, and crossed them behind the horse's hind legs.

As soon as he felt them touch him, he began to step forward, and as long as the strings were in contact he would keep moving, right into the box, as good as gold.

We had had a very busy day transporting horses and racing the boat, and when we returned to the field and were loading the first of our pair of horses I noticed a lady with a young daughter having some trouble in persuading her pony to go into its box. I offered to assist, and was told politely to go away, so off I went, delivered our first charge and came back to collect Fury.

The lady was by then in a state of some despair. The child was raging about in a foul temper, the pony was standing in front of his box and stubbornly determined not to go any further, and the mother was in tears, shouting at the child, the horse, the horsebox, and the few spectators left on the field.

"Look, missus," I said," I think I can get your pony into his box, if you will let me try."

"Oh, go on," she replied; "take him to hell if you want."

Further shrieks and wailing from the spoilt brat.

"Please take the child into your car, and I will have a shot at it."

Very smartly we tried the string trick, and in he went, first time.

I went round to the front and told her he was loaded, and she could depart now if she wanted. I also asked if she would like any assistance at her home field to get him out, but she had obvious plans for that event, and wanted no witnesses.

Funny thing, I never saw them at any later events after that.

The field where we kept Fury had become overgrazed, and we had to find another place for him. We soon found a field with other horses and came to an arrangement with the farmer for his board. This place was much nearer our house and more convenient for riding out, as one side was fenced by the Forestry Commission and there were pleasant walks and rides, well away from traffic.

I had been noticing that he was getting a bit thin, despite the fact that he was eating like the proverbial horse, and we took advice and asked the local vet to come along and have a look. He came by a couple of days later, and announced that the horse's teeth needed rasping. This was a new one on me, but he explained that, because the teeth had worn, he was not breaking up his oats and they were going straight through, not delivering any great nourishment. (Good for the farmer, though.)

An appointment was made for the Saturday morning, we went out and caught him, and tethered him by the fence, waiting for the vet to arrive. By sheer good chance, I was standing on the other side of the fence from the horse, and when the vet arrived he put on an old coat and came over, with one hand in his pocket.

After the usual salutations he said, "Catch a hold of his tongue."

"No bloody fear," I replied.

"Well," he came back, "I can't rasp his teeth with his tongue in the way. Just catch onto it, and hold it to one side."

Faced with this logic I was prepared to give it a go, and took a grip on his tongue. Great commotion, with the horse rearing up, and me still hanging on. Just as well I was on the safe side of the fence.

"Oh dear," he said; "looks like we'll need to put a snitch on him."

For those unaware of the finer points of animal husbandry, a snitch is a sort of a tourniquet, applied to the upper lip. Apparently the idea is to focus the horse's attention on a great pain in its upper lip, so that all manner of other cruel acts can be performed, without it really noticing them. That is the theory, anyway.

The vet applied the snitch, and I caught onto its tongue again.

Like a conjurer, a large rasp was produced from the vet's coat pocket, and he set to work with one hand on the rasp and the other on the snitch, and me holding the tongue. Fortunately, I had sent Susan off on an errand, so she was not present as a witness during this performance.

When one side of the animal's mouth was done I had to change my grip and catch his tongue on the other side, and the performance continued.

When the operation was completed to the vet's satisfaction he said, "Let go now," loosened the snitch, and at the same time released the tether.

The horse shied up in the air, hooves flailing, turned on the spot, and raced away to the other end of the field.

"That wasn't too bad, was it?" the vet said. "He'll be as right as rain in a couple of days, you'll see. I'll send you my bill at the end of the month."

Just then Susan arrived back, and she went off to catch Fury. He was a little bit more elusive than usual, but she soon had him, and brought him back to me at the fence. We were not friends for a couple of weeks after that, but I suppose he must have forgotten, or forgiven, and he would come to me eventually, when he saw I was becoming exasperated.

The next medical treatment due was worming. We bought some powder, with instructions to put it in with his oats in his bucket, but this was no use, as he sneezed deliberately into it and most of the powder was dispersed on the grass.

Next day we purchased a different brand of wormer. This one came in the form of a sort of syringe, not unlike those used for sealing gaps alongside windows in the house, about eight inches long, with a bore of about an inch.

The instructions were obviously written for experts: "Expel contents into horse's mouth by firm pressure on the rounded end." (Presumably they mean the tube's end, not the horse's.) Not a mention as to how you got the other end of the gadget into the horse's mouth.

We tried all ways – holding his nose, pulling his mane, putting a blanket over his eyes, and combinations of all three – and eventually he had to take a breath, and the deed was done. Until, that was, he coughed the whole tube-worth out over me.

The only way we managed to get any of the stuff into him was to mix his oats with treacle, and as he was about halfway down the bucket slip in a second dose, heavily smeared with treacle and loose oats.

Apart from these minor excitements, he was really a well-behaved and very healthy animal, and we were sorry to part with him when Susan grew out of him and needed a larger animal.

We found a good home for him, and set about selecting a bigger beast.

This time it was a white horse (officially, grey. The only white horse was associated with Banbury Cross, I am reliably informed; also, a brand of whisky.)

This creature was called Clint, and when we first saw him he had been kept in stables for some weeks, so that when he came out and was saddled up he was off like an arrow and galloped all round the field, with Susan hanging on, and rather enjoying herself. He and I got on quite well, as we had not come in contact with the vet and his rasp, nor had other previous painful association.

One evening, in the wintertime, the farmer telephoned to say that someone had skidded off the road into Clint's field, broken down the fence, and the three horses had escaped. We set out at once with two cars, looking for a white horse in a snowstorm, with six inches of snow on the ground and no idea which way he had gone, or even where the edge of the road was.

After about an hour of fruitless searching we agreed that it was pretty pointless. If he had any sense, he would have taken shelter in the woods, or even trotted down the road to the house, so we abandoned the search and went home. As soon as we got in, the farmer telephoned to say that the horses had all walked into his farmyard, and he would keep them overnight. What a relief! Next day we undertook repairs to the fence, and peace was restored, with the horses back in their own field.

At the time of the local Highland Games there was always a Pony Club gymkhana, and on the evening before these events the horse would be brought down to the house, scrubbed, polished, his hooves shone, his mane plaited, and his tail dressed, ready to stay in our garden overnight for a quick start in the morning.

We had gone to bed quite late, and were woken by a telephone ringing, at about five o'clock in the morning. This was a neighbour, complaining that his wife had been frightened out of her skin when a horse had put its head through the open window of their bedroom.

For a few moments I thought that this was really quite funny, but then I realised that it might be our horse, although how the hell it had got itself round to their bedroom window was more than I could fathom.

"OK," I said, "I'll be right round to collect him," and dressed quickly.

When I arrived at his house, going through the gardens at the rear, I found the horse chomping away on some garden plants, obviously quite happy, but somehow on the other side of the fence.

I had a quick look at our immediate neighbour's garden, and could see a neat set of hoof prints walking down between two rows of lettuce, with no obvious gaps where plants had been torn up, thank God.

The next trick was working out how to get us out of the back garden of this other house, and by following a path round I found a small garden gate, which led out onto the road.

Although the back gardens of our houses were very close, it was a considerable distance round the front, and I climbed up at the gate and mounted Clint, and was riding quietly back when I met some of the staff from the smelter, setting out on their early morning shift.

If they were surprised to see me in my pyjamas and overcoat, riding bareback round the estate at about 5:30, they never spoke of it again. We exchanged greetings and passed by.

Later in the day I mentioned to our neighbours that they had had a visitation from Clint in the night, and that, somehow, he had leapt over their fence into the other garden, and I hoped that he had not eaten any of their vegetables.

We went out and had a further inspection, and found not a leaf trampled on, nor any plants displaced, except for some flowers around Archie's window on the other side of the fence.

Some reparations were in order there, a box of chocolates changed hands, and I bought Archie a beer when next we met at the golf club.

The Pony Club meetings and gymkhanas were a lot of fun, not only for the children, but because they provided much-needed exercise for some of the parents, who had to run back and forth, picking up dropped eggs, carrying marker flags out and in, moving hurdles and jump stands, and having a good picnic at the end of it all, before the entertainment of loading up the horseboxes and going home again.

CHAPTER 136
Grangemouth Dock

The section of the work at Invergordon under my jurisdiction was coming to a close, and I was asked to transfer to a new contract at Grangemouth docks.

As there was not much other major work in Scotland in the near future, I agreed, and went down to Stirling to see about a house.

The work at Grangemouth was a very interesting project to replace the sea lock into the main dock, and would require some clever engineering and careful planning, one of the conditions being that the construction work should not interfere with the normal commercial operation of the dock.

I started work one Monday, having driven down from Invergordon and found lodgings in a local hotel.

The contract was expected to run for about four years, and as we had decided to move house we worked out a rough specification for our requirements.

It had to have at least four bedrooms, a garage, a garden, be near schools, have a pleasant outlook, be near a golf course, and not too far from the site.

We visited all the local house agents, and departed loaded down with the handouts of totally unsuitable properties, and for the next few days I worked on site during the day, and went round in the evening scoring these places out of our lists and collecting other properties every day.

Our specification had to be modified; the distance from the site was one of the first variables to change, and we then searched from Dollar in the north, Bridge of Allan, Stirling, Falkirk, Linlithgow, and along the south shore of the Forth as far as Edinburgh. Meanwhile, some others of the staff had rented a large house near Alloa, and I joined them there on a Monday-to-Friday basis, travelling back to Invergordon each weekend, with most weekday nights spent looking for a house.

I got wind of a farmhouse, near Bridge of Allan, which had all the other requirements on our list, but it was in rather poor repair – although, as they say, it had great potential.

459

Jean and Susan came down and were not unduly dismayed, and, as we were all getting pretty fed up with the weekend commuting, agreed to make an offer.

Rather to our surprise the offer was accepted, we put the Invergordon house on the market, and I moved into Bridge of Allan.

The survey of the farmhouse had indicated some worm, dry rot, death-watch beetle, and a host of other complaints, but we had anticipated the cost of the necessary treatment in our offer and were not unduly dismayed.

I arranged for the various contractors to come in and do their business, and moved into one of the back rooms while this was in progress.

The first day, when I returned in the evening, all the front windows had been removed and there was a whole gale blowing through the place. The kitchen was at the rear, and was not seriously affected, so I had a quick brew up and retired to my bedroom, to try to catch up on some work from the site.

For the next week I lived in half a house, until the windows had been restored, and I had to move into the front while the rear section was treated. This would not have been too bad, but the Rentokil merchants had been in and the entire place reeked of their poisonous liquid.

At last they were all clear, and I made a start on panelling the large kitchen, fitting some more modern equipment, connecting up the boiler to a new central heating system, running new electrics, and generally being the complete home handyman.

We had a few inquiries about the house in Invergordon, and were considering a contract, when I was called into the site office and asked to make a call to London.

After the usual pleasantries the question was posed: "How would you like to go out to Indonesia, and run our nickel exploration contract?"

I must confess that I had not been very happy about my part of the job in Grangemouth, and there was a bit of a personality clash between me, the site manager, and the general foreman, whose declared intention was to run the young engineers ragged.

We had exchanged words on this matter frequently, and I was not being supported in my arguments, so I agreed, on the spot, to come down to London and arrange to go out to Indonesia as soon as possible.

The site manager was a bit put out by this sudden posting away, but as this was a summons from on high there was not much he could do about it.

I think someone explained to him that the present project manager in Indonesia had been taken ill and had returned home for treatment, and as I had worked in the same department in London I was the most likely lad for the job.

There followed a frantic two weeks during which I cleared my desk at Grangemouth, leaving copious notes, visited London for a medical check, read up on the contract in Indonesia, arranged to rent out the house in Invergordon, arranged to sell the house in Bridge of Allan, complete with all the anti-fungus certificates, and my newly fitted kitchen, and packed up ready to depart for Singapore.

When I arrived in London I discovered that diplomatic relations were a bit strained between Indonesia and Singapore, and obtaining a visa and work permit meant staying for about three weeks in Singapore while the formalities were processed. It turned out that this stay would be invaluable, as I was able to see how the contract was set up, and meet a number of the site staff who came out from Indonesia on short leave.

The whole affair was quite complicated. We had a contract from the nickel company, in Canada. They had offices in New York, Sydney, London, Jakarta and Makasar (now Ujung Pandang), and a project manager on site in Sulawesi.

Our contract was to manage the exploration for nickel on the island of Sulawesi, including all accounting, servicing, paying of wages and salaries, shipping of supplies and equipment, licences, insurances, dealing with the local customs offices, and – in fact – everything except controlling the reporting of test samples, and operating the fleet of four helicopters and a small fixed-wing aircraft.

We had an office in Singapore with a small staff, who organised the shipment of supplies and equipment and took care of the documentation of staff travelling out and in, together with all the import and export formalities between Singapore and Jakarta.

In addition to these duties, they also looked after the needs of people on local leave, visits to the dentist, doctors, passports, visas

461

and work permits, and also provided a service of meeting such persons when they arrived, seeing that they had a place to stay, some cash, knew when they were expected back on site, and arranged local travelling to and from the airport.

CHAPTER 137
Singapore

I flew out from London and arrived in Singapore in the afternoon, completely disorientated by the time changes. I was met at the airfield and taken to a very quiet hotel, where I unpacked my bags, had a bath and fell fast asleep, to be awakened at about eight o'clock, local time, and taken out to dinner in a fashionable Russian restaurant, with dancing Cossacks, balalaikas, loads of garlic, and a lot of vodka.

As far as I was concerned it was still breakfast time, but I tried to keep awake and make intelligent conversation until I could get back to bed, and catch up on the hours lost in transit. Eventually the dinner ended, and I was taken back to the hotel, and told they would call for me in the morning. Apart from the violent indigestion and the lack of sleep, I felt reasonably fit in the morning, and after a couple of pints of fresh orange juice was ready to face the remainder of the day.

I was collected as arranged and taken down to the office, where my business in the East really began.

Our office in Singapore was in a couple of rooms in the Ming Hotel, one of the newer, large hotels that were springing up all round the island. This location was very convenient, the hotel was equipped with all the latest forms of telex, telephone and fax machines, the restaurants were open day and night, and there was always a spare bed available for members of the staff who arrived late or unexpectedly, or who were unable to fly out due to weather or other difficulties.

The office dealt with all the formalities of shipping goods and supplies to the site; we had a coaster on charter, which sailed every four weeks from Singapore, taking out food, drink, and all the equipment, spares and consumables required to maintain the operations in the field.

They also steered the site staff through the maze of work permits, health requirements, passport and visa formalities and police authorisation, to ensure that everyone had the necessary papers duly signed and stamped by the appropriate official in the correct sequence.

There was a state of tension between Singapore and Indonesia at a high level in the bureaucracies, although the individual officers were generally very obliging, and tried to hasten the process. Permits were issued in Singapore for 12-month periods, and took about three weeks to process, during which time the applicant had to be available to attend the various offices, and as a result of these procedures there were almost always one or two members resident in Singapore waiting for their approval.

In general, people on site were allowed leave on a regular basis; those working in small camps and outstations were given a few days off every month, while others who lived and worked in more permanent establishments had a week off every three months.

From the luxury of the hotel in Singapore this seemed a fairly generous arrangement, but after a spell in the bush it took on a more miserable aspect.

During my waiting period I met a number of the staff on leave, and visited various suppliers in the town, our shipowner, the Singapore customs office, our appointed doctor and dentist, various engineering companies, and representatives of contractor's plant manufacturers.

One evening I was invited to attend a film show by one of these companies, to be followed by a dinner in another large hotel.

The evening started with an introductory drink, and we all trooped in to see the film. It was the general run of such films, with machines seen running up and down hills, and tables of performances, and so on, and at the end the representative announced: "That was that, thank God. If you have any questions, please see me after the food."

We then sat down at my first Chinese meal, and were served a great number of small dishes, some very spicy indeed, and some looking like something the cat should not have bothered to bring in. The majority of those attending this banquet were Chinese engineers and it was all a lot of fun. Someone realised that I was a bit new to all this, and took me in hand, serving me particular items and suggesting I should decline others. At the first short lull in the proceedings the waiters rushed round with small glasses of spirit, and when everyone was ready the local president of the society made a short speech of thanks to the speaker, welcomed the guests, raised his glass, shouted 'yamsing', and drained it at a gulp.

Short applause, then all cried 'yamsing' – and there it was, gone.

No sooner had we sat down than the glasses were refilled by hovering waiters, who were obviously well drilled in this activity.

While the sight was returning to my eyes, which had somehow become misty, the speaker rose, thanked the president, and we were at the 'yamsinging' again.

This happened several times, with short addresses by different people, all ending with the ritual of gulping down the spirit and having the glasses instantly refilled. That ended the formal part of the evening; the groupings at the tables broke up, old friends renewed their acquaintanceships, some from earlier in the evening, and I was introduced to half the engineering population of the island.

This 'yamsing' business was potentially very dangerous; it seemed to be the trigger for the waiters to rush back and fill your glass again, and it was going on all round. Someone came along and took me in hand, we bade our thanks and farewells to the president and the speaker (more 'yamsings'), and were carefully placed in a taxi and sent back to our hotel. What a splendid party.

The next day was a Saturday, and spent in quiet contemplation.

No one who goes to Singapore should neglect a visit to the famous Raffles Hotel. The grounds were carefully tended, with magnificent palm trees set in lawns, and colourful flower beds bordering the paths. The hotel building was undergoing some repairs, but it was easy to imagine what it must have been like in its heyday.

The tone was lowered a bit by the presence of some American drilling contractor's staff, who were sprawling about in the lounge, wearing brightly coloured shirts and shorts, with no socks – would you believe? – and talking very loudly. We were recognised as being of a different order (poor engineers), and escorted to the dining room by a very dignified head waiter, where we had a most excellent dinner in very civilised surroundings.

On another occasion I was having lunch with one of our local suppliers at the Mandarin Hotel – another very large hotel, which had special lunches of Chinese dishes, prepared for Europeans, and catered for parties off the cruise liners, which were often in port.

We were sitting eating and chatting away in a quiet corner, when a man at the next table, who was trying to use a chopstick as a knife, managed to break it, with a loud crack. This was at one of those sudden silences that often occur in public places, and all eyes turned on the unfortunate man, while a waiter, who was passing nearby, fielded the broken chopstick and continued on his way, returning immediately with a replacement pair of chopsticks, to a round of spontaneous applause.

Our office manager was a member of the local sailing club. He had a small boat and we went out sailing on a couple of weekends. He was also a member of one or two other clubs, where he took me swimming and we played tennis, and generally I was kept amused during the period I was collecting my various documents. I tried to spend some time learning the Indonesian language and reading up on the history of the country, and also on the state of the contract, so that I would have some idea of what was going on when I eventually arrived on site.

I found that the client had an office in Jakarta, where it was necessary to stop overnight on passage, and also an office in Makasar, where we also had a representative. The main site office was at a place called Malili, about an hour and a half's flying time away, and there were other offices scattered round the island of Sulawesi.

The contract was to support the exploration operations for alluvial nickel, and this was carried out in three different ways. Before the war, the Dutch had carried out fairly extensive prospecting in two main areas, and we had 'outstations' at these locations, undertaking drilling operations using fairly large equipment, from semi-permanent camps. The second method was to go along the shoreline, taking samples of the material at the river mouths, and then going up the rivers where reasonable prospects were indicated, taking samples by a hand-carried drill. The third method was by travelling across country, taking samples to a regular pattern.

At long last all my documents were in order, and I was booked out to Jakarta on the morning plane, flying direct from Singapore, on Garuda, the Indonesian national carrier.

SECTION 3 – INDONESIA
General map of Indonesia

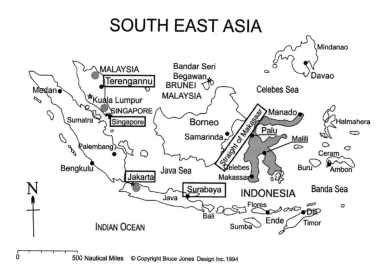

SOUTH EAST ASIA

Mindanao

MALAYSIA
Bandar Seri
Begawan
BRUNEI
Davao

Terengannu

MALAYSIA

Celebes Sea

Medan

Kuala Lumpur

SINGAPORE

Manado

Singapore

Borneo

Halmahera

Sumatra

Samarinda

Palu

Palembang

Malili

Ceram

Bengkulu

Java Sea

Celebes

Buru

Ambon

Jakarta

Makassar

N

Surabaya

INDONESIA

Banda Sea

Java

Flores

Bali

Dili

INDIAN OCEAN

Sumba

Ende

Timor

0   500 Nautical Miles   © Copyright Bruce Jones Design Inc.1994

467

CHAPTER 138
Indonesia

We had a smooth flight, and came in over the red-tiled roofs of Jakarta, where the houses were arranged in geometrical patterns along the banks of the many canals.

I had brought in all the mail for the staff, and a few items of special purchase: a couple of tennis rackets, some darts for the club, and a large selection of newspapers and periodicals. There were no problems at the customs – my passport and all the other bits of paper were in order – and I was soon through into the main waiting hall of the airport. Like all the other similar buildings I have passed through, this airport was being reconstructed, and was full of temporary fences and barricades.

Once through the building I was set upon by half a dozen porters and taxi touts, and had some trouble in making sure that all my gear was together and about to go into the one vehicle. Then came the paying out of handfuls of rupiahs before I could make my escape. I had been told to go straight to the Hotel Indonesia, where a room had been booked for me, and my earlier study of the language bore fruit.

"Hotel Indonesia," I ordered, and off we went at great speed through the traffic, which became thicker and thicker as we approached the city. There were many beaten-up taxis trundling around, a great mixture of vans and lorries, many pedal-powered rickshaws and about 10,000 cyclists, burdened with impossible loads. In addition to all the powered traffic, many hand-barrows were in use, and pedestrians were everywhere.

The hotel was an imposing new building in well-tended gardens, and I was welcomed in and taken up to my room, which had a beautiful view over the city and was furnished in dark wood; my attention was drawn to a notice concerning the limitations of the city water supply: baths should only be taken during specific hours, etc., etc.

I telephoned the offices of the client in town, and was advised that someone would be round to collect me shortly.

No arrangements had been made concerning my onward travel to Makasar, and then to site, and I supposed that this could be arranged later that day, from the client's offices. Soon enough a

468

driver came to collect me, and we were off again into the traffic, driving on the hooter all the time, until we arrived in a quiet street, and into a well-laid-out garden with offices at the rear.

After the introductions were completed I asked about further transport to the site, and they seemed to be in no great hurry. "Perhaps tomorrow, or the next day – we will wait and see; meantime, relax and enjoy Jakarta."

This seemed to be a bit of a hint, so I invited the two members of the reception committee round to the hotel for drinks – and perhaps dinner?

"What a splendid idea," they said; "we will just clear up here and be with you directly," and within a few minutes we were back in the traffic again.

The hotel had what was known as a 'happy hour' every evening, when drinks were much cheaper, and we arrived just in time to take the full advantage of it.

I suppose they must have been pretty regular customers, as they knew everyone in the room, and I was introduced as 'our new man in Malili'.

My further mastery of the language was confirmed when I ordered, "Three more beers, please," and that was what arrived.

My guests went off to change, and I retired to my room. So far, so good. Just in time for the official bath hour, but only cold water was on offer. As they say, there is nothing like a cold shower to get you going, however, and that was about the motto of the day.

Back in the bar for some light refreshment before dinner, I was astonished to see the extent of the menu, and took some time with my selection before my guests arrived.

We settled on pretty exotic fare, but it was well cooked and well presented, and we had a selection of local gamelan music playing some of the time, followed by a sort of jazz band. During the meal they said that I would be on the flight out to Makasar two days later, and they would call again tomorrow and take me round the various offices of the bank and the customs, etc.

In the morning we visited the customs office, which was full of men with gold-plated shoulders, carrying sheaves of paper and trying to impress one another, and then we went round to the bank.

The company account was with the American Express Bank, and when I went in I found that all the staff were Scotsmen. A good rapport was established immediately; I handed over my authority, we exchanged signatures and all that, and then it was time for lunch.

The bank was well within walking distance of the hotel, and we strolled round in the warm sunshine, dodging the cyclists and ignoring the entreaties of the rickshaw drivers to travel in special luxury in their three-wheeled contraptions.

They looked a most dangerous form of transport; there were two seats in the front over the two front wheels, and the driver sat at some distance back from the business end, in relative security.

I tried to find out about the banking arrangements at Makasar, but there was no Amex branch there. The local bank was not keen on the business of taking cheques, and the arrangement was that we should send a courier back to Jakarta each time we wanted some money, and he would deposit it in the local bank in Makasar, where we could draw on it as required, but with extended periods of notice and some other odd restrictions. It appeared that they had had some dealings with unscrupulous persons in the past.

The customs arrangements were somewhat obscure too. When our ship entered Indonesian waters the captain had to clear customs at Jakarta. He deposited the documents and sailed on to Makasar, about four day's steaming time away.

Sometimes the documents had arrived before him, sometimes he had to wait, and then the ship was inspected, the documents checked, the duties assessed, and a bill presented. Our man in Makasar had to collect this bill, argue if he thought there were any glaring errors, and then signal the site by radio so that a cheque could be prepared and sent off to Makasar.

If sufficient funds were not available in Makasar, someone had to go back to Jakarta, collect the cash, and deposit it in our account in the town.

Only when the cash was in the hands of the customs was the ship permitted to sail on to Malili – another two days' steaming – with the whole process to be repeated on the trip back to Singapore. It began to look as if the logistics and financial problems were likely to be greater than the engineering problems associated with this operation.

In London we had discussed the possibility of setting up an office in Jakarta, but the general feeling had been to leave all the negotiations between the client and the government in the hands of the client's staff in Jakarta. I believe that this was the best arrangement, and they were always very friendly and most helpful.

Next day I was off at the crack of dawn to the airfield, collected a lot of other baggage, parcels, mail, spare parts for machinery, etc., and was soon in the air on a Viscount of the local airline. Most of the time we flew over the sea, along the coast of Java, to Surabaya, where we refuelled, and then it was on again to Makasar, where we landed in a downpour.

CHAPTER 139
Makasar

On arrival at the airport buildings I was met by our man in Makasar, and together we sorted out all the stores and baggage into his car and set off to the town. The general informality of the airfield was a great comfort to me after all the regulations and formalities of so-called international airports, and the drive into town through paddy fields and small homesteads, with chickens in the fields and children playing, was very reassuring. The country had not long emerged from a period of civil strife and political disorder, and most people seemed to be happy to find that they had not been shot, and that the chances of their being shot had reduced to reasonable odds.

The town of Makasar itself was a curious relic of the Dutch colonial past. There was a splendid carriage drive along the seafront to the south of the main port area, with substantial old houses within walled gardens; nearer the town there were many tenement-type buildings with washing hanging on lines over the streets, and the industrial areas were much the same as anywhere else, with small workshops open to the street and people hammering away at all sorts of metalwork.

There was a fashionable shopping centre round the main square of the city area, but as Makasar was not on the tourist route the prices were fairly reasonable, and business seemed to be conducted in a civilised manner.

We drove down through the town to the office, which was in part of an old colonial house on the seafront, where I met the other staff, and was shown round the premises. Apart from the office there were three bedrooms available for up-country staff in transit, but these were all already taken up, and I was driven round to the nearby Queen's Hotel. This was another ex-colonial building, set in large grounds on the seafront, with the public rooms on the ground floor and the bedrooms on long galleries overhead.

My room was furnished in dark polished wood, and the slatted doors onto the front veranda were closed, making the place cool and dim. The bed was a massive structure, furnished with mosquito net, starched white linen sheets, four pillows at the headboard, and a fifth long pillow set out down the centre of the bed. From my earlier reading I guessed that this was what was called a Dutch wife, and was designed to absorb some of the sweat usually lost during the night.

The bathroom (?) adjoining was tiled throughout, with a wash handbasin with brightly polished brass taps, a conventional throne, and a sort of corner piece, where the tiles had been set out across a corner and the space behind was filled with cold water. There was also a sort of saucepan hung up on a hook nearby, and the theory was that you scooped the water out of the corner piece, and sluiced it over yourself with the saucepan. I learned later that, for a small extra charge, one of the houseboys would do the sluicing for you. Having tried out all these novel gadgets I went down for my dinner, into a large room set with a couple of dozen tables.

First I was invited into the bar, where I was handed the menu – a four-page printed affair, all in Indonesian – and sat down in a corner to make my choice. So far I had not been able to find anyone who spoke English, and so it was out with the phrase book to see what was what. By a brief flash of my Indonesian I was able to order a beer, and then sat trying to make sense out of the menu.

There were about three words on the menu that I could find in my phrase book, and they seemed most unlikely to be relevant, so – what the hell? – I selected a number of items on a random basis, and gave my order to the waiter by pointing them out with my finger. I must say he looked a bit surprised, but he went off, and some time later beckoned me to my table.

A lot of Indonesians, especially the more elderly ones, could speak Dutch, but they did not like to admit it and generally declined to speak it. The head waiter addressed me as 'meinheer', but when I replied in my best page-five Indonesian he was delighted, and broke into a torrent of Indonesian – which went straight over my head.

The waiter then appeared with a sort of barrow containing about a sack of rice in a heap, and an assortment of vegetables, crabs, shrimps, lumps of meat, bits of chicken, and strange-looking pieces of fish, together with many small bowls of spices and sauces and a collection of exotic fruit, together with a bottle of wine, compliments of the head waiter. I now understood why the waiter had seemed a bit surprised, as there was enough food there to feed a troop of dragoons. I struggled through as much of it as I could, and then took a short walk along the seafront to try to aid with the digestion.

Out in the bay, the sea was alight with the oil lamps of the shrimp fishermen.

Scattered all over were tall scaffold-like structures with a small hut on top. Each little cabin had an arrangement like outrigger poles extending outwards from the central platform, and a series of bright lights shining down into the water. Underneath they had spread a large net, and from time to time they winched the net up to the underside of the platform, unloaded the shrimps, and set it back again.

Other lights were from small boats fishing octopus with harpoons, and long poles fitted with nets, rather after the fashion of butterfly nets, and still other small boats were trawling with long lines. I walked about a mile along the promenade in the soft, warm darkness, exchanging greetings with the few people I met, and returned to the hotel feeling very contented and pleased to be in such a pleasant place.

In the morning it was back to the office, meet the customs, have my work permit endorsed by the local officials, and out to the airfield to meet our incoming flight. We had a small corner of the airfield to ourselves, with a shed for storing spares and tools, etc., and also a radio where we could hear our aircraft coming into the airfield radio zone. After calling the tower he advised us of his needs for refuelling, his passengers, and any special orders that he had been given from individual members of the staff in the field.

CHAPTER 140
First Days in Malili

We met the aircraft on the tarmac in front of our little shed, and I was introduced to the pilot and his passengers, who were going on to Singapore for a few days' leave. The pilot went over to file his flight plan and get the weather for our journey back while we saw to the refuelling and stowed my luggage and stuff in the rear. All very informal; we shook hands with the Makasar staff, climbed aboard, fired up the engine, and took off after a quick word with the tower.

The airfield in Makasar is on a long flat plain cultivated with paddy fields, shining green with rice, and the reflections of the sky in the water were sparkling in the sunlight. As we went north the ground changed to a more mountainous nature, with dense forests clinging to the sides of steep gorges. We swung away to the east, and crossed some steep limestone features, with thin grass on the tops and almost vertical walls falling into thickly wooded narrow valleys, and then came out onto the shore of the Gulf of Bone.

Here there was a fair bit of cultivation in progress, with small farms grazing sheep and goats, and a few tracks leading down into the little town of Palopo, where there was a small harbour and the seat of the provincial governor.

We continued on over the sea until we could see the little town of Malili, at the top of the Gulf of Bone, and our field camp, with its bright, corrugated aluminium roofs, sitting on the small hill behind the town.

Next thing we were landing on our own little airstrip, and meeting all the people who had turned out to greet me. After the introductions I announced that there would be a small celebration in the club that evening, and I was taken down to my bungalow, where the previous incumbent had lived.

This was a steel-framed structure, similar to those we had used in Iran, with a large lounge, two bedrooms, plus a spare room for guests, and kitchen and bathroom. The previous occupants had added a screened veranda and a small pool, about 12 feet square by four feet deep.

The houseboy and garden man came with the territory, and both appeared to be perfectly satisfactory; everything seemed to be in its place, and in good order, and both spoke about ten words of English.

The camp was on the top of a small hill, about 200 feet above the village and about half a mile away, and comprised a dozen or so houses, a bachelors' mess and quarters, a small school, a medical centre, the club with its tennis courts and small garden, and the main offices and administration buildings, which were set out in a square at the top of the hill.

There was a large modern laboratory where samples were analysed, a storehouse for samples, a general workshop, an aircraft workshop, radio room, stores buildings for equipment and materials, a shop, a small power station, water supply system, sewage system and the rudimentary facilities for the helicopter landing area, and the airstrip for the short take-off and landing aircraft.

Moored in the village at our little quay were a 50-foot launch, two other smaller launches and our supply barge, the *Bintang Selatin*. This craft was made up from a set of Bailey bridge pontoons, and had three large outboard motors mounted at one end.

I was taken for a brief tour of inspection in the afternoon, during which I met the village headman and the local chief of police, and friendly relations were established. In the evening we had a sociable knees-up in the club, and I went to bed – exhausted.

In the morning I had arranged to see all the local staff in my office, and was very glad to find that most of them spoke good English and seemed very keen to keep up the good work already accomplished in the area. During my visit to the workshops my Scottish sense of conservation was delighted to see people employed in rewinding electric motors and dynamos, and even more so when I found a couple of men extracting the nails from old packing cases and bashing them straight again.

The problems were that anything ordered on site had to come from afar. There were practically no stores facilities in Makasar, very few in Jakarta, and some in Singapore; all proprietary spares had to come from the UK or the USA, and this could take at least two months before delivery. Our site workshops were able to make most things; we had some first-class fitters and a turner who had come from the railway workshops in Makasar, and he was a real wizard. Absolutely nothing was thrown away, and there was a small storehouse full of obscure pieces of metal recovered from earlier vehicles and equipment.

The helicopter workshop was manned by two fully qualified aircraft mechanics and a radio mechanic, and they had all the latest tools and diagnostic equipment, so we had an excellent record of aircraft hours serviceable.

The fixed-wing pilot was also qualified to undertake some servicing of his plane, and was permitted to lift out and replace the engine at the designated service intervals.

I had a long meeting with the chief helicopter pilot, and tried to understand his problems on endurance, weather limitations, radio interference and black spots, and crew duties. We had five aircraft on station, and a total strength of eight pilots, of whom two were generally on leave away from site, and they worked a rota to try to keep one pilot with hours available to run any emergency services we might be called on to operate.

The shipping situation was quite complicated. The *Bintang Selatin* was used mainly as a ferry between our steamer, which had to anchor out in the bay, and the quay in the town, but it was also used to carry stores and machinery, fuel and soil samples back and forth between the various sites and the base, and sometimes to ferry oil in drums from the other side of the gulf to Malili.

When such a need arose and the *Bintang* was not available we sent a message by radio to the oil company office in Makasar, who passed the word to their man in Palopo, and next thing we knew was the arrival of a fleet of large outrigger canoes, each about 50 feet long, with their crews and families, livestock and household pets aboard, and a couple of dozen drums of oil secured to the outrigger poles. The arrival of these craft was the occasion for much singing and dancing in the village, and a great increase in local trade, goods being taken in and shipped out as if there was no tomorrow.

The two smaller launches were normally away in support of coastal operations, carrying men and equipment from one site to another, leapfrogging over the intermediate sites in passing, and we saw relatively little of them. The other vessel, the pride of the fleet, was a twin-engined motor yacht, and she was mainly used for carrying important guests across the gulf, or for boosting the support runs to some of the nearer outstations.

She was also called into use most Sundays, when we all went off to an island in the bay, swam in the warm sea, and lunched in a thatched house built on stilts.

Some time after my arrival the captain of the launch went off on leave, and his replacement was ordered to Palopo, to meet one of the ministers and bring him to site. I asked him to come into the office, and gave him his instructions concerning the minister and all that, and then asked him if he had been into Palopo before.

When he said that he had never been there I advised him of some of the navigation marks he would find, and where he should moor up, all done through an interpreter.

When I was sure that he understood he went out, and I heard him ask the clerk, "How does he know all about Palopo? Has he been there before?"

The clerk replied, "No, captain, he has not been there before, but he knows everything" – and off he went.

Of course, all this information was taken from the Admiralty pilot for the area, but he was not to know that. I had the clerk translate out the sailing directions and gave the captain a copy, and he was as pleased as Punch.

When he returned he came up to see me, and said, "Indeed, tuan, it was just as you had written out for me."

Another satisfied customer.

We had a bit of a bump on the *Bintang* when one of the engines caught on a rock in the entrance to the harbour. She managed to come on up the river all right, but when we lifted the engine off the steel casting that was the main cover for the drive to the propeller had been damaged, and no immediate spares were available.

This was potentially quite serious, as the main means of directional control was by turning the engine drives; with three engines it was simple, with two quite satisfactory, but with only one the craft became rather unmanoeuvrable, and was a bit of a hazard all round.

Our first call for a spare part was answered by a delivery date some two months away, and this was not satisfactory. A search of all our scrap heaps was also unhelpful, but I happened to be in Makasar during the week and found a section of cast-iron lamp-post, and an antique industrial gas bottle of about the correct diameter, and had them sent on back to site by return. When I arrived back the section of bottle was already turned down and fitted to the engine, and the bit of lamp-post was in the lathe being prepared as a standby spare.

The commissary supply problems on site were such that it would have been very difficult to run a shop where cash changed hands. The supplies were irregular and the costs dependent on the vagaries of the local markets, the weather, and the available space in the store.

The arrangement that had been developed was that all food was provided free to individual customers, and could be collected at normal shop hours. Drinks and tobacco were charged at duty-free rates to people living in the permanent camp, but outstation staff were provided with free beer and cigarettes as part of the 'hard lying' allowance (a naval expression derived from the rough life experienced by early destroyer crews). All refrigerated food came monthly on the ship from Singapore, in special containers, and some other food, such as fresh fish, vegetables, chickens, spices, etc., was bought locally or in Makasar, and taken into the camp stores for free issue on demand.

During periods of bad weather, when flying to Makasar was curtailed, everyone went on short rations as far as local produce was concerned, and from time to time there were problems with the refrigeration equipment and some food was defrosted and refrozen more than once. Despite all the normal hullabaloo about this practice, hunger is a splendid sauce, and no one ever came to any harm by eating such food.

For entertainment we tried to have a film show every week, and the client's manager had an early video machine, with tapes of the last three season's ice hockey matches in Canada. Most households had radios and gramophones, and there was always plenty of music.

CHAPTER 141
Family Reunion

After I had been on site for about a month news came through that the documentation for Jean and Susan had all been completed, and they were expected in Singapore during the following week. I arranged to travel to Singapore so that I would be there to meet them when they arrived, and set off for Makasar by our little aircraft.

We had been delayed by bad weather and I missed the connecting flight to Jakarta, so I stayed over in the office guest house in Makasar, but ate in the old Queen's Hotel, in company with one of the Indonesian office staff. He was most amused when I told him of my earlier meal in that establishment.

Next day it was on again to Jakarta, catching the early morning flight and connecting with an international flight into Singapore. I had booked into the Ladyhill Hotel, which had been built as a rest and recuperation establishment for troops from Korea, and it was a pleasant place to stay, with secluded gardens and a good swimming pool.

Jean and Susan arrived on time the following day, and we went straight back to the hotel for them to recover from the jet lag. In the morning the following day I went into the office, and found out that there would be a further delay of about a week in processing their papers, so we lived in comparative luxury for that time, me beavering away in the office and the family basking in the sunshine around the pool – at least for the first day, when Susan caught a bad case of sunburn, and had to remain indoors for most of the waiting period.

Eventually the papers were cleared and we set off again for Jakarta in a Boeing 707 of Qantas, and after clearing customs and all that set out for the Hotel Indonesia. We had practically the whole day to spend there, as there was no suitable flight out to Makasar that day, and we sat around the pool in the shade, selecting the dishes we would choose for our evening meal.

Off early in the morning again, landing in Makasar in time to fly straight on by our little aircraft directly to Malili, where we arrived safely in the afternoon, to a warm welcome by the whole camp. Jean was quite favourably impressed by the house and garden, although some comments were passed on the colour of the paintwork on the walls, and our first home-cooked meal together for some six weeks or so went down very well.

Next day Susan was enrolled in the school, which had about 20 pupils – English, Danish, Canadian, American, Indian, Indonesian and Dutch – staffed by one Irish teacher, who ran to an Australian curriculum. Jean met up again with the other ladies, and a routine was established for tea parties, 'sewing bees' (a meeting of ladies who brought along needles and threads, and small patches of material, to divert their attentions from the subjects of conversation), bridge afternoons, and other good works.

Whilst in Singapore I had bought a couple of tennis rackets, and we tried to play most evenings at the club.

We acquired the local habit of removing our shoes when visiting, mainly because of the mud that built up on the soles as soon as you stepped off the concrete paths, and it was always a clue as to where people were assembling when a collection of muddy shoes was observed outside a house.

For such visits most people had plastic shoes or sandals, and carried their more formal footwear in a bag.

As all the travelling shoes were very muddy, and generally of the same size (being the only size available), there was no guarantee that you would be able to find your own shoes on coming out. The same rules applied for umbrellas and plastic raincoats, each house having several spares of each in case of eventualities.

When they had settled in we went up to Soroaka for a weekend visit, and enjoyed the facilities of the guest house and a splendid mess building known as the 'Soroaka Hilton'. This was a wooden-floored structure with a thatched roof, set up about five feet over the water, and contained a kitchen and pantry, a lounge, and bar and dining areas. After duties it was the centre for swimming in the lake and other social occasions. I had borrowed a film from the Caterpillar Tractor Company, giving advice on how to avoid accidents at work with the heavy machines, and full of gory details about people being run over etc., and we arranged to show this at Soroaka during our visit.

I had to deliver a short address, in Indonesian, pointing out the error of some people's ways, but when the film was run each disaster was greeted with gales of laughter.

Maybe I delivered the wrong message.

The visit was regarded as a roaring success, and we promised to come back again soon.

And so back to Malili.

CHAPTER 142
The Routine

We were engaged in a comprehensive search for nickel on the client's territory.

The island of Sulawesi had been divided up into concessions, and our concession covered almost half the island, with the government controlling almost all of the remainder.

The Dutch had been conducting exploratory investigations during their occupation, and had found evidence of considerable deposits, but had not carried out complete studies either of quality or quantity on the areas they had examined, and there were many other areas where it looked as if further deposits might be located.

What we were finding was alluvial nickel, and it appeared as a green clay-like band underlying the surface, and normally just on top of the rock. The depths of the seams, which varied from several feet to a few inches, lay at depths of up to 50 feet below the surface.

Where nickel was suspected we had to drill trial holes, and take continuous samples of the material as it was excavated, and for this we had a number of specialised drilling machines, mounted on large tractors. We also had slightly smaller machines, which were fitted to smaller bulldozer-type tractors.

The principle of the sampling was to drive a double-walled tube into the ground by a small compressed-air-operated pile hammer, and to blow the displaced material up the centre of the tube by a blast of air down between the walls. The displaced material was collected in trays, and kept for sampling and analysis back at the main camp at Malili.

To keep these machines at work surveys had to be completed in advance, borehole locations set out and marked, access routes planned and prepared, food, fuel and sample bags had to be carried out and in, drilling records maintained, and the household needs of the staff and crews provided for.

There was one major nickel deposit about 40 miles away from Malili, where we had established another main camp, with a few bungalows for families and some bachelor quarters, and the 'Soroaka Hilton', which was the main centre for social activities, with film shows, frequent birthday celebrations, coming and going parties and entertainment for visiting firemen.

The lake was an enormous stretch of fresh water, at an altitude of about 3,000 feet. The swimming was excellent, and there were plenty of local fish.

The elevation of this camp made it a very attractive place to visit at weekends, to escape the heat and humidity of Malili, and, generally, water sports were undertaken by all and sundry.

We had a New Zealand engineer who was trying to make a large kite, in which he hoped to fly when towed behind one of our assault boats. These were aluminium vessels of about 20 feet length, with two large outboard engines, and could make about 40 knots on a good day.

His first attempt was to have this large kite fixed across his shoulders, the theory being that he would get up to flying speed crouched down on his water skis, and then lift off when he stood upright. The crouching down bit worked all right, but when he stood up the resistance was such that he nearly pulled his arms out. Back to the drawing board.

Next time he fastened his tow rope onto his chest. This was not a good idea.

When he lifted off he discovered that he had no effective means of control, the thing dipped to one side and dived into the water, folding up on impact, and – this time – causing grievous pain to the shoulders.

Try again.

Experiments continued, this time with the flying part directly attached to the ski board – a more promising line of attack, as he was able to bale out just before losing control, and suffered much less damage to his person in the process. The critical period when the board came out of the water, and all the forces altered, was always the stumbling block, however, and he went off on leave before being able to perfect his technique.

Not far from the camp were some submerged caves, and the swimming there was quite beautiful. The sun shone in under the water at the entrance, and the whole area was bathed in moving green light, picking out the stalactites dangling from the roof and the wavy shapes of the stalagmites rising up from the floor. There were some brightly coloured small fish always swimming in these caves, and we later learned that there was a steady market for them if they could be brought to Makasar or Singapore.

A slalom course had been set out just off the 'Hilton', and timed runs were conducted, on two skis, one ski, and on a flat piece of plywood about four foot square, with the boat pulling from the centre of one edge.

This was the most exciting event, as when the boat turned at full power you swung away out, travelling on the smallest patch of one corner of the board, probably at about 50 mph, before coming off. It was also quite painful.

Apart from the social events, working on the drill rigs was pretty tiresome. The weather was such that it rained almost every day for a couple of hours, so that everywhere was muddy and sticky, the heat was pretty intense, the flies awful, and the mud got everywhere.

The other major site was at Sua Sua, about halfway down the gulf, where we had five Europeans and about 100 men in total, drilling a pattern up the side of a river.

The camp there was a lot less opulent than at Soroaka; we had a mess hut, some locally built bungalows, thatch roofs, thatch half walls, uneven floors raised on short poles, and a storehouse. The men lived in similar huts about half a mile away.

This camp was right on the beach, on a narrow coastal plain about 50 yards wide, with the reef running off for about half a mile before the water deepened in the gulf. The site was supplied by helicopter as far as changing crews and mail, but all other supplies were either brought in by one of the launches, the *Bintang*, or by local purchase from the nearby villages.

There was no money required in these places; everything was done by barter.

Our cornflakes were supplied in large thin-walled boxes each about a metre cube.

These boxes were pretty well insect-proof, and, in good condition, each empty box was worth a goat. The normal rate of exchange for empty beer bottles was one chicken for two bottles. At times of stress, when the helicopter was unable to fly because of bad weather, this rate went up to five bottles per head. The enterprising driller loaded all his empties into one of the assault boats and took his custom to the next village down the coast, where a more reasonable rate prevailed, and came back with the boat full of live chickens. Build a hen house, quick.

I was visiting Sua Sua one day, and, after a tiresome climb up and down the river valley inspecting the roads and the machinery, came back to the camp, had a quick beer, a swim in the sea at my doorstep and a shower in warm fresh water, changed, and sat down to await dinner in the mess. It was a beautiful evening; we had a few drinks and turned in quite early.

Some time later I was awoken by a loud noise; the hut was shaking, the floor jumping up and down, the bed rattling about, and things were falling off the walls.

This was quite a serious earthquake. We all rushed out to the mess, and discussed the situation. First we had to calm down, and have a beer.

If the centre of the earthquake was behind us on the island, or on the other side of the island, we were probably safe enough where we were. This was a great comfort. From time to time the earth shook again, rattling the bottles in the bar, but not so violently as in the first shock.

If the centre of the disturbance was on our side of the island, out to sea, we could reasonably expect a tidal wave. Have another beer.

Various actions were proposed and discussed. It was low tide, and the reef was half a mile wide in front of the camp. The level of the ground under the mess was probably about 15 feet above the present sea level out on the reef. There was a flat area about 50 yards across between us and the first small hill behind us, and there were well-marked tracks up the hill for a couple of hundred feet at least.

Have another beer.

The general consensus was that we should remain in the mess, ready to rush up the hill at the first sound of heavy surf on the reef. People began to shift about their positions to be near the door, just in case. We tried to raise Malili on the wireless, but had no response; perhaps they had been caught up in the earthquake too. For the next few hours we sat on the veranda of the mess, engaged in desultory conversation, but with one ear cocked for the sound of roaring surf on the reef.

In the morning we were still there, but when we finally raised Malili on the wireless they said they were all right, and that the epicentre had been on the other side of the island. A tidal wave *had* come in, and there were reports of people being washed out to sea, and numerous houses destroyed. We offered to send in one of the helicopters, and the *Bintang*, as soon as she could be loaded up with supplies for the survivors, but the local authority said they had the matter under control and were able to cope, thanks very much.

I had words with the local foreman later, and he assured me that he knew that the epicentre was on the other side of the island, and after the first shake he and all his men had gone back to bed, quite unconcerned. I had another look at the hillside behind us, and was very glad we had not had to try to run up there in the dark.

Better have yet another beer.

Our concession boundary marched with that of the government nickel company, and we were both drilling pretty close to one another. The boundaries were marked out on the ground by small concrete plinths, and the first one, along the shore, was noted as being near the mouth of a small river.

We were a bit suspicious that someone had moved the boundary markers about 100 feet into what was our ground, and they were drilling away there, with obvious signs of success. We ran another check survey, and were pretty sure they had encroached on our ground, but it was their island, after all, and all I could do was report it to the client.

The results of the survey were sent off to Jakarta, and we heard no more.

CHAPTER 143
The Island Retreat

Some weekends, when the weather was suitable and the launch available, a picnic was arranged to a little island in the gulf, about ten miles out from the harbour at Malili.

A small house had been built on the island, where food could be served under the shade, and, normally, the launch with its crew and passengers would leave the harbour at about ten o'clock in the morning. Food was provided from the stores, and cooks and stewards were delighted to assist from the various households.

There was a small coral reef extending for about 100 feet practically all round, with one narrow passage just wide enough to allow the launch to come in and anchor in the shallows. As soon and the anchor was dropped, most passengers dived in and swam ashore, while the others supervised the landing of the supplies and then came ashore in the dinghy.

One day as I was swimming ashore I noticed a fish swimming ahead of me, quite slowly, and made a grab at it. It was a small shark, about 15 inches long, and must have been lost, or tired, as it made no serious efforts to evade me.

Elated by this success, I held it up out of the water, and shouted out, "Look, I've just caught a baby shark."

All the other swimmers reacted in unison, obviously believing, as indeed I did myself on second thoughts, that, where baby was, ma would not be far behind, and there was a concerted rush to get ashore and up the beach, or back onto the launch.

By this time I had thrown the fish onto the beach, and we made a cautious inspection.

It was indeed a small shark, complete with large dorsal fin and lots of teeth, and it was decided that it should be taken back and preserved as a trophy, to hang in the club, with a photograph entitled 'Man Catching Shark'.

This rather put the mockers on swimming about the reef until after lunch, but some intrepid persons went out looking for the rest of the brood.

The coral was most spectacular, with many different colours growing at different depths, and always a vast array of brightly coloured fish swimming nearby. There was an unofficial competition for finding the largest clam, and also the five-fingered conch shell, but these were normally found on the outer edge of the reef, where the water was suddenly several hundred fathoms deep, and quite scary, as there were frequently some large fish cruising about there.

Perhaps the most alarming were the barracuda, which appeared in large packs and took some interest in occasional swimmers, although I never heard of an attack.

They looked quite menacing, with their large eyes, underslung jaws and a great array of teeth, and they could travel very rapidly.

Another somewhat frightening creature was the sea snake. These particular fish were about ten or 12 feet long, yellow and black banded, and well camouflaged against the seaweed and coral background. They were generally fairly timid, but if come upon by surprise would rear up and swim off to one side. This caused violent neck-twisting on the part of the swimmer, as it was often difficult to keep both ends of the beast in view at the same time. This situation was usually compounded, as the snakes often kept company and several would rear up at one time.

As well as sea snakes, we had a bit of excitement one day when a long green snake was disturbed alongside our little house. It shot up a nearby palm tree, and stopped about 50 feet up, in amongst the leafy fronds at the top. We all turned out to see this piece of levitation, and suddenly it fell, or jumped off, and sort of flew down onto the roof of the house.

Great panic all round. Wives with children rushing out of the house; stewards quickly covering up the exposed food, with one eye on the roof; others trying to find long sticks to attempt to dislodge the snake from the roof; still others keeping well clear and going off for a swim in the shallows.

At the end of all this the snake was pronounced harmless and peace was restored, although no one was very keen to go in for lunch for some time.

Sometimes we would be visited by some of the local fishermen, and we often persuaded them to take us for a sail, back to the harbour mouth on our way home in the evening. Their canoes were dug-out tree trunks, with some splash boarding along the top, and another log fixed as an outrigger. They had a tall mast and a curious triangular sail, usually made out of old flour sacks, and were steered by the foot, working on a cranked tiller bar (both hands were too busy on other attachments).

Off the wind they had a great turn of speed, and any spare hands aboard were kept busy bailing, what with spray and the water coming in over and through the splashboards.

An additional hazard aboard was the number of large crayfish clattering and swimming about, and possibly the odd lobster. We made a point of keeping good relations with the fishermen, and, if we saw they were becalmed, often towed them into the harbour on our way home.

One evening, on our way back, one of our engines stopped, and when this was investigated we found that the silencer had broken, and water was coming in quite rapidly through the water jacket. A chain gang was set up with all the buckets on board. With all the spare passengers standing at the bow on one side, we managed to get the water level down enough to be able to bung up the exhaust pipe so that much less water was coming in. Contingency plans were laid to try to beach the boat should the other engine pack in, but we managed to chug back safely, and had the repairs completed that evening.

One of our major problems was a lack of equipment; we had bulldozers, tractors, drills, Land-Rovers, a couple of lorries, and our fleet of shipping, but no crane or other lifting device. What we had to do was to find a convenient tree, rig up a block and tackle, and use one of the tractors as a hauling engine. While we tried to improvise in a responsible manner, I feel that the Board of Trade or the factory inspector might have had some unkind comments to make, and would have banned a number of the tricks we had to apply.

How could they have issued a Safety Certificate for a tree?

CHAPTER 144
Other Distractions

Every few weeks we were advised of an official visit. These visitors ranged from the ambassadors and their parties from Canada, Australia or Great Britain, to ministers of the Indonesian government and their guests, to senior officials from the Department of Mines, the Labour Directorate or the Civil Aviation Authority, and – of course – to senior engineers and directors of all the participating companies.

We developed a scale of events appropriate to each potential visitor, and tried to work out a programme that we thought might interest them.

The first event was a welcome committee, where the visitors were greeted and introduced to the senior staff. They were then led away to their appointed accommodation, in one or other of the spare rooms in some of the houses, or the bachelor accommodation.

Next, a quick trip round the camp, visiting the offices, laboratory, hospital, school, workshops, stores and the helicopter office. Then a pause, in preparation for a big feast in the club, organised by the ladies and with some singing and dancing, and off to bed.

Next day, a quick flight to visit the main drilling site at Soroaka, lunch in the 'Hilton', then back to base, quick farewells, and flying off back to Makasar by helicopter.

Sometimes, due to bad weather, we were unable to unload our guests on the appointed day, and alternative sources of entertainment had to be found – such as a visit to the school in session, or a trip out to the island – to keep them out of our hair for the day, while the flying programme was reorganised.

The weather was a big factor in making the travel arrangements for these visits, as it was normally fine for flying out from the site in the early mornings, and sometimes late in the evening, but around midday there were often thunderclouds in the area, and these, even if they did not result in actual rainfall, caused violent interference to the radio communications.

Furthermore, we had flying programmes necessary for the operations, shifting men and materials between sites, collecting passengers going on and returning from leave, and keeping the pilots within the regulation flying hours limitations, and maintaining the routine servicing and inspections.

CHAPTER 145
The Weather

The island of Sulawesi has some very high mountains, which tend to modify the weather, causing rain to fall practically every day somewhere on the island. I think I can recall a period of about 15 days when we failed to record any precipitation, and this was most unusual.

Locally in Malili we could expect heavy showers most afternoons, and this restricted the flying arrangements, making for early starts and the frequent requirement for an aircraft to overnight in Makasar, to allow him to return the following day before the weather broke. Although the showers were heavy they only lasted for a couple of hours, laying the dust, keeping the forests green, and maintaining a plentiful supply of water in the rivers.

Quite frequently we had more severe storms, with violent thunder and lightning, and torrential rain accompanied by fierce gusts of wind, and such events usually continued into the nights.

We had a storm routine whereby a check was made on all the items of equipment likely to be blown away, such as aircraft, large doors on the stores and workshops, tin roofs on the houses, the fence round the tennis court, etc., and any loose items were secured by heavy ropes and wedges.

During our stay in Malili we had three exceptionally violent such storms. The first one occurred in the early afternoon shortly after my arrival. During the morning clouds began to build up all round, and all our flying was cancelled. The inter-site radio communications were very disturbed, and the humidity got worse and worse.

By noon the cloud cover was complete, and the wind dropped.

I had gone into the house for lunch, when suddenly there was a great blast of wind, causing all the trees and bushes in the garden to lean over, and the whole house creaked and groaned. A few seconds later there was a brilliant flash of lightning, followed immediately by a violent clap of thunder. I rushed round the house, closing all the windows and trying to secure the canvas screens along the veranda – and the rains came.

The noise on the roof was incredible; instant torrents of water were cascading off the roof, forming little watercourses in the garden, and the rain was bouncing off the road at the rear of the house about 18 inches high, making a sort of zone of invisibility and filling all the drains with red muddy water.

The thunder and lightning were almost continuous for about 20 minutes and then began to move away, with the noise coming progressively later after each flash, although the rain kept up its violent row on the tin roof. Half an hour later the rain stopped, the clouds broke up, and the sun came out, although the thunder could still be heard in the distance.

What a mess! A number of trees had been struck by lightning, and their broken branches were trailing in the road; some culverts had proved to be inadequate, and the water had washed out sections of road; all the concrete paths around the houses and offices were inches deep in red mud; and where windows had been left open a lot of rain had blown inside, causing some damage to furniture, clothing and equipment.

The generator had stopped working, probably because of a blown fuse, when the power cables had clashed together in the wind.

It was imperative that power was restored as soon as possible, as all our frozen food in the store was at risk, and I went across to the station, where I found the plant manager deep in diagnostic testing to try to find out where the trouble had been caused. He obviously had the situation well in hand, so I went round the houses to see if everyone was all right, and assure them that the power would be restored as quickly as possible.

Apart from the mud no serious damage had been caused, and we soon had repairs to the roads in progress in what turned out to be a fine evening.

The second noteworthy storm happened after dark, shortly after Jean and Susan had arrived.

We had just finished dinner and were sitting listening to the wireless when the lightning began. The interference became much worse, and we began to hear the rumble of thunder in the distance. Outside the sky was overcast, with not much wind, but the atmosphere was very oppressive and humid.

"I think we might be going to have a bit of a storm," I announced. "Close all the doors and windows, fasten the screens, and fill up some saucepans and things with clean water."

While this was going on I looked out a couple of torches and some candles and matches, and checked our emergency tilley lamp, and then I said I would go and have a look around to see that the windows in the offices were shut and that the other buildings were secure. The plant manager came in to say that he would go and check the workshops and generator house, and we should meet again in the bachelors' mess and organise what we could from there.

I went out with my torch into the wind, and crossed over to the office area. By then the lightning was pretty continuous, with the thunder getting louder and nearer, but still no rain. Just as I was closing the last window, the rain started. No light drizzle to begin with, just a complete deluge, and the thunder almost instantaneous with the lightning flashes.

I dashed across to the mess, and was happy to hear that all was well, all the doors and windows about the site secured, and everyone present.

There was a slight lull in the downpour so I ran back to the house, taking care to avoid any trees, or even small bushes, and arrived back breathless and very wet.

The intensity increased again, and there was one searing flash, followed by an almighty bang, and all the lights went out. Looking out, we could see that all the electrics in the camp had gone out, but some of the houses had candles going already.

The rain was still pounding on the roof, the thunder and lightning banging and flashing away, the window screens flapping against the mesh, but we appeared to be secure, with not much water coming in, and the storm seemed to be travelling on, although the rain was still a complete downpour.

In one prolonged flash I could see that one of our electricity supply poles had been brought down, just at the back of our house, and the wires were trailing across the road. By the time I had put my overcoat on I could see that Joe, the plant manager, and a couple of our men from the mess were there already, and they had put some sort of barricade across the road to prevent anyone from coming onto the cables unaware.

Bits of the road had washed out again, and everywhere was covered by an inch-thick layer of mud, which built up on your shoes as you walked, making things very difficult. We went round to the generator house and checked that the set was shut down, and then made a round of all the houses to reassure the people that we had the situation in hand (?).

At least we knew about it.

The rain had eased a bit, it was now only pouring down, and Joe and his assistants were already trying to bypass our pole and restore power by another circuit. (There was no chance of putting the old pole back, as it was split down the middle, and broken off just above ground level.)

About an hour later, still in the pouring rain, the loop was completed and the lights came on again; the storm had passed, and apart from the mud and some repairs to the roads no other serious damage had been caused.

The third serious storm happened late one Saturday, when there had been some party or other in progress in the club. Much the same sequence of events: very humid, no wind, lightning in the distance, some thunder, getting nearer, and suddenly – FLASH, BANG. All the lights went out, and the rains came. Most people left at once, running up the hill to the houses or the mess, while Joe and his assistants, who had come down in their Land-Rover, set off to see what the problem was with the electricity.

We arrived back at the house as they were coming up the hill and actually saw the lightning strike the power pole at the rear of the house. For an instant the pole, its stays and the power lines all lit up; then there was darkness, and in the next flash we could see the pole topple over and hang in the cables from the adjacent poles, right across the road, and in the path of the vehicle. Fortunately they must have seen it too, as they stopped rather abruptly, before running into the wires. At least they could see what had happened in the headlights of the car, and by the time I had gone out some sort of barricade had been erected on the other side of the damage.

We then went round to the generator house, and made the necessary changes to the switches to allow the supply to be restored by an alternative circuit.

When you hear that lightning never strikes twice in the same place, don't believe it.

The next few days were spent considering the electric supply layout, to shift the pole opposite our house. I was not too keen on this proposal, as, next to the pole, our roof was the nearest object at a

496

similar elevation, so a compromise was reached, by putting in a pole for the cables and a dummy pole to fool the lightning.

This seemed to work, too, as we had no more bother with that circuit, although the lights always went out in a thunderstorm.

CHAPTER 146
A local Wedding

For some time there had been talk of romance in the air in the office, and this was brought to a head by the announcement of a wedding between one of our Indonesian staff and a young lady from another part of the island.

Once the announcement had been made and a date fixed, the real business began. A temporary house was built near the village, and invitations sent out to attend the festivities. The preparations were extensive, as this was to be an important social and political event; the young man was a member of an important local family, and his bride was similarly well connected.

When the house was almost completed a large pit was excavated nearby, in preparation for the roasting of a buffalo, and suitable firewood was collected and stored, together with an assortment of utensils and other smaller cooking stoves.

The celebrations were arranged in three separate functions, and began with the bride and groom taking station together on a large elevated platform in the temporary house, which had been richly furnished for the occasion. He was dressed in silk jacket and long baggy trousers, and wore a tall turban, studded with diamonds and other precious stones, and several gold rings on both hands. His bride was resplendent in a gold-coloured silk sari type of dress, and wore long gold earrings set with precious stones. On her arms were many gold bracelets, and she wore a number of gold necklaces, also studded with jewels. Her headdress was a sort of coronet, also of gold, and set with many sparkling stones.

The whole house was beautifully set with flowers, and the guests were welcomed and introduced by the local headman.

There were three separate functions organised, and all ran on similar lines.

The first group to gather was drawn from the local male relatives of the couple, and the various members of the site staff and their wives. We attended outside the house, in a long thatched-roof ante-room that had been built that morning, and took off our shoes before going into the house. We then moved slowly round in procession, and were announced by a sort of major-domo person, who appeared to know everyone, and as we entered the wedding chamber we were shown where our gifts should be laid, and then went forward to offer our congratulations and all that.

We then moved out into another thatched house, which was set out with tables and chairs, cutlery, crockery and glasses galore, and sat down.

As soon as everybody was seated the major-domo announced that the bride's father would say a few words; meanwhile, stewards had been dashing round filling all the glasses with whatever liquor they may have had in hand. After his speech there was a toast, and an immediate refill all round. Next the groom's father said a few words, and more drinking. By this time everyone was feeling very merry and bright, and the food was served together with a general refill. I must say I can thoroughly recommend pit roasted buffalo, and all the other delicacies that were being served, and we had a really splendid meal.

The client's project manager was then prevailed upon to say a few words, and this went down very well, followed by more refills, toasts all round, and general jollification. I enquired as to where the happy couple were during all this jamboree, and was told that they had to remain in their house until all the other guests had passed through. While we had been feasting, the ladies of the families had been doing their processing, and they were taken out to an adjacent dining house, and the various speakers then went over there and performed again.

Just thinking about it, I could see that there might be a bit of a problem in collecting the right shoes when all this was over, as they had been dumped more or less indiscriminately at the entrance, and by then most people had eaten and drunk very well; also, it was getting quite dark.

There were one or two further speeches and toasts, and then the party began to disperse. Most people seemed to find a suitable pair of shoes to go home in, and the place quietened down for a couple of hours.

Later in the evening there was a further session for neighbours, children and guests from the village, with music from a bamboo band and some gamelan players, which we heard from afar.

Altogether a most enjoyable day.

CHAPTER 147
An Encounter with a Buffalo

During the earlier troubles in Indonesia there had been a campaign to disrupt the communications systems on the island, and practically all the major bridges had been destroyed, with all the others suffering badly from lack of maintenance and repair.

We were conducting a detailed survey for a proposed port and harbour, a new town, airfield and all services on the opposite side of the river at Malili, and I was selecting the site for the main bridge over the river. The river was about 200 feet wide between banks, and the previous bridge had been destroyed, with one abutment washed out completely. I went across the river by a dug-out canoe and asked the owner to meet me about a couple of hours later by the site of the old bridge, then proceeded along the river bank, in fairly thick elephant grass, along a narrow track probably used by fishermen.

This track sort of petered out and the going became very difficult, so I decided to try to get up to the top of the embankment, about 20 feet above me. Armed only with my notebook and my rolled umbrella I forced my way upwards, and came out suddenly, at eye level, with some open ground. Standing about two feet away was a large buffalo. It must have heard the commotion I was making forcing my way through and come along to investigate, but we both took off, me down the bank again, and the buffalo I could hear galloping along the top of the bank. I am not sure which of us had the bigger fright.

After what I took to be a suitable period of time (until my heartbeat had returned to a more normal rate), I crept up to the top again, and looked very carefully left and right at ground level. No strange creatures were in sight, so I climbed out onto what had been a road but had long since fallen into disrepair.

This had obviously been the old road that connected up with the bridge, and it was on top of a sort of bund – a raised earthen embankment – along the river bank, about five feet above the surrounding countryside, but overgrown with trees and bushes and so invisible from the air.

I walked along for about a mile or so, to see if there might be an alternative site for the new bridge, but decided that whoever had located the previous bridge had probably selected the best site, so I walked back again, not without frequent turnings around to ensure that my inquisitive buffalo was not following me.

The old road had been washed out where old culverts had been overwhelmed, and it was quite hard going at times, descending into deep gullies full of rustling noises, with frogs and suchlike croaking away. Just the sort of place where the large lizards, such as Komodo dragons, or even crocodiles, are likely to live.

When I arrived back at the bridge my canoe man was waiting for me, and asked in a very polite manner if I had had any trouble getting up the other bank. He had obviously seen my descent, and was suitably satisfied when I told him I had had to scare off a large buffalo.

Most impressed he was, in fact; and me with only my umbrella.

CHAPTER 148
A Death in the Camp

One night I was called out by the doctor, who advised me that one of the staff wives had died a few minutes earlier. She had been sick for a couple of days before, but it had not seemed to be very serious, and the doc had been satisfied with her progress. She had taken a turn for the worse in the afternoon, however, and despite all his attention had died in the evening, from a massive heart attack.

I spoke to the senior fitter and asked him to organise the making of a coffin in the workshops, and to call me when it was ready, and then went back with the doctor to the house, where Joe, her husband, was sitting up with the body. We suggested that he should come over to my house while the remains were being properly laid out, and we could discuss arrangements for the funeral. Joe said he wanted her to be taken home and buried in England, and the doc said that this could be arranged, and he would see to it in the morning.

During our discussion the fitter came in and said he was ready with the coffin, and we all went down to Joe's house and assisted in laying her out. Because of the delays that might occur in shipping the body back to England, the fitter suggested that it would be necessary to make and seal a lead covering for the coffin, and he took off to undertake this sad task.

Meanwhile we sat up, the doctor, Joe and I, all night, trying to foresee all the problems associated with getting the body back as quickly as possible, and all the local formalities that had to be completed before anything else.

In the morning we paid a visit to the local government resident, who was most kind and helpful, spoke to the Lutheran minister in Malili, who was also very kind, and arranged to hold a service in his little church.

We also advised the office in Makasar that we would be bringing the body in by helicopter that afternoon, and they should arrange immediate transhipment back to the UK.

We held our remembrance service in the church in the village, and returned to see the body loaded into the helicopter and flown off to Makasar. Joe had elected to remain in camp until the final arrangements had been made, and to clear up all his belongings from the house, and a rota was set up to keep him company at all times until he departed.

We heard later by wireless from the office that our coffin was too big to go into the service aircraft between Makasar and Jakarta, and they had instructed a local undertaker to make a new coffin of suitable dimensions and proceed as quickly as possible.

A few days later the arrangements were complete, and Joe flew out with his late wife and was met in London by the personnel officer from the company, who had made all the arrangements as requested by Joe.

Altogether a very sad and disturbing story, but we all felt that our efforts had been appreciated.

CHAPTER 149
A short Holiday

Some of the staff had been on station for a considerable time, and it was agreed that they should take some local leave. A particular problem had become evident, however, and it was not possible to complete the work in time if one particular chap went off on leave.

It was agreed that his wife and children should go off, and Jean and Susan also volunteered to go along to keep them company, so Ralph and I were left, abandoned, while the families tootled off all round Indonesia. They flew by our helicopter service to Makasar, on to Jakarta by the local airline, and then visited Bandung, took the train to Yogyakarta, and on to Suva Bay, visiting most of the tourist sites on the way, and then back to Makasar by the local airline.

According to all accounts they had a marvellous time, and came back refreshed and full of beans.

While they were away we managed to cause immense problems in the houses.

We had a small pool in our veranda, which I decided to have drained, cleaned and repainted, and this was more or less successful. The other house had a similar pool, but it was used as a fish garden, and there were a number of exotic fish and some lily-type plants resident in this pool.

The pool-cleaning mania had caught on, and one Sunday we carefully caught all the fish, transferred them to another tank, scooped out the plants and drained the pool. We gave it a good scrubbing, and washed it out several times to make sure that there was no bleach or scouring powder still present, and refilled it with clean water. We returned the fish without too much trouble, but some of the plants looked a bit tired, so we had some similar specimens brought down from Soroaka and put them in as well.

A couple of days later there was a bit of a panic. Most of the fish had died, and the remainder were looking pretty far gone. What had we done? How were we going to explain this away when the travellers returned? We arranged a consultation with the doctor, and he came over to have a look.

"Looks like a lack of something, or too much of something else, in the water," was his considered opinion. A lot of help that was.

Some other expert suggested that the new plants we had introduced were taking all the oxygen out of the water, so, straight off, out they came, and the water was changed again. No more fish died, and perhaps the remainder looked a little bit fresher.

Replacement fish were ordered up from Soroaka, and they were brought down in various vessels and tipped into the tank.

They all survived, but now we had too many fish, and would still have to explain that away. What about a short breeding season? Perhaps some were kangaroo-type fish, which carried their full-grown offspring about in a pouch, or something?

When the travellers returned they were so full of themselves, and their travels and adventures – the night sleeper to Surabaya, and going down inside a volcano – that they took no notice of the clean pool, and additional fish.

While they were away we had a small problem with the customs in Makasar.

They had altered the rate of duty on some items of equipment on the ship, and we had insufficient funds in the bank in Makasar to cover the revised bill. I sent off a telex from Makasar to our bank in Jakarta to prepare a draft of £120,000 worth of rupiahs, which I would arrange to have collected and taken back.

We had an almost immediate reply to the effect that the draft was ready, but would we mind sending in the odd signature with such requests in future?

This seemed a very reasonable request, and as the weather and the aircraft bookings were convenient I went back to Jakarta myself and collected all this cash. The actual volume of all these notes was considerable, and I had to pop out and purchase a suitcase to hold it all.

I returned to Makasar the same evening, with this suitcase full of used notes on my lap, not daring for an instant to let go of the handle. When I got off the plane in Makasar we went straight to the office and lodged all the cash in the office safe; next morning, while I was in transit back to the site, the office manager took the cash round to the bank, and we were able to pay the customs, the ship was cleared, and there was a sporting chance that we would be eating again when she arrived back off Malili and discharged her cargo.

CHAPTER 150
The bulk Sample

As our exploration continued small samples were taken and evaluated, but the time came for a large sample of the material to be collected and sent off to the processing plant in Canada.

We were required to set up a number of fenced areas and collect specific quantities of ore from particular sites, to be ready for shipping out on a large ocean-going vessel. This required a certain degree of organisation, as the sites where the deposits were located were scattered about the island, there were practically no roads suitable for heavy transport, and we had no vehicles anyway.

A plan was developed whereby we would build the roads where necessary, buy a number of lorries and a loading machine and have them delivered to Malili, set up a loading platform at Malili and a second one at Sua Sua, buy or hire a large barge and a tug and convert it for carrying ore, and start to make stockpiles of ore at Malili as soon as possible.

News came in that a large iron ore mine in Malaysia was closing down, and there was a lot of suitable equipment being sold off there, so the plant manager and I set off back to Singapore, hotfoot, after this equipment.

There was still some political disaffection between Singapore and Jakarta, and the telephone was very trying. First, you had to book an international call. This meant going to Makasar in the first place, and waiting until a slot was available, to make the booking. Next you were called to say that your call would be in so many hours' time. When the call came through, after much repetition of numbers and exchanges, no sooner had you started to speak than you were asked: "Have you finished yet?" This was repeated at about ten-second intervals, until frustration got the better of you and you sent a telex. Sometimes, this was almost as bad.

Anyway, we arrived and stopped overnight in Singapore, and were off to Malaysia in the morning, flying in a Fokker Friendship. This machine, in my opinion, was the perfect example of a 'heavier than air' contraption. It took forever to get off the ground, and climbed very slowly out over the Jahore Strait. Coffee was served, in plastic cups, and just as I was handed my ration the plane took a sickening drop. I managed, somehow, to stand up with the cup in my hand, and didn't spill a drop, but there were a lot of people jumping about with hot coffee on their laps.

We landed at Trengannau, near the mine, and were met by one of the engineers and taken up to a guest house. The mine was a massive operation. The ore was in a large piece of hillside, and they had opened up a huge crater, about a mile across, with haul roads descending into the pit, and still a good deal of traffic about.

We examined the fleet of machines and vehicles set out for inspection, and asked to see particulars of some of the more likely-looking items.

We went back to our guest house, and had a look around the compound. This place had carried a staff of about 100, and there were many bungalows and staff houses scattered all round the area. The clubhouse was a very fine building, now running to seed, but the meals were excellent.

Next day I went off to see some barges and tugboats that had been used to ferry the ore out to ships lying in the roads off the little river, and made a selection of possible purchases for our operations.

While we had been in Malaysia, news of our proposed operations had spread to Japan, and the accountant, who was in charge during my absence, was invited to Tokyo to see some new dumper lorries at a very reasonable price, so he took off just as we landed back and spent a fabulous week in Japan.

A few days after our return we received the instructions to proceed, and a date was given for the loading of the ship off Malili, so we really had to get a move on.

We undertook a road improvement programme between Malili and Soroaka, about 50 miles through the mountains, with steep hillsides and rushing torrents.

We had no pipes for making culverts, and had to improvise by using oil drums and heavy timbers, widening what had been a footpath not long beforehand into a road suitable for ten-ton trucks.

The Japanese trucks arrived in Makasar within a few weeks, and we arranged for our barge to bring them round to Malili, where we had set up the loading ramp, and soon the material was coming down from the various up-country locations and being set into separate pens, ready for loading out on the ship.

When the ship arrived we went out to see the captain, and inspected the arrangements he had made for keeping the material in separate piles.

Of course, he had made no arrangements at all, and we then had to see about making individual pens for each location and having them checked out against the ship's stability curves.

When we were ready for the first consignment the barge was towed out and a gang of labourers was engaged to load the ore into skips, which were lifted by the ship's derrick, lowered into the hold, tipped, and then moved by wheelbarrow into their proper place. All a very slow and tedious business.

On our first day of loading it became obvious that we would not clear the barge until after dark, and to assist in a speedy return we decided to set lights at critical turning points in the approach channel. I had selected the general location for each light, and set out in an assault boat with a dozen tilley lamps, all full and ready to go.

The first one or two were not too bad; we had to select a place on a tree where the light could be seen, and not set the tree afire. We beached the boat, my man climbed up the tree and made a suitable hanger, I lit the lamp and hauled it up.

By the fourth position it was getting dark, and I had not reckoned with the insects.

My man climbed up the tree and made his loop, but when I lit the lamp all the insects in Sulawesi were attracted, and it was very difficult to keep them out of my hair and shirt as we were hoisting the lantern up into its position.

Soon it was too dark to see where you were stepping – perhaps on a crocodile? – and we had to light the lamp before going ashore, and sometimes a second lamp to show us the way back to the boat. I must confess that we reduced the number of leading lights by about half, but reckoned that the passage was still well lit, and we had no problems bringing the barge back that night.

The loading continued round the clock, with different people going out with the lights each night. All came back in some state of dismay, bitten by strange insects and half scared to death by the odd splash in the water. Was it a fish, or a bird, or a croc – or perhaps even a giant octopus – attracted by the light?

After about three weeks of this continuous performance the ship was finally loaded; we had a dreadful dinner party on board, thrown by the Greek captain, washed down with pints of retsina and lots of ouzo, signed him off, and departed back to town.

He was still there the next day; probably couldn't get the fire to light in his boiler, or perhaps couldn't even find the boiler, as he had been in a pretty good state when we left. He got steam up later in the day, however, and we flew over him in the helicopter to wave him off – bon voyage to Canada!

He cleared customs in Makasar, and sent us a very nice case of Greek wine.

CHAPTER 151
The pilot Survey

As the investigations were going very well, and lots of suitable nickel was being discovered, it was decided that a marine survey should be undertaken throughout the gulf to chart the channels for large ore-carriers, and to establish suitable mooring areas etc.

We were advised that an Australian company would be in our area shortly, with a survey vessel, and we should make contact and assist as necessary. We heard on the airport radio that they were coming our way, and kept a lookout for them from our helicopter traffic, and soon we were talking to them on our own wireless net.

They thought that they would be in our vicinity by the weekend, and we invited them all ashore for a tennis match, and other forms of relaxation.

During our flying over the gulf we frequently saw whales swimming, close to the surface, blowing away just like Moby Dick. The water in the gulf was very deep, but there were numerous coral heads that rose up out of the depths and were only just submerged, and we could see these very clearly from the air.

When our Australian visitors arrived we asked if they would like an aerial view of the scenery, and going looking for whales, and the captain came along on the first trip. He must have passed very close to a number of those coral heads without picking them up on his survey, and when we landed he sent off a message saying that he had better stay with us for another couple of weeks and survey them all before he ran into one on the way home. This was agreed, and we had the pleasure of their company a bit longer.

CHAPTER 152
The new Road

As the results of our exploration and evaluation of the deposits grew, it became necessary to consider the support systems that would be required to sustain the exploitation of the minerals. The basis of the study was to consider the development of a number of opencast mining sites around the Soroaka area, a new road, or perhaps railroad, to the sea near Malili, a new town, complete with airfield, harbour, and all services, and a power supply system, possibly based on the hydroelectric potential of the main river out of Lake Soroaka. Initial flying surveys had identified the possible locations for the town site, and some earlier work had been completed on the hydroelectric possibilities.

There was an existing footpath that followed the river for about 30 of the 50 miles between Malili and our present base at Soroaka, and we elected to try to see if this route could be made passable for vehicles, at least as a basic means of access for conducting a proper survey. Having explored most of the footpath by driving in as far as possible from each end, and landing in the middle from our helicopter, I was of the opinion that a suitable route could be found, and in a rash moment wagered that I would drive the length in my Land-Rover within three weeks' time. The bet was taken up, and a crate of champagne was the agreed prize.

My Land-Rover was prepared with the addition of a sort of roof-rack, which was loaded with long timber boards, various lengths of rope, blocks and tackle, hand winches, picks and shovels, some ground anchors, and the plan was prepared.

The intention was that some work was required at the Malili end, where the river had reached some more gently rolling landscapes, with fairly deep gullies, and a machine was put to work to cut its way along between survey marks we had established in advance. This machine was to push fill into the valleys and get itself across onto the easier stretches of hillside, while a second machine would follow on, making temporary culverts as it progressed and running supplies of fuel to the leading machine.

Somewhere about the middle of the new line we made a temporary access, following the line of a tributary of the main river, until we arrived at the footpath, and then turned to continue along the footpath towards the Malili end. From this junction we forced a passage along the footpath towards Soroaka, widening the track as required, making temporary bridges with our planks, lifting them behind us, and pulling ourselves through many swampy sections, with the vehicle sitting on our boards and being pulled by our winches from tree to tree, or other suitable anchorages as we could find them.

We progressed at the rate of about three miles per day, driving in on another vehicle, walking to the Land-Rover and fighting our way onwards, leaving the vehicle at the end of the day, walking back to our other car, and then back to base.

The going on this section improved, and we were soon able to run over reasonably open ground, and quickly connected with the original route. So far so good.

The work continued at both ends of the remaining section, with diversions on account of bees' nests, snakes, very large trees and frequent heavy downpours, but progress continued. It was agreed that we had driven the upper section, and that was written off as completed and not needing to be run again (thank God), but time was running out and the more difficult section remained to be traversed.

We were now able to drive all but about three miles of the centre section from each end, but on this section the path ran along the steep side slope of the river valley, with the river about 300 feet below. The slope was practically bare rock, with no obvious trees for anchorages, nor any – for that matter – to catch you before you reached the river, if you came off.

Our machines did what they could and scratched the rocky hillside, making the footpath a little bit wider, with the diggings falling away down the slope, ending with a splash into the water away down below. Then they came to a particularly hard section of rock, and were unable to do very much for us over a gap of about 200 feet.

Time was catching us up, and our only chance was to try to make ourselves a movable bridge over this section, with the high side wheels on the original rock, and the other side wheels on our planks, which we had to set up on small piers of assorted rocks and stones, partly dug into the hillside to give a sporting chance of a foundation. As we moved slowly forward we had to set ground anchors ahead of us and run lines with our winches, just as preventers, in case one of our piers decided to slip, and for the last 30 or so feet we proceeded with extreme caution, until we were once more on the slightly wider section that had been prepared previously.

Once over this rather alarming section we thought we had cracked it, and drove on for a couple of miles, even getting into third gear from time to time, until we came to a large valley that had a wide river running between banks about three feet deep. The machines had made a ramp into and out of the water, which was about a foot deep, but the ramp had washed out, and before we knew it the front wheels had fallen down onto the river bed while the rear wheels were still on the bank, leaving the vehicle sitting on its nose and with the crew squashed up against the windscreen.

No serious damage appeared to have been done, the engine having been switched off as we went over. We found a suitable tree, set up our winches, and wound away until we had pulled the vehicle level again, with the water up to the underside of the floor. An inspection of the far bank revealed a similar situation; we were faced with a five-foot vertical bank, full of tree roots, stones and heavy clay.

This was too much for our simple plank trick, and we had to dismantle the scaffolding of our roof-rack to make up a temporary support, and reduce the apparent gradient before we could winch ourselves up the bank and onto the track again. By this time news of our progress had been carried back to the camp, and a welcome committee came out to meet us, just in time to see our miraculous deliverance from the river. We then drove back in some style, although very wet and muddy, and had a splendid party in the club in the evening.

Next day we were advised that, due to a sudden fall in the price of nickel worldwide, and some changes in the tax regimes whereby exploration was no longer a tax-free activity, our exploration contract was to be terminated, and we were told we should start to arrange for the equipment to be recovered into a central area and for the operations to be placed on a care and maintenance basis. All the site staff were to be repatriated, the local staff paid off, and the final account prepared and agreed.

This was bad news all round, as we had believed that we were in on the start of an important and exciting development, and had done the initial hard graft – all of which was likely now to be scrapped.

CHAPTER 153
Closing down again

The next few weeks were busy with the details of closing down and sending people off home, and at last it was our turn to go. There was a party in the club with the remainder of the camp dwellers, during which we were presented with a large wooden carving of a fisherman. As we had nowhere to pack this figure we had to have a large box made up overnight, and as they were at it we asked them to make it a bit bigger, so that we could take home our veranda furniture – a couple of long chairs made in rattan.

Next day, after an emotional send-off, we flew back to Makasar, and then on to Jakarta, where I still had some business to complete. We then set off on a short holiday, stopping for a few days at Yogyakarta, and then Bali, where we stayed in a little beach hotel in a thatched room, right on the beach. Most of the tourists who came to Bali were unable to speak Indonesian, so we were made specially welcome, and had an enjoyable and restful few days, admiring the native carvings and being entertained by performances of various village dancing, and the recitation of long obscure tales of the Ramiannah complete with actions, dragons, and music.

We returned to Jakarta for a couple of days, and then went back to Singapore, flying through a violent thunderstorm on the way, with the plane lurching about all over the sky.

In Singapore we had some more closing down to see about, and managed to visit a very interesting exhibition of shellfish, which had been established in an old barrack block by two ex-accountants who had taken up diving in their spare time; they had now set up a recognised collection, and were trading in shells worldwide.

They ran evening sessions that included a trip round their collection, a lecture on certain specimens, including how they had been found, and an excellent supper afterwards. They both agreed that there was not much money in it, but it was a much more satisfying sort of life than fiddling other people's accounts. What a nice way to make a living.

We had intended on going home via Bangkok and over the North Pole, but as we felt we had had enough of travelling and hotels we went straight back to London, where a car was waiting for us, and we set out on our tour of visitations: Devon, Glasgow and home to Invergordon.

Our house had been rented out, and when we returned everything was in good order, the grass had been cut, and the garden was looking fine.

While in London I had been asked to come back in a month, and this was soon up. I called in at the office and was asked to go on another contract on a nuclear power station, and I agreed, but just as I was leaving to go home to break the news I was called back and told that our man in Invergordon had gone sick, and asked if I would care to soldier on there until something else came up.

There was really very little option, and I said, "Certainly, I'll make a start tomorrow." Much better to avoid any more shifting houses than was really necessary, and Invergordon was a nice place to stay – convenient for schools, sailing, golf, hang-gliding and all our friends.

SECTION 4 – THE OIL BUSINESS

CHAPTER 154
Invergordon again

When I returned to the site I found that a number of different jobs were running and that I had about 50 men to look after, finalising the commissioning and handover of the smelter, some new work concerned with additional fume treatment facilities, various tidy-up and minor repair works, and general jobbing contractors' items of maintenance, repainting, clearing drains, altering fences, etc., etc.

All this was quite interesting, and very time-consuming, and I was surprised when I was asked to look after another contract, building some small extensions to the existing harbour works in the town.

Some time later the big construction programme of building rigs for the North Sea oil got under way, and I was asked to take an interest in activities in the north of Scotland, and some short time later I went on an expedition around the north and west coasts looking for suitable sites for rig construction.

This work started with a desk study, looking for places with readily accessible deep water, a safe construction basin area, and a safe passage out with a minimum depth of about 30 fathoms. We were basically interested in promoting our concrete design of structure, which was built to float out of a shallow basin, be submerged to the floating depth, and have all the top structure fitted within sheltered water before being sunk to the towing draft and being towed out to its final location, where it would be settled on the seabed and start work.

Due to limitations of draft, the only suitable places for deep-water rigs were confined to the Inner Sound, off Skye, but modifications to the design, and requirements for shallower water exploration, opened up some alternative areas, notably Campbelltown and in the Cromarty Firth.

A colleague and I set out to visit the selected sites, and we had a grand tour of the north and west coast, from Loch Eriboll to the Clyde estuary, all undertaken in the dead of winter, with snow, sleet, gales, icy rain, fog and only a few nice days.

We came across the Rest and Be Thankful (a high pass in the hills between Loch Long and Loch Fyne, famous for an interesting hill climb motoring event) in a fearsome blizzard, with visibility reduced to about five feet – when you could see that far. We were creeping along, following a car in front, when suddenly a policeman appeared out of the snow, carrying a bright light.

"Follow me, and keep closed up," he said, and marched off.

We passed another policeman, who was going to hold the traffic behind us to allow vehicles coming the other way to pass, and drove slowly onwards. About quarter of a mile on we saw the lights of cars waiting to let us pass, and thanked our kindly guide for his assistance. The last thing we saw of him he was marshalling the traffic ready to go back again. Aren't our policemen wonderful? After that, we had only to dig ourselves out of a couple of minor drifts, and were nearly home and dry.

As a result of our survey we selected two possible sites in the west, and one in the Cromarty Firth, very handy for Invergordon, and then conducted more serious investigations and land ownership negotiations. The company came to an agreement with the landowner in the Cromarty Firth, and more detailed surveys were conducted, including drilling trial holes in the seabed from a small barge normally used for maintenance in the Caledonian Canal.

We had advised all the various interested parties of our operations and proposals, and I was asked to deal with a comment from the Migrant Birds Society, who alleged that the foreshore where we intended to site our basin was an important stopping place for several species of birds, and at that time of the year they would be resting on passage in large numbers.

I had been out on the site every day for about a month and had seen only the regular winter resident birds, and as it had been snowing pretty frequently I could see that there was no great number of new footprints on the shore and the site was obviously not the place the migrants were landing on, and – as far as I could ascertain from some of the local worthies – never had been. I went along to their office in Inverness and discussed the matter, producing a number of photographs as evidence, but they maintained their case, even after admitting that none of their staff had ever been out on our bit of foreshore. So much for the experts.

Another of the sites we had selected was at Drumbuie, in Wester Ross, where the railway ran close to the shore; there was a sheltered little basin, access to the Sound of Rassay, a clear passage out, and a small village of about ten houses. This was an ideal site, and we believed that it could be isolated from the village by the presence of the railway line, which ran on an embankment across the head of the small bay.

The land belonged to the National Trust, and at first they seemed to be favourably disposed towards the development.

A special train was arranged one Saturday, for members to attend and see for themselves what we were proposing, and how it would affect the locality. We attended, suitably dressed in oilskin coats and sea boots (as it was a fine west coast day, with driving rain and half a gale blowing) to point out the various landmarks and positions of our proposed site works.

One of the major arguments against our proposal was the visual impact of the works, whatever that might be. The general ignorant comment was that it might spoil the view of the far shore of the loch, to which the only sensible answer was that with 1,200 men working there the view would receive a lot more looking at then than it ever had before. Surely a very reasonable argument?

After the site visit a general meeting had been arranged in the village hall in Kyle of Lochalsh, and the spokesmen for the Trust still seemed to be generally in favour, as was most of the local opinion, and we imagined that all would be well.

There was a change of policy by the Trust, however, who now opposed our application, and a public inquiry was called, to be held in one of the nearby hotels.

This performance lasted for about five months, and ended up by the Trust refusing to allow for the lease or purchase of the site, and the scheme was then abandoned.

Meanwhile, at another site on the opposite side of the water, a site was approved without delay for another contractor, who started work fairly soon thereafter, making an enormous hole in the mountain, in full view of Drumbuie, and running heavy vehicles over the local roads, despite having suggested that he would never dream of such vandalism; and – worst of all – working on the sabbath.

Put not your trust in public inquiries.

In the meantime my other work continued to progress, the handover of the smelter was completed, the little additions to the harbour grew apace, and I was asked to make regular visits to Dounreay, where we were completing some additional work on the power station site. These extra jobs kept me pretty busy, but I could see that they were all coming to an end, and I was a bit concerned as to where I might be asked to go to next.

There was further work in prospect in the nuclear power station construction business, but most of the stations were in outlandish places, where the sailing was strictly limited, and as we had managed to buy a small sailing boat with our earnings in Indonesia this was a factor in any move we might have to make. In addition, Susan was doing very well in the school, and we did not want her studies disturbed any more, if this could be arranged.

CHAPTER 155
Other Activities in Invergordon

Our equestrian activities broadened out, and we began to attend meetings of the Pony Club all round the area, dragging the horse in its box from Loch Ness to Dornoch, attending agricultural and county shows, eventing, dressaging, jumping, and falling off, getting horseshoes fitted, and generally having a good time, but it did interfere with the sailing in the summertime.

One of our neighbours had taken up hang-gliding, and I was invited along to have a go.

I had a little experience of flying in a proper glider, and had often flown the site aircraft, under instruction, in Indonesia, but this was something different.

On our first outing we climbed up a long hill above the fabrication yard at Nigg on the North Sutor, one of the two rocky headlands guarding the Cromarty Firth. The wind was about 15 to 20 knots – fine for our machine, I was told – and then I was given a demonstration by our neighbour, who had obviously been at this game before. He arranged himself into the correct position while I shouted out the wind speed from a small hand-held anemometer, and, next thing, he was off. He made about 20 feet above the ground, and flew gracefully down for about half a mile, alighting without any seeming effort in the middle of a large field. We rushed down and assisted to recover the machine and carry it back to the launching point, where it was my turn now.

The wind had dropped slightly, and I stood, poised and ready to fly, when he shouted: "15 knots."

Fourteen and 14½ kept coming up, but no 15, so I was advised to run gently downhill until I lifted off. I took a couple of steps, and I was up, leaving the ground about ten feet below, and travelling quite fast down the hill. I tried a gentle turn, first one way, then the other, but I was not really high enough, and feared that I might catch the wing-tip. Then I looked ahead.

I was losing height, and a hedge was coming up rapidly. My eyes were streaming tears from the wind. If I can get my feet down, and give a bit of a jump, I thought, I can probably clear the hedge, the long thorns of which were now clearly visible. I made a sudden dip, got my feet on the ground, and gave a mighty leap – and there I was, soaring clear over the hedge with about three feet to spare.

The next obstacle was a barbed wire fence, but I reckoned that I had used up my luck for that trip, and put down, with a bit of a bump, some distance short.

From then on I was hooked, and had a flight whenever it was offered.

Our next expedition was to a hill on the side of Dornoch Firth, which rose steeply out of a gently sloping, heather-clad, boggy moor. When we arrived several other gliders were flying, and conditions were excellent, with about 20 knots blowing up the slope.

We climbed up, assembled the machine, and the first flight was away, just by stepping off a small outcrop projecting from the hill. The slope and the wind were such that it was possible to soar on the updraught, and after about five minutes the wind dropped slightly, and the first flight ended on the heather at the foot of the steep slope.

We recovered the machine, climbed back up, and I was ready to go. I could feel my body lifting as I adjusted the angle of the bar, and suddenly I was away. No conscious effort, just lifting off like a bird, and holding my height, about 50 feet off the face of the hill, I drifted along first one way, then the other, though slowly losing height as the slope eased.

To look at, a field of heather appears to be an even surface. Well, this is so, but there is no guarantee that the surface of the ground is also level. This I found out the hard way. When I came in to land, I stalled the wing in the approved manner, and dropped a foot or so onto the surface – well, one foot only, as I had landed on the side of a large rock, with one foot on the top, my knee bent up to my chin, and the other foot some distance down and scrabbling for a grip. No great damage was sustained, but by the time I arrived home I had a very stiff leg.

We went out quite often, and I had a number of good hops, taking great care to try to land on bare rock, or places with no heather, and certainly no fir trees. This hang-gliding is like old-fashioned ski-ing; it is hard work going up, marvellous when you get there, and bloody dangerous on the way down – but great fun all the same.

CHAPTER 156
Commuting to London

As a direct result of my involvement with the Drumbuie public inquiry, when that exercise came to a dismal failure I was asked to report to London, where work was just starting on the management contract for the Thistle oilfield. The major design work had been completed, and construction work was being farmed out to a number of international contractors, with postings available to Spain, France and the Netherlands, as well as in the United Kingdom.

I had arrived just too late to be sent to one or other of these exotic locations, and was drafted into the logistics department. This organisation dealt with the provision of materials, equipment and services throughout the entire scope of the work, from ordering bulk steel plates at the start of a construction contract to arranging the shipping and delivery to the offshore location, and included the establishment of staging bases and supply bases.

I was selected to assist in the preparation of 'contingency plans', and we set out to consider individual possible disasters that might upset the programme of the work, then to put these events into some sort of order of possibility, and finally to suggest alternative schemes that might minimise the effect of such calamities and their consequent cost implications. This was something of a novel concept as far as I was concerned, having spent most of my working life coping with numerous minor calamities on a day-to-day basis.

Here was a chance to get into the 'big time'.

We were allocated areas of responsibility, and sat down to consider the possible worst eventualities.

We soon discovered that there were a number of headings that could be considered as pretty serious, and the work was further subdivided into individual events, which were then strung together to form a series of catastrophes – and talk yourself out of that lot.

The easiest cases to imagine were everyday events, such as a loaded barge breaking loose at sea, or in the channel, and this could be countered by having a standby tug in attendance. A more serious case was to lose one large item of cargo on passage, and arrangements had to be made to rearrange the construction schedule to allow for the late delivery of a replacement section. As there were something like 30 large individual packages, which were designed to be landed on the platform in a particular sequence, there was plenty of scope for argument there.

Political considerations were also taken into account – the effect of delays caused by industrial action, or rather inaction, for example, and alternative plans were considered to allow for the removal of partially completed items from one yard in one country to another facility elsewhere.

We spent many hours dreaming up accidents, following their consequences through the programme, and suggesting means of reducing their effects. Internally, we had review meetings where these matters were discussed, and our solutions were either accepted or rejected, and we were told to go off and think again.

Every so often we had to make a presentation to the general management, and this was followed by a presentation to the client. These affairs were known as a 'dog and pony show', and required us to appear in our best boots, complete with our schedule of disasters, possible remedies and the consequential costs, and to display our wares, assisted by the magic of modern technology, before a very critical audience.

I had attended, as a spectator, such performances on other aspects of the work, and was more than a little apprehensive before our first presentation. It seemed that the objective of the inquisition was to rattle the confidence of the presenters from the start, by asking seemingly unconnected questions for which no answers had been prepared, thus leading the unfortunate presenters into a state of confusion and apparent ignorance or incompetence. We determined that we would try to be one jump ahead of them this time, and had thought out a comprehensive series of calamities, with some sort of plan, in outline, conjured up to get around the problems.

On the day of our initial encounter with the management within the construction group we collected our documents and went in to bat. The whole force of the bureaucracy was in attendance, and we were introduced and asked to proceed.

The chief logistics engineer/manager opened with a general picture of the duties of his department, and somewhat testily, I thought, was advised to hurry up and get on with it. In the spirit of 'getting your retaliation in first' he presented the case for the worst possible disaster.

"Consider please, gentlemen, the case of the client running out of cash."

There was a deathly hush.

Perhaps we had hit on something we should not have known about?

Perhaps there was such a possibility in the immediate offing?

There was a bit of huffing and puffing, and it was declared that this was beyond our remit, and would we please proceed with our other proposals? We sailed through the other items with hardly a serious question being asked, and were able to provide reasonable answers to all the questions posed.

After the presentation we were taken aside and instructed that this subject must not be raised in our forthcoming meeting with the client, and various other taboos were imposed. Apart from that, we were congratulated on a good effort, and told to: "Keep up the good work."

As the work in the field progressed some of our contingency plans were activated, and arrangements made for some of the others to be brought to a state of readiness.

Our other work in support of the operations continued, with contracts for towing, provision of vessels, establishment of supply bases at Peterhead and Bergen in Norway, and the provision of staff and office facilities at these locations.

The Thistle project was one of the earliest large offshore installations to be designed and managed in the United Kingdom, and we were determined to make a success of it. The design called for the large platform to be built on its side, towed out to the location and upended, then sunk into its correct position, and a secondary large steel frame set level on top of the legs of the platform, and a series of steel boxes landed on this frame. These boxes, called modules, contained the equipment necessary to make up large-diameter steel piles and then drive them into the seabed, and they also contained limited living and office accommodation.

When the piling had been completed, and the weather had settled down for the so-called British Summer Time, these modules would be removed and replaced by the permanent modules, which contained all the oil-drilling and -processing equipment, the living quarters, workshops, stores, generation equipment and the helicopter deck.

The programme was arranged so that, while the piledriving was continuing, the final modules would be barged into forward holding locations, and the barges for collecting the piling modules would be held in Bergen.

I arranged that I would be present offshore during the piling operations, and also that I would be in Bergen attending to the forward base until the piling barges had collected their loads, and would then return offshore during the installation of the permanent modules; after a short visit to Peterhead, setting up the base there, I proceeded to sea to observe the piling.

CHAPTER 157
Hotels and Aeroplanes

During this period I spent most weekdays in London, and had been booked into various hotels in the Hounslow area or around the airport. Some of these places were very pleasant, but others were pretty dismal, with poor food and limited facilities. In one large hotel near the airport we had a couple of fire alarms to enliven the winter evenings.

On one occasion I had gone to bed, and was awoken by someone coming into my room. "What the hell are you doing still lying asleep?" he demanded. "Don't you realise that the place is on fire, and the alarm has been going for the last ten minutes? We only discovered you were missing when we called the roll."

I dressed quickly and went out into the corridor, which was full of smoke and still had bells ringing, and was ushered down into the hall and then outside onto the forecourt.

By then two fire engines had arrived, and firemen were scrambling into a room on the first floor, having first broken the large window. Dense clouds of smoke came belching out, followed by a smouldering mattress and some charred bedclothes. When the firemen had declared the place safe again, we returned to the coffee shop and had a free brew, and heard all the gossip.

It appeared that some Indian gentleman had fallen asleep while smoking in bed, and had rushed out in his nightshirt, closing the door behind him. There was a bit of doubt as to whether he was unable or unwilling to report the fire, but it was some time later before the alarm was raised, allowing the fire to get a good hold.

There was a certain amount of speculation as to who had come out of whose rooms, and the funny clothes they were wearing, but I suppose that travellers and aircrew are generally assumed to behave in rather an immoral manner, and we all had some coffee and a bit of a laugh about it. The poor man who had started the fire was somewhat distraught, as it appeared that he had lost his money, passport and most of his clothes, and what was not lost was decidedly damp.

The second fire where I was present was in a large hotel in Manchester, where I was attending a visit to a steelwork contractor. I arrived in the hotel in the late evening and went up to my room, changed, and descended to the dining room, where a further meeting over dinner had been arranged.

We selected our food from the menu, and just before the soup was delivered the fire alarm was sounded, and we were ordered out into the street. Those of us experienced in these matters gulped down our drinks before leaving the table, while some others who were halfway through their meal tried to cram as much food as possible into their mouths and pockets before departing.

There were wisps of smoke evident as we left the dining room – could have been someone burning the toast, I suppose – but we were herded out, and stood around outside for about 20 minutes, in good time to witness the arrival of the fire brigade, who descended into the bowels of the building. At the end of this period we were allowed back into the main lounge, where coffee and free drinks were being served, and much later still we returned to the dining room, where normal service was resumed.

Much later, in the middle of the night, the alarm was sounded again, and most people turned out very quickly; this time we were allowed to remain inside the building – just as well, too, as it was pouring rain (as it often does in Manchester) and not many people had brought suitable overcoats and shoes with them. The fire brigade were in attendance again, and after a careful search announced that, this time, it had been a false alarm, and we could all go back to bed. No free drinks and coffee here, as the bar and catering staff had all gone home.

Fires in hotels are scary enough, but one day, coming back to Glasgow from Aberdeen by air, in a small aeroplane, the flight was going smoothly until we arrived over Glasgow, and the captain announced that we would be landing in a few minutes, so could we put out cigarettes and get belted up?

We circled the field several times, and I could see no traffic on the taxi tracks or runways, or other aircraft in the sky, but there was a distinct smell of smoke in the cabin. Next thing we saw was the airport fire department turning out of their shed, and, followed by a fleet of other cars and ambulances, stream off to the end of the runway.

The captain then announced that he was sorry for the short delay, but he had been having a bit of bother putting out a small fire under his dashboard, but that it seemed to be out now and we would proceed with our final approach and touchdown. When we landed we were followed down the runway by all these vehicles, and taxied to a stop well away from other aircraft and buildings.

As we climbed into a bus, I could see that some of the fire crew seemed to be slightly miffed that they had not been allowed to squirt us with their hoses, and the first aid merchants had lost their chance to wrap everyone up in bandages. It was a bit of a comfort to see that they were all there, and seemingly ready for action the next time.

During my travelling back and forth between London and Inverness we had numerous delays caused by fog at Inverness, strikes by the baggage-handlers, strikes by the fuel tanker drivers, and other minor malfunctions in the aircraft on the ground. I must say, they always got us through, although once or twice we had to go by bus for the last 100 or so miles, and they were always generous at handing out refreshments.

One evening, after we had boarded in London, the captain announced that there would be a slight delay, and the steward would be coming round directly. My companion had just returned from some exhausting trip into central Europe, where he had been plagued by delays, and had only just managed to catch this plane. He complained bitterly to the steward, who called the captain down, and we had a pleasant chat, with refreshment, until the trouble was cleared, and we prepared to take off.

Shortly after we were airborne the captain regretted to announce that the firemen at Inverness had gone off duty, and we would now have to land at the RAF base at Kinloss, but we weren't to worry as a bus would take us on to Inverness, and arrangements had been made to keep the terminal open at Inverness, and look after anyone coming to meet passengers off the plane. More refreshment was served. A great pity this, as I had left my car at Inverness, and still had about 40 miles to drive from there. We landed at Kinloss and were taken by a service bus to the gates, where we transferred to a chartered vehicle, and set out back to the airfield. By this time it was about eleven o'clock.

As we approached Nairn one of the passengers realised that there was a dance on at the golf club, so the bus was diverted there, we all disembarked and joined briefly in the dancing, and then it was back aboard again and on to Inverness, where some people had some explaining to do concerning our late arrival.

On another evening we departed more or less on time, and arrived overhead Inverness in thick driving snow. The captain made three attempts to land, but the weather was just too bad, and we turned back and landed at Glasgow. We were told that we could either stay in the hotel overnight and catch the midday plane on the following day, or go on by coach, which would leave in about half an hour. Most people settled on the coach trip, and we set off into the snow and rain.

The majority of the passengers were attending some marketing function in Aviemore, and the bus had to go round to half a dozen different hotels, unloading passengers and their baggage, before we were able to continue on to Inverness, through thick snow, eventually arriving at about 7:00 in the morning.

The next task was to dig the car out of a large drift, and then try to escape from the car park, which some foolish person had locked up the previous evening. This was not too difficult, as the snow had filled in a bit of a ditch around the open end of the park, and as it was frozen hard we all drove across without any trouble. Very much one up on the National Car Park company, which had arrived only recently in Inverness, where parking had been free since the Wright brothers first flew.

Going back to London after that weekend we had some more entertainment. The airport at Inverness had been closed the previous day, and the aircraft had landed at Kinloss and stopped there overnight. When I arrived at Inverness I was put on a bus that would take passengers out to Kinloss to join the plane there. We changed buses at the gate and went out across the field in deep snow, and were loaded onto the aircraft without delay.

The captain then announced that he was unable to take off because of snow and ice on the wings, and he had called for the station staff to come and assist. Some time later we could see two men walking across the field, one carrying a long ladder, the other a sort of squeegee on a long pole. This was the best the RAF could do for us, backed up with all the latest gadgets and high-tech equipment my income tax could buy. Primitive it might have been, but it certainly worked, and we were soon off and away, flying by the tourist route on a beautiful clear morning, with deep snow all round.

When I was working offshore the programme was that we flew out on a Tuesday from Aberdeen by helicopter to the rig. As we had an early start in the morning we were booked into one or other of the new hotels around the city and stayed overnight, usually arriving in time for dinner in the evening. I had booked in and gone down to the dining room, and was ordering my meal when there was some excitement at the next table. The diner had ordered some kind of flambé, and the waiter and his assistant came along with a trolley and began the preparations for serving this dish.

When they came to light the flame it got a bit out of hand, and went searing up to the ceiling, where it set off the automatic fire spray.

These devices work on the principle of an expanding plastic ring, which, when heated, is forced off the nozzle, and the deluge descends. Fortunately there is a delay of a few seconds between the ring pinging off and the water coming out, and the waiter, obviously well practised at this drill, leapt up onto the table and put his thumb over the nozzle.

Meanwhile the fire alarm had sounded, but as we could all see where the trouble was nobody moved, except to eat up quickly, if they thought about it. The head waiter came in and advised us to remain seated, and a few moments later the Aberdeen fire department came in, dressed overall in oilskins and sea boots, etc., bearing extinguishers, axes, respirators and portable radio sets. By this time a spare ring had been found and replaced, and the waiter who had done his Pieter and the dyke act climbed down to a round of applause. It seemed that this was not an uncommon event, and provided a bit of harmless entertainment for the guests and a turnout fee to the firemen. The hotel probably had a special requirement for athletic waiters, trained specifically for vaulting onto tables.

The organisation for the movement of personnel to and from the rig was quite considerable, when you consider that there were a great number of variables to be taken into account.

The first call would come from the platform, indicating that so many men would be coming ashore and so many others would be required. The general arrangement was that people would not be released from the platform until their replacement had arrived, and frequently, due to late arrival in Aberdeen, the returning passengers went on a later flight than they had earlier expected. In the fields served from Shetland this introduced a further complication, because of the differing capacities of the aircraft and helicopters.

Hotel facilities were pretty limited in these early days, and there were some interesting places to stay. The great worry for those on the homeward trip was the helicopter journey, and the expectation that the bars would be open, and for the outward bound it meant another night's drinking ashore before a fortnight's forced abstinence.

CHAPTER 158
Out at Sea

My first experience of offshore working was at the start of the loading of the piling modules onto the platform. The tower had been landed in its correct position and the substructure framework had been lifted on by a very large crane, operating from a barge moored alongside. This framework had to be set to a level, by jacking and packing spacing pieces onto the legs of the platform, and the modules were arranged on their barges so that they could all be lifted onto the frame and skidded into their final position without excessive movements of the crane barge.

The staff carrying out this operation all lived either on the crane barge or on another barge, called an accommodation vessel, also moored alongside and with a lifting bridge connection to the platform. The piling modules were lifted onto the platform and set into their correct position, and the crane barge departed.

The piles were made up out of six-foot diameter steel tubes, welded together and driven into the seabed around the legs of the platform by a large piling hammer. In some cases, where the underground conditions required, the material from the inside of the piles was drilled out, and the piles later knocked down by the hammer.

Staff came out and returned by helicopter to Shetland, and then by charter flight to Aberdeen. Persons stranded in Shetland by late arrival or early departure were lodged in a local hotel, which had bunk beds all round the rooms and could put up about 200 at a pinch.

The work offshore was noisy, dirty, and continuous; staff worked on two 12-hour shifts, changing from days to nights after one week, and changing opposite numbers every week. This made for continuity in some things but utter confusion in others, and it took some time to settle down into a sensible changeover pattern. Just to make things more interesting there were frequent extensions due to bad weather, with people being held on the platform waiting to go home, and others waiting at Aberdeen or Shetland ready to come out.

I shared a cabin with my opposite number, but this was fine, as he was sleeping while I was on duty and vice versa, so there wasn't any trouble. The only times we met were when handing over at the end of the shift. The barge was quite comfortable, with an excellent dining room and a cinema, which had the only comfortable seats aboard and which had three film shows each day. The problem was that the piling operations continued, cargo was being loaded and landed, and during the daytime helicopters were landing and taking off, all making a fearful row and acting as special sound effects to most of the films.

We had a photocopying machine in the office, which required a pool of developing liquid, and when the barge was loading or unloading the movement of the crane tipped the liquid to one end or the other, so that a special skill was required to hold the machine level as the barge tilted.

During the winter we had some fairly heavy weather, with 30- or 40-foot waves not infrequent. In one storm the connecting access bridge was lifted and the barge backed off about 100 yards, pulling out on its anchors. During the night one of the anchor chains broke. There was a violent bang, and a sudden change in the movement. The barge was bobbing up and down a few feet when all the tackle was secure, but with one anchor gone she slewed around and all the other chains went relatively slack, allowing for a great deal more lurching about.

The crew went into action at once and began heaving up on the other cables until some sort of equal tension was restored, but we were a bit apprehensive if another chain should go, and we might slew into the platform. The standby tug was called in, and he took a line aboard just in case, but the weather eased a bit later in the night and all was well; a spare anchor was rigged, and run out as soon as the sea allowed, and we returned to our station.

For some time we had a wave-rider buoy set out about half a mile away, and this had a mechanism for measuring the height and direction of each wave, and sent back a continuous record to a radio display on the bridge of the barge. It had been fairly rough for about a week, with very strong gusts sufficient to keep the helicopters away, and the waves were a regular 35 feet, with the odd one nearer 50. I happened to be in the bridge when the recorder reported a 75-foot wave; a message was put out on the Tannoy to expect something special and hang on tight, and it was upon us.

We could see the water rising around the legs of the platform; up and up it came, covering the lower staging of the support structure, and bursting around the undersides of the modules. Then it came to us, and the barge began to rise up like a lift. From the bridge we were normally on a level with the upper deck of the top modules, and the helicopter landing platform, but now we were about 20 feet above it and could see the deck laid out before us, and even people pointing up to us as we crested the wave. There was a lot of rushing of the waters, and some banging about down below, and then we subsided, back to our previous level. We must have stretched the anchor cables, however, or pulled the anchors in, as we had a lot more movement now and were lurching about quite wildly.

While the crew shortened up on the anchor cables we dashed around our stations to see if any serious damage had been sustained, but apart from a few bent handrails, and the loss of some walkways at the lower levels, everything seemed to be in reasonable order. After that, all other waves were pretty tame, and we became quite blasé about anything less than 50 knots, and 50-foot waves were just something to be remarked upon.

All the time we were on station we had an old trawler in attendance as a standby boat, and during these stormy times it was amazing to see how she rode the waves, frequently disappearing from view altogether as she rolled between the seas. These ships were out for about four weeks at a time, and towards the end of their stint were pretty short of fresh supplies. Sometimes we tried to drop a plastic bag of fresh bread, and the newspapers and suchlike, as they steamed close in. Frequently it landed on board, and we had a grateful vote of thanks from the skipper; sometimes it fell in the sea, and if they were quick they could get it on board before the sea got into it, but as one old skipper said, "Your new bread, even a bit salty, is better than the stuff our cook makes, any day."

The platform was supplied by a number of supply boats, operating out of Peterhead, bringing food, materials, pieces of machinery, and equipment, and they usually tried to have one boat alongside loading or discharging at all times. We heard on the radio that one particular supply boat had a case of some contagious disease aboard, and as soon as he approached everyone on deck shouted for him to 'go away', or words to that effect. When the next shift came on duty, not only did they say 'go away', but: "Keep away to leeward of us." There were some signals exchanged between the port

535

health authority and the ship, and eventually he put back to Peterhead, and we did not see him again.

We had two large, steel fabricated clamps, affectionately referred to as 'rocket launchers', for holding the piles during assembly, which had been designed in London but were out of favour with one of the alternating platform managers. As soon as he came aboard he would order these item taken ashore, and as soon as his alternate arrived he would summon them up. One of the supply boat captains once asked me why we were using so many of these things, as practically every trip he brought them out. I put it down to contingencies.

CHAPTER 159
Bergen

The piling programme was drawing to a close and my offshore work was declining, so I was asked to attend some meetings in London, where discussions were in progress for the establishment of a staging post for modules in Bergen and Haugesund, in Norway. This having been settled in principle, I went out to Bergen to supervise the final arrangements.

The general intention was that the barges carrying the operational modules would be towed from their manufacturing sites in Spain and France, and held in convenient safe anchorages until the weather would permit the final voyage to the platform. The piling modules would be lifted off and sent back to Bergen, where arrangements could be made for their further disposal, and the best use could be made of the tugs already on charter.

It was my job to see that all went well with the security of the barges whilst in Norway, and to arrange for maintenance of the mechanical and electrical equipment aboard. Through our agents, the Bergen Line, I arranged contracts for all these operations, and moved into a flat on the edge of town.

The news came through that the first barge was on its way, and I alerted the various contractors and authorities.

In Norway the working of overtime is very strictly controlled, and I had to make special arrangements for an emergency call-out squad – mainly shipyard workers. The workshops manager suggested that it would be very difficult to obtain permission, and even more difficult to ensure that the men would be available, as most of them left town each weekend and spent the time at their cottages in the mountains, or on the fjords. With the able assistance of our agent, we obtained the directions for each of the emergency squad's holiday homes, ranging from 'the first house on the left when the road runs out' to 'the house with the red roof past the three with black roofs down by the water'.

In the event, the first barge, a 400-footer carrying six modules, came into the Inner Sound under tow by the ocean-going tug. I went out with the two harbour tugs and saw the pilot put aboard the ocean-going tug, and then I boarded the barge with a couple of men from the harbour gang. As we sailed majestically up the channel

537

towards Bergen harbour I went below and checked the levels of ballast in each of the tanks, tried out the main pumping engines, and laid out the lines for mooring the barge when we should arrive alongside.

I suddenly noticed that we seemed to be overtaking the towing tug. What had happened was that he was steaming into a very strong tidal stream, and the barge was travelling towards him on a heavy eddy in the opposite direction. We tried to raise him on our hand-held radios by passing messages through the harbour tugs, who then called him direct, but by the time he realised what was happening the tow line had gone slack and the barge was heading for the shore, not very far away.

He managed to get some weight onto the line, and the barge began to turn back into the channel, but it was too late. From the bow of the barge the rock on the shore was about ten feet away, and it was impossible to see the water's edge below us due to the curvature of the hull.

There was no doubt that we were about to run into Norway, and I shouted out to the crew to get into the middle of the deck and hold on tight. I think it was my tone rather than my command of Norwegian that moved them, but we all dashed inboard and caught hold of something substantial. It seemed a long time before anything happened, but there was a loud screeching sound of tortured metal and the barge gave a lurch, but continued to move in the right direction.

Next thing we knew was an absolute deluge of water coming off the roof of the module behind us. The water had obviously ponded on the flat roof, and when the barge grounded the water spilt over the leading edge and soaked us all.

We counted heads – all were present and very damp – and we watched as the stern of the barge swung closer and closer to the rock, then there was another screech and lurch, and we were clear.

We were now heading for some rocks on the other shore, however. By this time we had passed a line to one of the harbour tugs and he was helping to pull us straight, the ocean tug had shortened his line, and we proceeded with a bit more control and managed to avoid any more contact with Norway.

We could hear the noise of water coming into one of the forward ballast tanks, and after a quick check to see that we were still reasonably level I went down and ran up the main pumping engine, so that we could trim the attitude as required, then went back to the bow, and we tried to undo the manhole cover into the forward tank space.

The first bolt that was loosened gave out a violent hissing noise.

Time to pause and consider

Could this mean that the ingress of water had built up such a pressure, and so quickly?

If so it must be a large hole, and the air pressure might be helping in keeping the water level down.

The noise of incoming water seemed no less, so I reasoned that it was probably quite a small leak, spraying the odd jets against the other steel plates of the hull, and decided that we should open the manhole. As we slackened the other bolts the rush of escaping air reduced, and when we came to lift the cover the pressure had all escaped.

There was a bit of water in the floor of the compartment, with jets of incoming spray coming in along one wall of the tank space, and a couple of the steel supports were a bit buckled, but we were in no danger of immediate sinking. I radioed ashore to summon our emergency turnout squad (of course, this happened at about seven o'clock on a Friday evening) and went along to the stern, where we tried to open another manhole in the stern compartment.

Once again we were frightened by a violent rush of air, but decided to take one bolt out and see what happened. After a few minutes the draught reduced, and we were able to open the entire manhole without much trouble. There was a little water on the floor, but nothing serious, and no more water seemed to be coming in, so we closed it down again.

Thinking about this later, the pressure must have been caused by the expansion of the air in the compartments, since they had been shut down in the Netherlands and had travelled up in bright sunlight for ten days, but it was a bit of a worry when first encountered.

In the meantime I had been in touch with the shipyard, who had alerted the Lloyd's surveyor, and our working party were already mustered at the quay with a full set of equipment. I spoke to the harbour master and got his agreement to trim the barge by the stern, so that we could see the damage above the water, so after a quick check of the stability conditions and the security of the modules I fired up the pumps to transfer some ballast to the stern. As a consequence of this operation, more rainwater was discharged from the module roofs onto the unsuspecting dock workers who were making us fast to the quay, causing much amusement in the ranks of the emergency squad.

Soon the barge was swarming with men with lights, welding and cutting equipment, jacks and crowbars, and after an initial inspection by the surveyor a plan of repairs was drawn up and the materials called up from the yard stock. I then had to prepare a damage report, and make a guess as to how long it would take to complete the repairs and have the barge ready for sea again, and get this off to London as soon as possible. It was then about 5:00 in the morning, sun shining, birds singing, and me still in my wet clothes from the previous night's ducking.

I managed to get some sort of story off on the telex, and went home to a hot bath and some sleep. No sooner had my head touched the pillow than the telephone rang. Someone must have been out early in London, and I had to explain that everything seemed to be in hand, and that I would call again later when there was some reliable information to report.

Later in the day I went round to the shipyard, where the manager was very complimentary on our efforts to turn out the emergency squad and was busy preparing a programme of work for the completion of the repairs, even authorising special overtime to help things along. In the event the repairs were completed in about ten days, the barge restored to an even keel and ready for sea in time for the original programme event.

We stood by, waiting on the return of the piling module barges, and when ordered took the first barge out into the channel with the harbour tugs, transferred the tow from the ocean tug to the module barge, and returned to harbour with the piling module barge. This equipment had completed its work at the platform, and had just been dumped onto the deck of the barge, and only notionally fastened down. The shipyard had to make these fastenings much more secure for the trip back to Teesside, where the loads were to be discharged for scrap and disposal.

We repeated this process for the other four barges parked in Bergen without much excitement.

Just before these events there had been some industrial palaver in France, and two barges were taken out of the north of France, incomplete, and towed to Haugesund, a small town between Bergen and Stavanger where a contract had been agreed for the local shipyard to complete the outstanding work on the modules on these two barges.

I was asked to go down and supervise the entry of these barges into the port, and arrange for the necessary berthing facilities, customs documentation and all that sort of performance, and then continue with my presence as available to see that all was going well.

I flew down to Haugesund on a stormy morning, and went out in the pilot boat with a couple of men to make sure that everything was ready for coming alongside and to slip the ocean-going tug when we were within the designated harbour area – as usual, in a storm of sleet and snow, and with half a gale blowing. There was some calamity in progress inside the harbour, and we were asked to stand off and wait.

Some short time later our tug was ordered away, and we had to drop the barge's anchor and wait out until the tug was available again. By some good chance I had practiced lowering the anchor on this type of vessel before, so apart from being covered in red rust from the cable, which quickly settled on all my wet clothes, this operation went off without any snags, and we swung at anchor for about five hours before being recovered by the tug, heaving in our anchor chain and going alongside in the shipyard.

To accommodate the additional workforce to enable the barges to be ready on their required date, the shipyard had chartered one of the Baltic car ferries, and a cabin had been reserved for me on one of the upper decks. I handed our agent all the papers and asked him to see them through, had a long hot shower, and slept for the next 12 hours.

Next day we had to collect the second barge, but this time there was no trouble, the weather was fine, the ocean tug brought us alongside, all the papers were in order, and I was back on the ferry in time for tea. I returned to Bergen in the morning, and arranged to visit Haugesund at least once a week until the barges were ready to sail.

According to our programme, it seemed likely that there would be a short period in Bergen when no traffic was expected, and apart from the routine running of engines, and general inspections and maintenance, I was not going to be too busy.

I arranged to have Jean and Susan come out to Bergen, and had borrowed a small summerhouse on one of the fjords for a couple of weeks. Prior to their arrival the weather had been pretty changeable (Norwegian for 'bloody awful'), and I was a bit concerned as to what they would do, marooned out in the woods all day when I was off in Bergen or Haugesund.

The day they arrived the summer came, and we had a marvellous fortnight: not a drop of rain, bright sunshine all day, and friendly neighbours not very far away. I managed a few days off between visits, and we drove all round the Bergen area, and up to the ski centre at Voss. The countryside was quite magnificent, with steep forested hills, full of bright summer flowers, set sharply against the deep blue of the waters in the fjords, and not a cloud in the sky. The owner of the cottage came across one day and said that his daughter had lived in the house for a month last year, and it had rained every day, and most days since.

The house had a small boat as part of the equipment, and we paddled about, or went on trips on the outboard engine round some of the nearby small islands. One weekend, when the sun was very hot, we went out in the little boat and rowed gently round one of the small islands nearby. The water was crystal clear, the air warm, and we were all dressed for swimming. We selected what we took to be a sheltered little bay, and tied up to a nearby tree. After some discussion, we persuaded Susan to dive in – after all, she was the youngest. She came up absolutely gasping from the cold; the water had only that day run off the nearby glacier, or so she maintained, and we had to haul her back aboard and towel her down to restore the circulation. That put an end to our swimming activities.

The house also boasted a long fishing net, and we set this out every evening just below the house. Every morning, before I set off for the office, I pulled the net in and cleared all the seaweed we had caught overnight. During the period we managed to catch one small black fish, which looked utterly inedible, and it was returned to the sea, still wriggling.

Jean and Susan returned home, and I went back to my flat in Bergen, continuing my visits up and down the coast to Haugesund, and the date for shipping out the barges approached.

I had returned to Invergordon on one of my weekends off, and had made arrangements to fly back to Bergen, taking advantage of the new service to there from Edinburgh. I booked my ticked from Inverness to Edinburgh, and on to Bergen through the local travel agent, and arrived at Inverness airport, booked in my bag, and we took off for Edinburgh.

This was the first day the new section of Edinburgh airport was open, and when we landed we taxied to one of the stands. The pilot advised that we would have to wait until one of the ground crew put a prop under the tail of the plane, as there was a danger of it tipping up as we all got out to leave by the rear exit. We sat about for a few minutes, with no apparent action, and the pilot climbed out at the front, only to discover that someone had put the prop in place. (Unless it had been dangling out all the way down from Inverness.)

He came round, opened the door, and led us round to the luggage compartment.

Surprise, surprise. The luggage hold was empty.

"Well," he said, "we had better get into the building and see what the hell has been going on."

We trooped after him all along the frontage of the new buildings. All the doors were closed, except one, which was full of customs persons, and they ordered us out as we had no business in that room. We continued around the outside, and came upon a door with passengers coming out, bound for London, or some such place.

Despite protests from some officials we barged in, went through this room, and came out in a spacious hall, at the departures end of the building, and followed the pilot along to the arrivals end. Here we found some other officials, who were very glad to see us as they had just discovered an abandoned aircraft out on the stand.

The pilot handed us over to these people and went off to report his arrival formally, and we asked about our baggage.

"No baggage has come in from that flight," we were informed. "Did you see it actually being loaded? Perhaps it is still in Inverness?"

One of the passengers could remember seeing her bags loaded into the hold.

We explained that we had been delayed out on the apron, and there were no bags left in the aircraft when we left it.

The pilot came back, full of regrets, and suggested that we should make out a lost baggage claim, and handed us out a handful of forms. As we were filling them in, I noticed a few bags going past on a conveyor belt, along the rear of the building, and thought I recognised my bag in the pile.

I dashed along the row of desks, disturbing the staff as I went and calling out for someone to stop the conveyor before it took the luggage out through the front of the building. I ran down a few more desks, jumped over the baggage-weighing machine, and grabbed my bag off the moving belt. This caused a bit of a security alert, but at least the belt stopped, and all the other passengers came down and took their bags off the line.

This caused an even greater security alert. "What were all these people doing, jumping over weighing machines and entering a forbidden area?"

Thank goodness, the pilot arrived and explained that we were only collecting our bags, which had somehow been transferred into the building while we had still been in the aircraft, and said, "Try to be a bit more reasonable about all this."

Somewhat mollified by this explanation we were all shepherded back to our station, our baggage tags examined, and the luggage handed out officially.

During all this kerfuffle the Tannoy system had been calling for a Mr Brown, the communications engineer, to report to the control room.

I went along to the desk for the flight to Bergen, but when I asked for my ticket, which had been ordered via Inverness through Norwich, to be collected in Edinburgh, I was advised that it had not arrived. "There is some trouble with the communications system," was the excuse.

"Can't you get onto Norwich, and get them to authorise you to give me a replacement ticket? The plane is due to leave in half an hour," I said.

"There is some trouble with the communications system, we are unable to talk out of Edinburgh; don't you hear, they are calling for Mr Brown, all the time?"

"Well, I have already paid for my ticket, and had it confirmed by telex to Inverness, so what are you going to do? I must be on that aircraft; there are people waiting for me at the other end."

"You must have a ticket before you can board the aircraft," she said, becoming all official.

"Please get me the managing director of this outfit," I asked.

Once more came the call for Mr Brown.

She had me there. The managing director was in Norwich.

There was nothing else for it but to buy another ticket, and register my complaints by letter at a later date.

I had a very polite letter back by return, explaining that on the first day of the opening of the new building some regrettable mistakes had become evident in the communications system, but hopefully all was now in good order, and he hoped for my custom in the future, and enclosed the cost of my original ticket and included a flight bag to keep me sweet.

Next time I flew by this route, everything went like clockwork.

I later learnt that what had happened on our original trip was that a new man had come out as soon as we had landed, put the prop in position, unloaded the baggage, and returned to the terminal. Because we had taken such a long time to find our way in, the bags had been recirculated, and would have come round again in a few minutes, had we waited.

Nothing like insisting on our rights to petty officialdom, especially when it concerns our luggage.

CHAPTER 160
More Barge Palaver in Bergen

We had been expecting a large barge with the oil-loading tower aboard to come into harbour, and learnt that it had been having some excitement on the way across. It had broken loose from the tug in a storm, and had been in some danger of bumping into another platform at sea, so the tug had pushed it clear by the simple expedient of ramming repeatedly into the side of it. Some time later they had managed to get a line aboard and resume the tow, and when they arrived in the harbour I had to go round and make an inspection of the damage to the barge, and also to see if the tower had been damaged on the passage.

The barge was all bashed in along one side, and was listing about five degrees.

I went down into the internal tanks and found them all securc and undamaged, so there was plenty of buoyancy still available, and by pumping some water into the undamaged side the trim was restored. The fastenings on the tower, which was lying horizontally on the deck, seemed to be intact, and there was no obvious damage on the structure itself.

I reported this back to London, and suggested that the barge was fit to sail down the inside channel to its rendezvous with a special lifting ship, and that this should be confirmed by a surveyor before we moved. This was accepted, the surveyor appeared and declared himself satisfied about the condition of the barge for the inside passage, and we set off the same day. The tower was unloaded and the barge returned to Bergen for repair, while the work on the tower continued in the fjord.

The tug had suffered a small bit of damage during this passage, and I was asked to muster another tug.

Through the good offices of the Bergen Line we selected a suitable tug, and had a surveyor appointed to carry out the on-hire inspection. He telephoned the office to say that he was satisfied, and asked if I would come down and sign the handover certificate.

When I arrived at the dock he was waiting for me in the bridge of the tug with a sheaf of papers in his hand, and he invited me to have a quick look round in his company, to make sure that everything was in order. He was a most meticulous old gentleman, as

I had found out on his earlier inspections, but very earnest. "That will not be necessary," I said, "but there seems to be something missing."

He became very agitated, and before he could come to any mischief I asked, "Where is the steering wheel?"

There was a moment's stunned silence; then he smiled quietly.

"Yes," he said, "you really had me going there, but you can see, in my report, that there are two independent controls for the steering, worked by hydraulics, one mounted on the table in the wheelhouse, and the other on a trailing lead, which you can carry about to the best vantage point," and he pointed to the small joystick on the table, and a similar gadget attached to a long coil of cable.

The work on the tower continued from the lifting ship, and I was monitoring the weather conditions that were laid down governing the tow out to the field. We seemed to be in a flow of minor depressions coming in across the North Sea, and these brought an increase in wind, rain and sleet, and reduced visibility.

I camped out in the local Met office in Bergen and compared information with them on a daily basis.

The Met officer was in a bit of a clove hitch; if he forecast good weather and it turned nasty, he had a load if irate local fishermen parading about in his office, and when he added the odd Beaufort number to his forecast, and the sun shone, he had the same people in complaining that they had missed a fine day for fishing.

We were governed by his forecasts, and were not allowed to sail if the wind was more than force five and rising, but we drew up our own charts and were able to predict with reasonable certainty that, as soon as the barometer began to rise as a front went through, the wind would drop, and it would not reach the force five limit for a couple of days at least as a result of the next depression. After long arguments with London we got them to agree with our appreciation, and the Bergen Met's increased number, and on the next front passing we were off, and had a fine passage through the inner channels, and clear out to sea, before the next front arrived.

After that I had to close up the office, see the remaining barges off to Teesside, and leave my flat in Bergen to fly back to Invergordon and London again.

CHAPTER 161
More Commuting offshore

When I returned to the London office I was engaged in preparing schemes for the demolition and removal of offshore structures. This was very interesting work, and among other things suggested that it might be sensible to consider the ultimate fate of these objects during the original design stage.

The majority of offshore platforms are based on steel towers, piled into the seabed, and to remove them requires special techniques, as when cut into sections to facilitate handling the sections are generally quite unstable and liable to collapse during lifting or transit, and also tend to have long legs sticking out below, which makes life difficult when trying to load them onto a flat-decked barge. The machinery and living quarters, which are set onto the upper levels, also present a problem, as they have been built into the structure and contribute to its strength, so that when one or more sections are removed it might cause the remaining sections to fall over.

We worked away, and produced possible solutions for each type of structure, and then tried our hand at estimating the probable costs and possible salvage values. No matter how we considered it, it always came to a great deal of money.

I was then available to go on a pipeline route survey, connecting two existing platforms in the northern North Sea, and after a week ashore studying the situation, and selecting possible routes that would cause least interference with existing pipelines, set off for Aberdeen, where the survey ship was lying.

The ship had previously been the lighthouse ship for that august body the Commissioners of the Northern Lighthouse Trust, known affectionately as Aurora, Bora and Alice, and had been fitted out to a high specification to permit these gentlemen to visit their lighthouses in great comfort and some splendour, and had originally been named *Hesperus*.

She had been replaced for the commissioners by a more modern vessel, and now, renamed *Esperus*, served as a general survey ship and a floating laboratory for underwater operations. I went aboard and was shown into the owner's suite, as I was acting as the client's representative on this voyage. This was a massive room, with its own shower compartment, fitted out in dark mahogany and rosewood and containing a large double bed set against one wall, facing a large hanging wardrobe with heavy mirrors, and on the other side was an ornate dressing table, also with mirrors, and numerous drawers and small compartments.

I stowed my gear and went down to the saloon, where I met the chief surveyor and a couple of his assistants. The cook was not yet aboard, so we decided to go ashore for a meal and departed for a local hostelry on the quayside. Here we met most of the remainder of the crew, and had a pleasant evening, but when we returned aboard we heard that the weather forecast was very bad, and we were not likely to sail the next day. At this news, most of the crew departed back ashore, but the surveyor and I sat in the saloon having some quiet refreshment well into the night.

In the morning it was blowing hard, and the general opinion was that we did the right thing by staying in port, but the next forecast was for an improvement, and we prepared to sail on the evening tide. The word must have gone round, but all afternoon various tired and overhung members of the crew crept back aboard, and it seemed possible that we might even be served dinner before we sailed.

We set off out of Aberdeen harbour and headed north, seeing the town lights dip below the horizon, and into the empty sea.

After dinner we had a small conference and the survey crew began checking their equipment to comply with the depths of water we would be working in, and also taking satellite readings to confirm the navigation equipment was in order. We would not be on station until late afternoon the following day, and everybody turned in for an early night.

When we arrived on station we launched the towed equipment and began the survey work. Basically, we sailed on a series of straight lines, taking readings of the depth by echo sounder, scanning sideways by side-scan sonar, and taking electronic soundings of the seabed to see how deep the sand cover was over the underlying gravel. The results came out on a series of paper prints, which indicated the position, time and heading of each set of readings, and these were then pieced together in the ship's office to provide an overall picture of the seabed.

On our first line we discovered a wreck, right on our proposed track, and this caused a minor change in plan and a small diversion, but we continued steaming up and down, extending the breadth of the survey as required.

This routine was disturbed by the arrival of a gale, which developed into a storm, and we had to recover the towed equipment and sail off in a generally southerly direction, at very slow speed. The waves built up until we were pitching quite heavily, but the movement was not uncomfortable, although requiring both hands to hold on while moving about, and all the office work was brought to a standstill.

Still we continued south, but the captain began to be concerned that we might run into Germany, and announced that he hoped that we would be turning round fairly soon.

I had gone to bed, and was lying against the wall, when he executed this manoeuvre, and it was rather alarming. I found myself pinned to the wall, the wardrobe doors flew open, and all my carefully hung-up clothes swung out towards me; next the doors slammed shut and I was sliding across the steeply sloping bed, vainly trying to find something to get a grip on.

There was a mighty crash from the dining saloon, just through my wall, and the sound of dishes and cutlery crashing about on the deck; we rolled back again, with the wardrobe doors performing again, and the dressing table then joined in the act. All the drawers flew open, discharging my socks and underwear onto the carpet, the mirrors swung out, and the doors of all the little compartments burst open, and my boots slid across the deck and under the bed. With the next roll, which was not nearly so violent, the dressing table returned to normal, my boots travelled out from under the bed, and the wardrobe remained shut.

The broken glass noises continued from through the wall, and I thought I had better see what I could do to salvage some of the crockery, so dressing quickly and putting on my boots I went across to the dining saloon. What a mess! The ship was now steaming gently into the storm, but from time to time we ran into an extra large wave and stopped dead, thus making the fielding of the remaining few cups and plates quite tricky. I was joined by the mate, who had fallen out of his bunk (receiving only the odd bruise), muttering that: "He might have told us he was going to turn," and together we salvaged four cups and a couple of plates from the dinner service. All the other fine crested crockery was grinding about in the pantry, mixed up with cutlery, sugar, tea, salt and marmalade.

The cook was called out and began to try to clear the place up, and I went up to the bridge to see how we were doing.

"Sorry about that," the captain said; "I thought I saw a calm patch, and just put her about. It got you out of bed, if nothing else."

Later in the morning we heard the tales of the others in the crew, how they had been pitched out of bed, or attacked by doors or loose gear, but no one seemed to be badly damaged. Quite miraculous, really.

We continued steaming gently north until we were more or less back on station, but the weather continued bad and we decided to return to Aberdeen and wait out the storm in some shelter, and effect some repairs to the survey equipment, which seemed to have been damaged during the night.

This called for another turn, and we had plenty of warning and performed it in the daylight, but it was still quite extreme. As we had been warned, and managed to secure most of the remaining items in the saloon, no further damage was sustained, however, and we returned sedately to port, where, as soon as we were tied up, the whole crew abandoned ship for the tavern in the town.

Someone must have been working hard, as in the morning the galley was restocked with another set of crockery and the stores replenished and the equipment repaired, and we were off again before noon. This time the weather was quiet, and we had no trouble with the survey, our lines coming out very clearly, and firm indications of the positions of other pipelines in the area. We completed a few additional measurements and then set course back to Aberdeen.

When we arrived it was early morning, and I went into the survey office and spent a couple of days completing the paperwork before returning to London.

There were some snide remarks about the survey having taken slightly longer than was originally estimated, and: "How was the duty-free?" but no one volunteered for the next offshore survey in a nearby piece of the sea.

CHAPTER 162
Still more Commuting

When I returned from the sea a further large construction project was being considered, and I was seconded to the staff, working in Berkeley Square in the client's head office preparing the initial estimates and programmes of work.

Some time later we moved into a much larger office in the West End of London, and work continued from there.

Initially we invited various construction companies to meet with us, and set out our overall requirements and then asked them for their proposals. Meanwhile the designs were being completed and orders placed for the bulk materials, and then tenders were prepared and offered to selected companies. These operations required a deal of travelling and visiting constructors' yards, suppliers' premises, and making preliminary visits to heavy-lift ship contractors, to see how our assemblies would fit into their projected programmes and what limitations we would need to recognise by way of weights and dimensions for individual parts to suit their special equipment.

Since the first platforms were installed the size and capacity of offshore lift ships and their cranes had increased considerably. On our first platform, the heaviest section was about 400 tons, and in all there were 27 individual packages. Because of the numbers of units, all needing to be connected up offshore, requiring a large labour force and taking a long time, it had been realised that larger modules would require less time spent offshore on connecting up, and heavier lift ships had been developed to match this need.

One problem that this philosophy revealed was that each module built ashore would be bigger and more complete before it was loaded out, and would require a larger shore staff to supervise it, as the intention was to have as much work completed on shore before load-out, and all the necessary parts for connection into adjacent modules loaded and prepared as far as possible.

We established sites in the north-east of England, France and Spain, and the main platform was to be built at Ardersier, near Inverness, which was very convenient for me. I was able to travel daily by car to the site, and lived at home for about a year.

Initially I worked on the construction of the main platform, and then transferred to the construction of the generation plant modules,

552

also at Ardersier.

We had a number of American engineers working on site, and at the start of the contract they all lived in hotels in Nairn or Inverness, but as they settled down, and their wives and families arrived, they rented local houses in the area.

One day when I was in one of the American's offices he placed a telephone call to his local solicitor to finalise on a house, and when he was put through a very excited young lady came on, shouting: "Get off this line."

He replied politely that he was trying to connect with a Mr Black, on that number, but was again told to: "Get off."

He hung up as requested, and I suggested that he must have been put through to the hotline on the nearby NATO airbase at Kinloss, and they were probably scrambling at this very moment. No such excitement ensued, however, and next time he was put through to his solicitor, and all was well.

We had a visit one day from a photographer from the *National Geographic* magazine, who was working on an article to demonstrate the effects of the oil construction business on the local amenities. I had a calendar on my wall with pictures of yachts racing, and one or two spectacular scenes with the yacht heeled right over and some confusion evident on board.

He had been brought in to me because I happened to have a boat, and I was asked if we could arrange a trip for him, so that he could take pictures of the boat with the rig construction in the background. I explained that, to take pictures of my boat sailing, we would need another boat but that this could be arranged, and we set up an expedition for the Saturday morning. He would be collected at Invergordon harbour by a colleague who owned a large speedboat, and we, having sailed off earlier, would meet up with them off Ardersier. It turned out to be a fine morning but without a lot of wind, and it took some explanation to advise him that we were unable to sail straight over there, directly into the wind, but had to tack about from side to side.

We got into position and he took a few pictures, and then came over and asked if we could tip the boat over a bit, like the pictures on the calendar; "A little more dramatic, like." We had to explain that it was a function of the wind, but that if he waited around the wind would blow sometime, not necessarily that day, and we might give him a bit of drama. This was not satisfactory, and he went off at great speed to Cromarty, where they spent the rest of the day having some refreshment, and I believe they had a little difficulty when they eventually returned to the harbour. Apparently the steps on the harbour wall kept moving.

The construction of the generator module continued apace, but due to a number of alterations and additional features the overall weight increased, the programme stretched, and the cost escalated – a series of events not uncommon in construction activities. I was asked to come down to London to attend a meeting, where the problems would be discussed and the necessary action proposed.

I attended in my best boots and spurs, and we all met in a large conference room about ten minutes before the scheduled time. The room had a large oval table, with chairs set around, and other chairs spread around against the walls. Soon everybody was present, drinking coffee out of plastic cups, and the chairman was wandering about exchanging greetings with all and sundry. I then noticed that there was a movement in the background by some of the participants.

Each section head was trying to keep near the chairman, and the camp followers of each section head were positioning themselves behind their masters. Suddenly the chairman sat down, and there was a muted rush to get the most favoured seats at the central table, while a lot of pushing and shoving was going on by the minions behind.

The meeting started and various individuals reported on their progress, calling for details from their supporters behind, and passing the questions back when they were unable to answer themselves. The meeting had been in progress for about 20 minutes when one individual got up and announced that he had thought that this was a finance meeting, and he had no idea why he had been summoned, and withdrew with his four followers.

I had prepared lots of drawings and details of the orders calling for the alterations and additions, and a new estimated cost and completion date, and when my time came to speak all I had to do was to assure the meeting that the revised structure would still be able to be lifted into position and would be completed before the target date. I sat through the remainder of the meeting and heard some pretty wild explanations offered, and serious questions being asked, but generally, as far as the construction was concerned, everything seemed to be in reasonably good order.

When the first two generators were completed we ran a trial test, setting up a large tank of water and heating up some elements, making clouds of steam and a lot of noise. Just like a rather large electric kettle, really. This test completed, we loaded the engines onto a barge and sailed it to Teesside, where the electrical control room module had been built. I went down and spent a few days setting up the machinery and observing the tests, as it was to be my duty to see the generators and switch room installed on the platform when it went out to sea.

At this time there had been an edict issued that all staff should undergo offshore survival training, and I was booked on a course at Aberdeen.

The course lasted a week and comprised lectures in the college, practical operations of various types of lifeboat, launching and recovering, and a day spent lurching around in Aberdeen Bay, with each candidate driving the lifeboat and the others picking up a dummy body, which weighed about half a ton and was in need of some repair having been speared by the boat-hook on almost every occasion.

Other lectures on first aid, resuscitation, hypothermia, drowning and general seasickness were followed by exercises swimming in the pool, righting a life-raft, and escaping from a helicopter in the water.

For the first exercise we all climbed into a mock helicopter, which was swung out over the water. It was lowered, quite rapidly, into the water, and the instructions were to take a deep breath as the water level rose, put one hand on the release buckle of the seat belt and the other hand towards the nearest exit, and when the motion ceased, or you could count to seven, release the belt and pull yourself out by your other hand, through the exit. This was not too difficult, although it is surprising how quickly some people can count up to seven.

The next task was to evacuate from the helicopter after it had overturned in the water, and the same drill was advised. This time, to add a bit of realism, the thing was dropped from about ten feet, in free fall, and when it struck the water it inverted immediately.

This sorted the men from the goats, or whatever.

There was a distinct shortage of time to take a deep breath, and when the thing rolled over bubbles of air from under the seats blew up your nose, causing some discomfort, to say the least. Then there was a general rush to escape from the nearest exit. Upside down, unless you remembered to set your hand in the correct direction before immersion, this could be a bit confusing. Also, some people neglected to release their belts and had a bit of a struggle, and others must have had an electronic calculator in their brain when they counted to seven, and were out before their time. The attending divers collared these members, and they had to go round again. If nothing else, these exercises made people realise how easy it is to become disorientated, even in clear water in daylight.

I was impressed by the care and attention the instructors gave to non-swimmers. They took them at the edge of the pool and slowly encouraged them to put their heads under while still holding onto the rail, and before long had them submerging like the rest of us.

Swimming with a life jacket on was pretty exhausting, but at least you would not sink. Climbing into a rubber life-raft with a life jacket was extremely difficult, as the buoyancy across your chest reduced the effective length of your arms, and it was a considerable pull-up to lift yourself over the edge. This procedure was made even more difficult by the instructors sluicing you in the face with a hosepipe.

At the end of the course we all left praying that we would never be required to put this experience into practice.

We attended another course on fire-fighting, which was very realistic, with smoke and flames, and red-hot steel walls and floors. We were put into a smoke-filled building and told to get out, but to collect two other men from somewhere in the building on our way. These two 'men' were life-sized dummies, also weighing about half a ton, and we had to find them and bring one down from the top floor and the other from the second floor. There were firemen wearing breathing apparatus and with radio communication inside to see that nothing terrible happened to us.

We found the first body and were having trouble lowering him down the access ladder My colleague let go without warning and the body fell onto one of the firemen, who was coming to our assistance from below. He was not amused.

Another of our party discovered a rivet hole in the floor and found that he could breathe fresh air through it, and was reluctant to move. The fireman designated him as a casualty, in need of removal and being taken down another ladder. Having seen what had happened to the dummy our casualty made a miraculous recovery, and he was downstairs and out of the building like a shot.

At the end of this course we had a few small blisters, a lot of coughing, the hair scorched off the back of our hands, and a few eyebrows singed – but a serious respect for smoke and flames as well. Since then, I have always been conscious of the need to seek out fire escapes in buildings, and the location of extinguishers and hose reels.

The structure was floated out, upended and sunk in position, and I went out with the generation modules and saw them lifted on, into position, and stood by as they were hooked up and commissioned.

On my first night on the lift ship I had been allocated a cabin that was underneath the ship's galley. The barge with the modules had not yet arrived alongside, and I turned in early. There was suddenly the most peculiar noise, for all the world like a woodpecker, working on metal. It continued intermittently for about half an hour, and it troubled me enough to go and investigate. What it had been was one of the cooks chopping up cucumber, on a metal table, right above my head. At least I was not located below the workshops, where they had some man hammering plates all night, and then dropping them on the steel deck.

There was another small workshop on the platform, and they were preparing to run a test on a section of pipeline under the sea. The plan for this test was to plug the far end of the pipeline, displace all the sea water with fresh water to which a dye had been added, and then pressurise the line by a pump on the platform. A submarine inspection would then be conducted, and any leaks that occurred would show up as fluorescence when illuminated by an ultraviolet lamp.

The work was well advanced, and the dye mixture was being prepared on the upper deck. This required the concentrated dye to be mixed from barrels into a tank, and while this work was in hand some slight spillage must have occurred. One unfortunate man, who was working on a bolt-threading machine on the lower deck, must have been standing under the drips, and when he went off shift and had a shower was horrified to discover that his hair had turned a bright purple shade of red.

He rushed into the medic, fearing that he was suffering from some dread disease, and was not comforted by the professional reply that it was outside his experience, but did not seem to be causing much obvious damage.

It was not until some time later that the penny dropped, and a connection was made between the pipeline dye and the turner's hair, but by this time all sorts of rumours were afoot. Informed medical opinion suggested that no great harm had been done, and that the colour would fade with time, but this was not much relief to the poor man, who was dreading going home with his peculiar colour of hair. He went ashore a few days later, but when next I saw him, some three weeks on, the colour had certainly faded a bit, though it still looked rather strange.

Talking of people with strange hair, one crew change brought out some new members of the catering and ship's service staff, one of whom was a Jamaican.

Normally he worked in the galley, and wore a tall white hat most of the time, but when off duty he went hatless and boasted a huge mass of dark, curly hair like an inverted beehive. This caused only mild amusement among the 'bears', as the construction crews liked to be called, until our man went to the cinema one evening, and was politely advised that the best seat for him was in the back row, where his hair would not obscure the view of the screen for other long-suffering sleepers.

We also had a number of Spanish workmen on board, some of whom had arrived with the modules from Spain, and who had then been taken on for their specialised knowledge of the modules and the peculiarities of Spanish design and construction techniques. They were billeted in a number of Portakabins set together in a block, with their own showers and toilet facilities, and pretty well kept themselves to themselves.

One evening there was a call to the medic from this block, and he went down to see what the trouble was. It seemed that these people had been making their own demon vino, or tiger juice, and had been fermenting fruit and potatoes, producing wood alcohol, and going blind in the process. The block was isolated and questions asked all round the platform if there were any other obvious sufferers, but no one would admit to it. In the morning the residents were taken ashore to be dried out, fitted with white sticks and Alsatian dogs, and a new batch of Spanish workmen arrived.

New regulations were brought into service concerning the reception and departure of helicopters from the platform, the radio drill was perked up, and the fire attendants were obliged to wear shiny fireproof aluminium suits. This gear was greatly fancied by the Spaniards, who volunteered to a man to act as fire hydrant attendants, and they were subjected to hours of training in the evenings, sluicing water jets all over the heli-deck.

The radio procedure was also a bit of a joke, modelled rather on the antics of the NASA communication system, and full of weird phrases. One that comes to mind, and was often heard, went as follows.

Platform: "Hello, Thistle tower to alpha tango two – over."

Helicopter: "Alpha tango two to Thistle tower, loud and clear, expect to be landing in five minutes – over."

Platform: "Thistle tower, roger (followed by weather report, wind speed, visibility, etc.) – over."

Helicopter: "Alpha tango two, roger – over."

Platform radio operator to assistant: "Nip out and see if you can see him."

Assistant, returning from quick trip outside: "Yes, he's about a mile away."

Platform: "Thistle tower to alpha tango two, we have you on visual – over."

Great roaring noise and fierce blast of wind, and the helicopter lands.

No mention of having him 'on audio'.

Normally, what happened next was the landing crew opened the doors and unloaded the baggage, while the passengers filed off the heli-deck, passing the shore-going crew in the ante-room below, the machine was refuelled, and the new passengers and their baggage embarked and were off again within a couple of minutes.

In these early times there was no requirement for the wearing of survival suits, and the flight back to Shetland was generally pretty uninteresting – mostly just sea.

From the airfield in Shetland the flight back to Aberdeen was in the relative comfort of a DC-3, with ever-smiling hostesses and coffee and biscuits.

On one occasion the flight was delayed out of Aberdeen, and we had a group of tired and disgruntled 'bears' waiting around in the Shetland terminal. When the aircraft eventually arrived the captain apologised for the delay. He said a stone had bounced up on landing at Aberdeen and they had had to patch the tailplane, and had to wait until the dope had cured. A likely tale, or so we thought. But when we went out to board, there it was: a small canvas patch, about the size of an envelope, stuck on the fabric of the tailplane.

During one of my periods offshore, when we were at an early stage of the hook-up operations, the platform was not equipped for normal occupation, and the staff went over by an access bridge for their shift on duty. This bridge crossed between the platform and our accommodation vessel, and was arranged to move backwards and forwards on the platform to allow for the wave action effect on the vessel. It also carried fuel, water and telephone lines.

During a severe storm it was decided that the bridge should be lifted and the accommodation vessel moved off to a safe distance on her anchors, and the signal to evacuate the platform was sounded.

The orders were that all men on board the platform should come back across to the accommodation vessel, and report in. When a roll-call was taken one of the men in my section was reported as missing, and I returned to the platform to look for him, in case he had suffered an accident on his way down to the bridge. One other engineer also came across the bridge, which by now was travelling to the full extent of its carriage, and he stood by with a radio while I completed my search. During this period the bridge came off its carriage, and fell down between the platform and the vessel, which immediately moved away, leaving the two of us stranded on the platform.

There was no sign of my man, and it turned out that he had defeated the system of booking in and out, and was already safely on board the vessel.

Part of the abandonment routine was to shut down the temporary generators before leaving, and this had not been done, so our first task was to close down one set to conserve fuel, just in case we were left aboard for a few days. This accomplished, we looked out the emergency rations and sleeping bags, and selected a comfortable cabin for our stay. When we were satisfied that we would not starve, and had somewhere warm and dry to sleep, we prospected about for reading matter to while away the idle hours, but only found some operating manuals for various exotic pieces of equipment. We returned to the bridge carriage platform and waved to our colleagues, who were then about 100 yards away, and spoke to them on our radio to let them know we were in good order, and quite happy.

The storm continued with heavy seas and high winds and showed no signs of abating, so we retired to the manager's office and sea cabin, where we managed to get some music out of the radio, and heard the midnight weather forecast, predicting continuing gales in our sea area.

In the morning we inspected the helicopter deck, which was all cluttered up with large crates and boxes and was no place to land a helicopter, but if desperation really set in I suppose we could have been winched up. Anyway, the weather was so bad that all flying had been cancelled.

Later in the day the wind had gone down a bit, and the vessel began to come back to her normal station, but it was still too rough to attempt to land the bridge, and we sat out another day in the manager's cabin. During the afternoon it was decided that we should be passed a hot meal, and after due warning, and taking cover, a hand line was shot across and we made fast the end very quickly. We then pulled across a much heavier line, and set up a messenger on an endless rope, running on a pulley wheel, which could travel across the heavy burden line. (Incidentally, the original hand line that was shot across makes excellent flag halyards for my boat.) The galley outshone themselves and we had an excellent dinner sent across, all piping hot, which we took up to our residence and ate while listening to the wireless.

The weather was still too bad for reconnecting the bridge, and we spent a further day and night aboard in relative comfort, quite undisturbed by movement, while the accommodation vessel heaved about in the sea.

Next morning the vessel came close in and we were within reach of the crane, which could just manage to spot his basket onto the landing place of the bridge. Although the vessel was not coming any nearer, because of the high sea it was still rolling quite heavily, and the head of the crane, which was about 80 feet out from the vessel, was going up and down about six or seven feet with each wave.

The basket used for these operations comprises a heavy rubber ring, about eight feet in diameter, with a canvas floor, and a number of rope supports, each about eight feet long, spread about the circumference and connected to a central point; this in turn is connected through a long rubber section, which then fits onto the crane hook. In the hands of a skilled crane-driver it can be set down accurately, even onto the heaving deck of a supply boat, and by use of the elastic section of the system the support ropes can be kept reasonably taut, the driver paying out and heaving in as necessary.

In our case we had a very limited area to land the basket, and there was always the possibility that the support ropes, if they went slack, could snag on some projecting parts of the platform.

We were perfectly happy to remain on board and assist in re-rigging the bridge when the weather abated, but it had been decided that we must be taken off, so after a short argument I elected to be the last person aboard, and my colleague prepared himself for the ordeal. As far as I was concerned this was a good idea, as I might learn how best to get onto the basket when my turn came.

After a bit of shouting and signalling the basket came down for him, and he calmly stepped aboard and was whisked away, out over the sea and onto the deck of the vessel.

When my turn came the basket landed, I stepped aboard, and the support ropes went slack. I wrapped my arms about them, and the next thing was they all came tight and I was off, swinging across the gap and landing on the deck of the barge.

Next day there was a bit of an enquiry as to why I had gone back aboard after the abandonment signal had sounded, but my explanation of going back to check that my man had not been left aboard was accepted, and he got a bit of a blast for not reporting in when he had come back in the first instance.

This event brought home to me the difficulties of ensuring that the abandonment signal was heard and understood by everyone aboard, and that areas where it was unlikely that the signal had been heard had to be physically checked and cleared by some responsible person.

A couple of days later the storm had abated, the bridge connection was restored, and normal work continued, with the helideck cleared for emergency landings.

My section of the work was drawing to a close, and I returned to Invergordon for a spell of leave, prior to returning to the commuting routine back to London.

CHAPTER 163
A Visit to Shetland

After a couple of weeks in the London office I was asked to do a trip to Shetland. The objectives were to survey a place called Basta Voe as a possible site for a staging post for module barges and a possible forward supply base, and also to look at some massive peat deposits, with a view to using peat as a fuel for power stations and as a material for absorbing oil in the case of oil spills and harbour pollution.

I arranged to fly up to Shetland, and arrived at Sumburgh airport in the south of the mainland in the early afternoon.

I was met by a Captain Mason, who had an interest in the island of Yell and was trying both to advance the development of Basta Voe and to promote the use of the peat as a material for mopping up oil spills. We drove across the mainland, then travelled by ferry to Yell, where arrangements had been made for my accommodation in a local guest house.

This was a Victorian house set at the crest of a small hill, and open to all the winds that blow. It was owned and operated by two elderly ladies, who were very kind and tried extremely hard to make my stay as comfortable as possible.

I spent a couple of days walking about, collecting samples of peat and studying maps and charts of the areas, working out suitable locations for moorings, storage and support buildings, possible quay sites, and access roads for the export of large volumes of peat, or factory sites for the processing of peat into a suitable state for absorbing oil spillages.

Peat is about 90 per cent water, and the natural drying process of stacking small slabs in an open structure and letting the wind blow through is somewhat lengthy. There are industrial processes that mine the peat using large tractor-drawn ploughs, and these machines turn out short cylinders of peat, known in the trade as 'elephant turds' (a pretty apt description), and these are then collected and dried in a special drying house. The energy required to dry out the peat is frequently derived from a peat-fired power station's waste heat; thus the economics of peat drying are difficult to evaluate.

Having completed my field studies I arranged for a taxi to take me back to Sumburgh, and was advised that I would need to leave

Yell at about 6:00 a.m. to catch the ferries in time for the midday plane back to Aberdeen. The taxi arrived and we set out in good style for the journey. The drive across Yell to the ferry was uneventful, and our quick transit of the sound to Toft was enlivened by a display of gulls diving for fish alongside the ferry. When we docked I noticed frantic activity amongst the local small boats to get out as soon as possible.

The road ran down the island towards Lerwick, calling to mind an earlier expedition I had made to Shetland during my national service (this being the furthest place from Salisbury that we could reach on our official travel warrants without extra costs).

When we arrived in Lerwick I noticed that the mailboat was docked alongside, and was dressed overall. I remarked to the driver that they must have known I was coming into town, what with all the bunting up, and asked where the band was. He crouched down to look up at the display, and in the fraction of a second necessary for this observation the car in front stopped suddenly, and we ran into the back of it.

We jumped out to see the extent of the damage and to remonstrate with the inconsiderate driver of the car in front, who turned out to be the local butcher's wife.

We had put a bit of a dent in the back bumper of her car, but not much else, and the front of the taxi was a bit disturbed, but nothing too serious. The lady insisted that we should follow her round to her husband's shop, so that he could see the damage and we could explain that it had not been her fault, so we followed on behind her and noticed a funny noise coming from under the bonnet of the taxi.

We turned into a narrow side street towards the shop.

"My God, look at that!" the driver observed.

The butcher, a very large man, was busy sharpening a meat cleaver, and looked up in surprise at the unexpected arrival of his wife.

There followed a short discussion, and due contrition was expressed, formalities were exchanged, and on the excuse that we had to be away quickly – or I might miss my flight – we were on our way again without further comment and continued on the road to Sumburgh. I left my name and address with the taxi-driver, in case I was wanted as a witness, paid the man off, and went into the terminal building. As usual it was full of oil workers who had come in from the rigs, but due to mist and high winds there was a delay in flights from Aberdeen, and the natives were becoming increasingly restive.

I had been booked on the scheduled BEA flight to Aberdeen, and it was also subject to delays, and was being postponed by an hour every hour. I tried to scrounge a passage on one of the oil company flights, but they were equally delayed, and at one o'clock the announcement was made that there would be no further flying that day, and passengers could either wait in the hotel at Sumburgh in the hope of a flight the next day, or try to go back to Lerwick and find a passage on the mailboat. Furthermore, the airport would now close and all the facilities would shut their shops.

I had been up and about since before five o'clock, and had had a pretty frugal breakfast, so by now – about one o'clock – the starvation had set in, and there was not a bite to be had in the airport. I telephoned the steamer booking office and booked a berth, and was asked to report in as soon as possible and confirm the booking, so I set off to find a bus or taxi to take me back to Lerwick. By this time the wind had really got up, and there was a big sea running, so I hoped that I would get a reasonable berth on the steamer, and also a reasonable dinner.

I suppose it was about four o'clock when we arrived back in town, and I went aboard the ferry, confirmed my booking, and was ushered down to my cabin. The ship was due to sail at 6:00, and the dining saloon opened as soon as we had cleared the harbour, as did the bar, and as the rain and wind outside were by now pretty fierce the thought of a walk up into the town for a sandwich or a bar of chocolate was discouraging, to say the least, so I had a walk around the ship while we were still tied up alongside. One of the crew said it had been pretty rough last night coming over and the weather had not improved during the day, so he was expecting another wild passage.

On the dot of 6:00 we set off from Lerwick, and as soon as we had cleared the harbour the ship began to pitch and roll in the heavy swell. I stood inside on the lower promenade deck and watched her put her bow into a couple of large seas, which sent green water all along the foredeck and great curtains of spray halfway up the mast.

I went down to the dining saloon in a series of lurches and rushes, and arrived in time to observe a general migration.

We had fallen off the side of a large wave, and the diners already in position, and busy with their soup, had been disturbed. All the soup, plates and cutlery and a few of the passengers had ended up in a messy heap in one corner. This scene had so upset the remaining passengers that they also were abandoning any thought of eating, and were assembled at the foot of the staircase, waiting their chance for a run up the stairs as the motion allowed.

I found a seat at the weather end of a table, and wrapped my leg around the leg and waited on the service. One other hardy, or hungry, soul appeared, and became similarly ensconced at another table, and the steward came out, pleased to see some customers, while some other minions cleared away the mess on the deck.

We had a splendid dinner, enlivened by the violent lurches and crashes as the ship fought her way south, and engaged in a silent competition to appear unconcerned, a performance much appreciated by the staff.

Afterwards we retired to the bar, where there was a similar lack of attendance by customers, and after a couple of drams to settle the stomach retired to our cabins.

My bunk was set athwartships, so I was not in much danger of being rolled out, but the rolling had some effect: one moment my head was being pressed against the headboard, the next my feet were pressing hard on the bulkhead. None of this motion was sufficient to keep me awake, and when I woke up in the morning we were steaming past Fraserburgh, just in time for breakfast before docking in Aberdeen.

The next trick was to find a taxi to take me to the airfield, where I had left my car, and then back to Invergordon for a quiet weekend, making out my reports and preparing for the week in London.

CHAPTER 164
Further Work in London

When I arrived in the office there was some excitement going, as we had landed another large contract for work in the North Sea and my reports were filed away somewhere, and I was set to work on the preparation for the construction of another large offshore platform.

The process of development is constrained by all manner of regulations and many practical considerations, many of which are based on prejudice, previous experience, or sheer bloody-mindedness – opposing some other person's ideas.

The size of the platform is determined by the water depth, the estimated capacity of the field, and the extent of the production equipment required to separate gas from oil and water. Further considerations are concerned with the method of exporting the product, whether by direct pipelines, or to a loading buoy, or to a moored tanker, which has a large storage capacity.

Our task was to prepare possible schemes, with cost and programme estimates, and make comparisons to select the best solution at that time, with a 'technology factor' that could be added to account for possible developments and refinements in equipment that might be expected to come into being before the platform commenced production. This factor was brought into play to discourage some of the more outlandish suggestions.

One chap who joined the staff had been a helicopter maintenance engineer, and he now specialised in risk assessment. He was the most infuriating person in his car, as he ran a full pre-flight check of all the fittings and instruments before even firing up the engine, and then waited until all the instruments were indicating that everything was working before moving off.

Some of us used to go for a quick game of squash at lunchtime, and if you went with this man you were always late arriving. Even playing squash I am sure he counted the number of strings on his racket before starting to play. It is comforting to think of people like him looking after your aircraft with such dedication, but he makes for a bit of a pain if you are stuck behind him in the car park.

I must confess to being a little bit careless in some aspects of motoring. Normally during this period I left my own car in Inverness, and hired a car in London during the week. This gave me great experience in driving different models of car, all of which seemed to have a different arrangement of the gears, light switches, indicators, windscreen wipers and other accessories from the previous model.

One Monday I had parked as usual in the office car park, which was on a sort of two-level concrete structure, and I was on the upper floor. It had started to rain during the afternoon, and when I came down to the car in the dark the whole vehicle was covered by large blobs of water. I got in, fired up the engine, and began to move down the ramp, putting on the lights and windscreen wipers as I went.

There was the most awful screeching noise, and my half of the windscreen continued to be covered in large raindrops, although the other half was being cleared. As I continued to roll down the ramp I tried all the switches I could find, but still the screen failed to clear, and the noise continued. I clapped on the brakes and got out to find that the rubber washer part of the wiper had been removed, and that the noise was coming from the small metal bracket on the arm scratching against the window glass, and it had already scored a groove. All I could do then was to change over the wiper part from the other side and put a small bit of rag under the bracket there to prevent the noise from continuing. When I telephoned the hire company in the morning they did not seem a lot bothered, and said this sort of thing happened all the time.

This phase of the work was taking some time to settle, and in the meantime we had landed another contract to manage the construction of a special support vessel for one of the major oil companies, to be built in Finland, and I was asked to report to Aberdeen with a view to going out to Finland to supervise the construction. I was given copies of the contract and piles of drawings, and told to go away into a corner and study them for a couple of weeks.

I had rather been hoping for an assignment in somewhere warm, such as Greece or Indonesia, but Finland sounded different, and it was only for a couple of years at most, so off I went as bidden.

CHAPTER 165
The Finnish Adventure

In the Aberdeen office we met the proposed construction team, the naval architect, the chief engineer and the construction staff, and after a number of difficult negotiations with the client and the shipyard in Finland the contract was eventually signed, and we went out.

On our first visit we were taken round the yard, met the staff, and made some preliminary arrangements for accommodation for our staff in flats and apartments in the town, organised cars, the bank, postal services, and all the other preliminary details for our stay in Finland. Our first visit was in the summertime and the country looked quite attractive, with small, rolling hills covered with coniferous trees, small open fields around little farmhouses, and almost continuous daylight.

We stayed in a large modern hotel not far from the yard during this visit, and had a few entertaining experiences, including my first ever visit to a proper sauna. We were told, of course, that this was not a proper sauna, as such establishments are situated on a beach somewhere, and are only effective when the saunee, or whatever, is expected to rush, naked, over the ice and jump into a hole. The skill at this point is to allow his shoes to become frozen into the surface around the hole, so that he has a lot of trouble getting back to the hut (that is, if he has not died from heart failure in the meantime).

On our last evening in the hotel it was decreed the night of eating freshwater prawns, and these were served up as a *spécialité de la maison*. We felt that, as prospective residents in the country, we had better have a go at these, and ordered accordingly. When they arrived, spread all round the periphery of a large plate, on a bed of chipped ice, they looked marvellous, and we enquired as to what the appropriate drink to accompany this fine fare was.

Meanwhile we had been fitted out in a waterproof bib and apron, and were pronounced ready for the food.

The wine waiter indicated something from his list and we ordered on his suggestion. The prawns were magnificent, and the pale amber-coloured drink went very well with them. Perhaps we should have known that there was a reason for serving it in small glasses, but it was very tasty and was having an excellent effect on us.

Next morning seemed a bit hazy, and we were taken back to town and put aboard our aircraft for Helsinki, where we changed planes for London. On the journey they came round with the leaflet describing the goodies available on board, and my colleague pointed out some expensive perfume, which looked exactly like our drinks of the previous evening. Come to think of it, the costs were quite comparable too.

After a busy two or three weeks between home, London and Aberdeen I was off on semi-permanent attachment to Finland.

CHAPTER 166
Resident in Finland

For the first few days I lived in the large hotel near the yard, and then I moved into a flat in an apartment block in the town, some eight miles away.

My first place was in the top floor of an old building, overlooking the square in the centre of the town, and it was served by an antique open-framed elevator, which creaked its way up and down its shaft, with loud whirring noises coming from the motor room just above my bedside and clattering noises as it passed each floor. I suppose this was to assure you that, when the wires broke, you would only fall down to the next floor.

I lived there for about a month, as that was the term of the lease, and then moved to a modern flat not very far from the town centre. This was a much better proposition, and although I did not have the same view the car park nearby was fitted with a heating switch for each parking place, so that it would defrost the seats and generally warm the car up for 20 minutes before you needed to come down.

The block also had a sauna and swimming pool, available to tenants one evening per week or by arrangement. We had people scattered all round the town, mostly in flats with a sauna, and it was possible to boil yourself away almost every night of the week, by judicious visiting. A practice not to be recommended.

The demon drink had a powerful hold on many parts of the Finnish scene. On the other side of the harbour there was an open space between the road and the old quayside, about 100 yards wide and about half a mile long. This space had been used previously for stacking timber, and there were photographs in the town showing the harbour full of sailing vessels loading timber from this area. The trade had ceased about 20 years ago and the area had reverted to grass, which the town kept reasonably short during the summer months.

I went over to look at some fishing boats one evening, and was horrified to find that this park area was covered by inert drunken men, sprawled about all over the place. It seems that they were all on the spree before the end of the summer, and it happened each year over a few weekends in the early autumn.

Later in the season we went out to the local hotel for dinner, and observed the dancing. Towards the end of the evening the bar became very crowded, and people were standing three or four deep, waiting to be served. This press of bodies had the effect of keeping some people on their feet, as when the crowd parted very often there would be a body left lying on the floor. The drinkers generally treated these bodies with some respect (you never know, you might be the next one), but the hotel staff gave them short shrift and dragged them out and downstairs, parking them in the yard, where the temperatures went down to about 30° below. There seemed to be some arrangement whereby taxis took these bodies home before they froze solid, as they were generally back again, and up to the same tricks, the following weekend. Statistics suggest that the number of early deaths among males in Finland is amongst the highest in Europe, and perhaps the collecting service is not quite as good as it might be.

We were established in a temporary office in the yard, and conducted our business from there. The main sections of the vessel were being constructed at a number of other locations around the country, and as they were completed they were delivered to the yard and the vessel began to take shape. There was a seasonal factor to be taken into account, as during the winter the yard became iced in, with sea ice forming up to seven or eight feet thick in the enclosed bays. This, incidentally, provided a track for heavy vehicle racing, with the track marked out by small branches of fir trees. It was quite impressive to see these huge eight-wheelers going round the course, sliding broadside on around the corners, and throwing up clouds of chipped ice from their winter tyres.

A further winter pastime was fishing through the ice. The secret here was to go out while the ice was pretty thin, and bore a hole about a foot across through the ice.

The next move was to lower a float through the hole, and then pay out a length of line behind it, allowing the float to be carried away on the tide. Then at some suitable distance a second hole was cut, and the float retrieved. A further line was attached on the surface with baited hooks and a few sinkers, and this was then passed down under the ice and the ends made fast. After a few hours the baited line was recovered, usually with a good few suitable fish attached. With a certain amount of care and attention this pair of holes could be used many times, the main problems being that the surface line frequently froze into the surface along the top, or the hole froze over.

Towards the end of the winter, about the middle of May (!), the ice became pretty thin, and frequent drownings were announced when fishers fell through while attempting a final run for the season.

In the autumn, before the frost really set in, there was the elk-shooting season.

At this time half the male population set out with guns and large knives to shoot elk. This was a dangerous time for everyone, as the shooters were liable to fire at anything that moved -- frequently at other hunters, but also at innocent pedestrians out for a walk, or even fulfilling a call of nature behind a tree at the roadside.

All the major roads seemed to be across the migration trails of the elk, and there were frequent stretches of the road, up to a mile long in some sections, officially designated as 'elk crossings'.

If you collided with, or were run down by, an elk on one of these areas, your insurance was declared invalid and you were liable to prosecution under the game regulations. The advice given out to visitors was that, if you hit an elk, you must stop the next vehicle, who would always tow your car (and the carcass of the elk) to beyond the designated crossing area, one condition being that half the carcass was taken by the driver of the other vehicle.

During the hunting season it was advisable to proceed with extreme caution through these designated sections. If you kept blowing your hooter to warn the shooters of your presence they became infuriated, and were likely to deliver a blast to frighten you off. On the other hand, if you drove along quietly there was a sporting chance that they would frighten some elk, which would come out onto the road and gallop along the carriageway, thus presenting a much better target for all the hunters in the vicinity, and probably bringing down enfilade fire on the poor motorist.

These hunters were out purely for the meat, and were not bothered if the animal had been run down by a car or a lorry, but if it moved they shot at it, and when it was down they cut it up and carried off the edible pieces.

Another native sport was cross-country ski-ing. In the winter, when the snow lay all around, every road was fringed with earnest young men with funny hats running on their skis, and in the so-called summertime the same people could be seen out on elongated roller skates, still with their funny hats, poling themselves along the verges.

We visited one of the famous ski jumps before the snow fell, and climbed up a series of ladders and steps to the top launching platform. This thing had been built into a hillside, and I suppose we were about 100 feet directly above the ground at our feet, but the take-off point was another 200 feet below on the track, and there was no view of the ground out over the jump for another 200 feet or so, then another 100 yards before the stopping area.

We returned in the wintertime, when the jumping was in full swing, with coaches standing by at a number of smaller jumps, encouraging children to have a go, and more experienced jumpers being given instruction at various levels of the full-sized tower. Probably the most useful advice would be to shut your eyes before leaping.

Some of us from the yard joined in the cross-country ski-ing, and I bought a ski bicycle – basically a pair of short skis fitted on a frame, with handlebars and footrests. It went well downhill, but was a bit of an effort to push uphill and seemed to have a will of its own when negotiating ski trails in the forest, where it appeared to imagine it could pass one ski on either side of trees bordering the path.

The work was progressing well when there was a bad accident to another vessel in the North Sea, where a pontoon broke away and a number of men were drowned.

The resultant enquiry called for some changes in the design rules for all new structures, and some extensive delays were caused to our construction while these changes were being implemented.

A new programme of work was agreed and the modifications put in hand, and to mark this change in the contract we were invited out by some of the senior staff in the yard to a party and sauna in the hotel.

It was all well arranged; we drove back to our billets, washed and changed, collected our swimming gear, and were collected and taken off to the hotel.

We assembled in the bar and had a few drinks to warm us up, so to speak, and then went into the sauna. There seems to be a stupid competition in all these places to find out who will collapse first from heatstroke, or steam in the lungs or whatever, but in these races

I am a non-contestant and I remained on the floor, where it was not too hot and I felt it might be doing me no great harm.

Some chap from the top shelf decided that he had had enough, and dashed out, into the swimming pool and dived in. He must have misjudged the depth, or slipped or something, but the next thing we saw was the hotel manager fishing him out of the water just by the door to the changing room. He had cracked his head on the bottom, and was bleeding like a stuck pig. We wrapped his head in towels and came to a decision that we should take him to the local hospital, so we all got dressed, carried him out to the van, and set off back into town.

There were a lot of words being spoken in Finnish in the back, and when we arrived at the hospital it appeared to have been decided that I should be the one to take him into the building, on the premise that I might assist in having this matter accepted as a pure accident and not taken as a self-inflicted wound – and, in any case, the authorities would try to impress the foreign visitor as to how they dealt with casualties after hours. I was not at all convinced by these arguments but felt that we could not have our colleague lying about in the back of the van all night – and him still in his bathing suit, his head swathed in bloodstained towels – so I agreed to give it a try.

I went in and asked if they had a wheelchair for my friend, and was instantly surrounded by the matron and her staff. After some language problems I managed to get the message across that I was not the one in trouble, and they descended on our friend, loaded him into the wheelchair and took him inside. Within a few minutes a doctor came out and said our friend would survive, but his colleague was just putting a couple of stitches in his head, and how did I enjoy being in Finland, and would I like a cup of tea or coffee?

Shortly thereafter our friend was brought out, now wrapped in hospital blankets and with a large bandage like a turban on his head. By this time the others in the party had come into the hospital and were happily chatting up the nurses, but when the matron appeared she gave them a great blast and ordered them all home at once, and our damaged friend was ordered to bed for two days.

When we got back into the van there was another long conversation in Finnish, and the outcome was that, as I had done so well at the hospital, I was elected to take our friend home to his house and explain the circumstances to his wife, again in the fond hope that she would not go off too loudly in front of a visitor. This, I thought, was going a bit too far. For all I knew his wife did not speak English, while I understood about ten words of Finnish and could get by with some difficulty in Norwegian, which can be understood by most Swedish-speakers, even though I had been told that I had a dreadful Bergen accent; all in all, the situation was pretty fraught.

After some further persuasion I agreed to take him in, and we struggled up the path to his house with some difficulty and rang the doorbell. A charming young lady appeared, and before I could explain the situation she had dashed back into the house shouting for her mother to come quickly: "Father has been damaged!"

Another lady came out, and together we half carried our friend into the house and laid him out on a settee. His wife spoke excellent English, and I tried to explain what had happened: that he had been to the hospital, and been passed fit for going home, but should stay in bed for a couple of days, and I had to be going as there were others involved.

She came to the door with me, thanked me profusely, and took careful note of the others in the van, greeting each by name, and returned to her house.

"Thank God that is over; you have no idea how that woman can react sometimes," someone remarked. "Let us now return to the hotel for dinner."

So, back we went, and after reassuring the hotel manager that all was well, and under control, we sat down to a great Finnish feast, complemented by that amber liquid we had previously consumed with our prawns on our earlier visit.

Our damaged friend stayed off work for about a week, and returned full of beans, saying that the only quick way to attract attention in the hospital was to be taken in by a visitor. He seemed to have squared things away with his wife, as I was invited over for dinner a few weeks later, and had a very pleasant evening.

The language in Finland is quite different from any other European language; in fact, I have heard it said that there are about ten common words between Finnish, Hungarian and Romanian, and that is the lot.

As a non-Hungarian- or Romanian-speaker, this is not a lot of help.

On one of my earliest visits to the main square in town one evening, when all the shops were shuttered up, I found it almost impossible to identify the merchandise available from the title of the name on the shop front. The only one I managed to guess correctly was the instrumentarium, which, of course, was an optician, who also sold cameras, telescopes, microscopes and – for all I knew – periscopes for submarines.

In all fairness, some of the shops had merely the owners' names on the frontage, so this gave no clue whatever.

Another example of the difficulty was evident from the notices of occupancy of the cabins on board Finnish ships. The general practice was to have the names set out in Swedish and Finnish. From the Swedish, it was always possible to make an intelligent guess – kapitain was captain, for example – but the Finnish equivalent was some unpronounceable selection of many consonants, mainly 'j's' and double-'t's', which gave no clue whatever.

The only solution was to carry the handbook, and if they still failed to understand speak more slowly, and a bit LOUDER.

The price of drink was prohibitive, and it was incumbent on all visitors to bring duty-free liquor to the resident staff. Some people took this to excess, and on one memorable occasion one of our crew was stopped on arrival. He passed as 'Nothing to Declare', but was asked to open one of his bags, to reveal four bottles of spirits. Another bag was opened to disclose four more bottles, and these eight were subjected to duty, making them marginally cheaper than the local prices. Our friend was still quite happy, as he had another eight bottles in the remainder of his baggage, so he was well ahead of the game.

My section of the work was drawing towards completion, and I made my arrangements to return to Scotland, for a spell of leave before taking up my next assignment, which was to be concerned with drilling for gas in Yorkshire

CHAPTER 167
Yorkshire Gas

After my leave I went down to London to find out about the next phase of my career.

The company had taken over as operator on some gas fields in Yorkshire, and drilling was already under way. I was given a pile of drawings and told to go off and think about the requirements for establishing a drilling rig in various possible sites around Maltby, Pickering, and in the Dales.

The general areas had already been selected, but the particular sites were as yet uncertain. This is a common state of affairs in all exploration activities; you never know what you will find until you have drilled, and, as often as not, the results raise other questions as to where you should go looking next.

On top of the technical problems in selecting particular drilling locations, with so many different potential fields available, the priorities kept changing, due to political and environmental considerations, the dates of individual county council planning committee meetings, discussions with the possible landowners, disturbance of crops, and the hunting season.

The drilling operation is basically a very simple and brutal process, and comes in three main parts. First, you bash a hole into the ground and fit a pipe around it.

Next you grind out a deeper hole, using a rotating drill bit, which is turned by a turntable at the drilling deck floor. Finally, the material ground up in the hole is carried up to the surface by means of a fluid called 'drilling mud'. This material has three main purposes.

Firstly, being fairly dense, it allows the ground-up material to be carried back to the surface. Secondly, it helps to support the wall of the hole, which has not yet been sleeved off. Thirdly, again because of its density, it creates a pressure at the foot of the hole sufficient to prevent any oil or gas bursting out to the surface. Where there is a so-called gusher, it is because the drilling engineer has miscalculated the necessary density of the drilling mud.

When the drilling mud carrying the ground-up material from the bottom of the hole arrives at the surface it is screened, and the reusable material is returned to the mud tanks, while the brought-up material is dumped in a holding pond, with samples being taken at frequent and regular intervals. These samples are then dried and examined under a microscope, and logged, with any special indications noted.

Sudden changes in the type of material being brought up may require alterations to the density of the mud mixture, and if any traces of hydrocarbons are identified the rig is put on conditions of extreme caution, and progress continued.

At the first signs of any hydrocarbons in the mud a message is dispatched to the head office, and the site is soon invaded by all manner of experts, who restrict progress and get in everybody's way.

There is also a veil of secrecy thrown around the situation, as a good result might have some effect on the share price, but any sensible observer would need only to look at the local hotel register, or count the number of expensive motor cars arriving on site, to see that something was afoot, and a study of the bar takings where the drillers congregated would also render interesting indications of the prospects.

When the hole has reached its target depth, and has been lined with a steel pipe, this pipe is perforated by the use of a small bomb at the appropriate depth, and then, by reducing the weight of the drilling mud, the pressure of the oil or gas displaces the mud from the pipe and comes to the surface, under control.

Various tests are then conducted to estimate the possible flow from the well, and a system of flares is established to burn off any excess pressure. At the end of all this palaver the well is closed down, the rig dismantled, the site cleared, and the whole caboodle moves on to the next site, where the performance is repeated.

As there is no fixed period that can be estimated for the completion of a hole, it is necessary to have the next site selected and prepared in good time, and this is sometimes quite difficult.

The selection of the actual optimum location for the well is determined from a study of the available geological records. These can be a combination of the results of nearby drilling logs and the results of seismic surveys, which are obtained by causing small explosions at the surface and detecting the echoes of these explosions from various different strata underground.

This position is then plotted on an Ordnance Survey map, and, in the nature of things, is frequently in impossible locations, such as the middle of a lake or pond, or across a railway line, or at the intersection of different boundaries, be they county council, National Trust, local farmers, school boards, the Forestry Commission, and a number of others. All of these must be approached for their approval, and then the process carried through the appropriate county council – a process that can take many months.

Against these conditions, it was necessary to have a number of potential sites and alternatives in varying states of readiness, and this all took a large portion of my time.

We had considerable success on one particular hole during my stay in Yorkshire.

The site was being drilled on behalf of a small oil company, and the owner was a regular visitor. We began to get good indications of substantial quantities of hydrocarbons in the screenings, and news of this must have leaked out to the outside world. The owner appeared on site with his whole family and announced that they had just become millionaires that morning, due to movements in the company stock – and would we all like to join them for lunch at a nearby restaurant?

My colleague, the drilling engineer, was unable to leave the site at this critical stage of the operation, so we departed in good spirits, promising to bring him back a 'doggy bag', and some champagne to wash it down.

When we were returned, after an excellent lunch, it was considered that I should not attempt to drive back to my hotel, and I stretched out in the spare room in the drilling engineer's caravan and slept the sleep of the just.

Next day, bright and early, the well 'came in'. This process was a formal means of displaying that there were hydrocarbons in the well. The density of the mud was reduced until the pressure of gas had displaced it all, and the gas was evident at the well-head. The well-head was then closed down by operating a series of valves, known as a 'Christmas tree', and then testing began, probably causing a further vast increase in the share price.

The situation concerning other sites was now in a reasonable state, and I was recalled to London and asked to pack up my section of the operation, with a view to starting work on another oil platform construction project, based on the Glasgow office of the oil company.

CHAPTER 168
Moving House

Throughout this later period we had maintained our house in Invergordon, with me commuting between home and the various locations. Sometimes it was not too bad travelling daily to Ardersier, near Inverness, but Grangemouth was a bit too far and Yorkshire even further.

When I was appointed to the contract in Glasgow we decided that we should move down to the Glasgow area. There were a number of circumstances that made this move an attractive proposition. For a start, Susan was now at university in Edinburgh, and we had sold the horse (amidst some tears). The drive down the A9 was pretty tiresome, and often difficult in the wintertime, although it was being improved by opening bypasses round most of the villages and small towns on the route.

The possibilities of further work in the north seemed to be receding, and a more central base looked like a better proposition. On top of all that, I was getting pretty fed up with spending a lot of my off-duty hours driving up and down the country at great expense, and being away from home.

Having taken this momentous decision, I found some reasonable lodgings near the office, and began house-hunting once again. First consideration: can we afford it? This was governed to some extent by the price we could expect for our Invergordon house, and this in turn was suddenly reduced by the closure of the smelter, putting a number of similar houses on the market.

The next consideration was the convenience of transport into the city, so we drew a 30-mile-radius circle on the map and examined the facilities.

Next we discarded any places less than a couple of miles from the sea, and from this much-reduced area I set out to collect brochures from all the advertised agents.

The next job was to make a flying visit to these places in the evenings, to try to establish the kind of property we might expect to be able to afford.

As a result of these visits we had to revise our expectations, and we prepared a shorter list of the properties that we thought might be worth a proper visit of inspection.

I made a flying visit to confirm that at least the overall description was reasonably accurate, according to house agents' terms of reference – e.g. 'stns thrw fr bch' did not refer to a bloody great tree in the garden – and after a further revision of the lists Jean came down and we spent a few evenings and a couple of weekends on a tour of inspections.

Some of the places were very nice but suffered from one or more unsuitable factors, e.g. no garage, too many large trees about, with consequent swarms of midges, small garden, enormous garden, proximity to boarding kennels, backing onto a motorway or railway line, north-facing frontage, possibility of inundation by tidal wave, or greenhouse effect, liability to subsidence from old mining workings, no room in the garden for the boat, etc., etc.

After a few visits we had settled for a fine house near Skelmorlie, but each time we visited the agent he produced another likely property, which in turn had to be visited, and judged against the previous selection.

This purchase fell through, and as we were running out of time regarding the sale of the house in Invergordon something had to be done.

Another suitable place came on the market near Helensburgh, and after one quick visit we agreed on the price and set the wheels in motion for an early move.

Apart from the legal formalities, and the fact that I was still living in lodgings and travelling at weekends, our move was complicated by the fact that the boat was in winter storage, so fitting out, launching and mooring had to be completed before we could make the passage down to the Clyde, and a suitable mooring had to be assured for our arrival.

Arrangements were made for a bridging loan to be made available through the company for the short period between moving in and the completion of the sale of our Invergordon house, but when the day approached it seemed that no one was available or competent to sign these documents, and I had to dash around Glasgow making alternative arrangements before the deed could be completed.

This left a bad taste in my mouth, and some hard words were spoken. All was sorted out in the end, however, with profuse apologies, and all was nearly forgiven.

We packed up the house in Invergordon with many regrets, as it was a very pleasant place to live: convenient to schools, the golf club, the sailing club, shops, charming neighbours, and a fabulous view up the Cromarty Firth.

One point that caused near-panic on the day of our departure was the disclosure of a large unpainted patch on the wall of the lounge, which had been covered by a large bookcase that I had neglected to move during an earlier fit of decoration.

There was a brief discussion as to whether a quick slap of paint was required, but this activity was ruled out on the grounds of their being no paint left and the brushes being all packed up.

I had been collecting odd lengths of timber off the beach from time to time, and had taken all this material out and stacked it neatly behind the garage, ready for collection on a further visit.

We set off, following the pantechnicon in two cars loaded up to the eyebrows, the arrangement being that I would go on ahead to collect the keys and have the gas and electricity turned on before the arrival of the other car and all our worldly goods.

When I arrived, and had opened up the house waiting on the arrival of the other car, I walked round the garden, and one of our new neighbours came over.

"I hope you are not another of those gardening fiends," he remarked, nodding towards another garden across the road, which was beautifully laid out with a border of daffodils in perfect alignment. I was able to assure him that he need have no worries on that score.

CHAPTER 169
Offshore again

At the start I was concerned with the design of the main jacket, or platform legs, and spent most of my time in the office, or on brief visits to suppliers and construction sites, and latterly went offshore when the fabrication was completed to assist in the hook-up and commissioning.

I went out with the installation crew, and was present during the lift-on of the various modules from their ferry barges. This time the installation vessel was an enormous barge with two huge cranes over one end, and a very sophisticated ballast system that could pump water across and along the barge at great speed to keep the deck level during lifting and slewing operations.

The weather was pretty settled while this work was continuing, and as it drew to completion the hook-up and commissioning crew were assembled and began to be ferried out to the accommodation barge. This vessel arrived from Norway, and we were transferred by helicopter while she was still steaming around the area, waiting for the lifting barge to clear her anchors.

These barges were typically dual-purpose vessels, large pontoons carrying six cylindrical legs, with living quarters, storage area and helicopter deck, and also available to be converted into mobile drilling platforms.

In service as an accommodation vessel she would be ballasted down and moored by a pattern of eight anchors, with sufficient scope to manoeuvre close up to the platform, and this particular vessel had a special hydraulically-operated bridge to maintain access to the adjacent platform.

We were still steaming about, waiting for the weather to settle down before we set the anchors, and the movement was quite lively. The new hands of the commissioning department were all mustered and taken on a tour of the barge, and then called to their first progress meeting. They had claimed a corner of one of the large office spaces and were all sitting round a table, being advised of their duties and responsibilities, keenly taking notes, and asking intelligent questions. One of their number, at sea for the first time, was observed to remove his safety helmet, bend over and be violently sick into it, and then conceal it under his table until the end of the lecture. Very commendable.

Until we were properly moored alongside the platform there was not a lot we could do, so we had frequent fire drills and lifeboat drills to keep people on their toes.

The accommodation was divided between two-man and four-man cabins, and we heard of one poor fellow, a newcomer to the offshore business, who was rebuked by his fellow occupants for not wearing his life jacket in bed. They all solemnly appeared, dressed up in their life jackets – hard cork, square things – climbed into their beds, and drew the curtains. Immediately inside, they undressed and went off to sleep.

In the morning the process was reversed, and they all appeared dressed in life jackets as they got up, but the poor newcomer had not slept a wink all night, and appeared all stiff and crooked. News of this soon got round and it was decided to play the same trick on some of the later arrivals, but as far as I heard nobody took the hook, and, anyway, by then the shift system was in operation, with only two men in residence at any one time.

The assembly work continued, and soon the drilling operations were started on one end of the platform while work was continuing on the completion of the other sections of the facilities: the accommodation units, administration offices, and other equipment that would be used to separate the oil and gas from the water that would come up when drilling was completed.

This work required a lot of welding, and as the fire-detecting equipment had already been installed we had numerous fire alarms arising from the sensitivity of the visual sensors. Apart from disturbing the progress of the work, these alarms were personally a great nuisance. My cabin on the barge was two decks down from the main deck, and each day I had to get up, go down one set of stairs to the mess, return to my cabin (up another set of stairs), collect my outside boots and overalls, go up two sets of stairs to the main deck, up another set onto the access bridge, and then up another 100 feet to my office. From there I had to go down to whatever level the work was in progress, and then back up to the office.

These trips were necessary each mealtime, and again at the handover period at the end of the shift, so that I estimated that in a normal two-week spell offshore I had climbed the equivalent of more than halfway up Everest.

By Sod's Law most of the fire alarms sounded just when I had reached the highest point of my circuit, and I had to descend at great speed to my lifeboat station, which was on the lowest deck. Conversely, if I was on the barge when the alarm sounded, I had to dress quickly and climb up to the main deck to my station on the barge.

At the end of a spell of two weeks offshore I was both very fit and quite worn out, and after the first couple of nights' sleep ashore at home ready for all manner of gardening, golfing and house repairing activities that might present themselves.

We were graced by a visit from a very senior official who had some rather peculiar requirements. His minion came out a few days in advance to prepare the way for the lord, so to speak. It appeared that he liked his tea from a silver service, and this was provided at some great expense. It also appeared that he had fallen out with one of the construction companies that had build the accommodation module, and we were ordered to remove a small brass plate at the entrance referring to the constructors.

Inside the accommodation there were a number of paintings that had been painted by the local schoolchildren, and there were small plaques by each picture stating the name of the artist and the fact that it had been selected by the constructor for display in the module.

It seemed that the very mention of the constructor's name might bring forth a heart attack, and we were instructed to remove all these pictures during the visit.

When the great man arrived he was taken to the installation manager's office, then to the dining room, which had been swept and polished until it shone, had his cup of tea from the silver service, and then climbed back up to the helicopter deck and departed, closely followed by his assistant bearing the silver service.

We had some pretty bad weather during the winter, with fog and ice affecting the flying routine. This caused some disruption in the manning, and led to extended periods at sea, and also periods of standby in Aberdeen. The system was that, generally, flights ran every weekday, and there was a change-out pattern whereby people offshore were released only when their replacement had arrived on board. Staff were expected to report to the helicopter office in Aberdeen early in the morning, and if there was no flying they were stood down until the following morning, but staff due to go out the following day were also expected to report in, so there was often a shortage of beds in the hotels around Aberdeen on these occasions.

We came in late one evening from offshore and there was a rush to organise a hire car. At that time there were three main agencies with hire desks in the airport foyer, and, what with the sudden rush of potential customers and the inadequacy of the staff and organisation, some mix-up resulted in customers from one company going off in another company's cars. What seemed to have happened was that someone was told: "It's a red Escort, and the keys are inside," and made off with the first red car they came across. The next customers found another car with keys in the switch, and set off in hot pursuit. The next man came back to the desk complaining that he could not find his car, despite having a bunch of papers, and then it was realised what had happened. At least two cars were off on the road with incorrect insurance and hire documents, and liable to a nasty surprise when they tried to hand their cars in to the wrong hire company in the morning. Common sense seemed to have prevailed, as they were given polite directions to the correct depot, and all turned out to be well. The telephone can sometimes be a force for good.

With the main structure complete, and all the modules operating, my part in the job came to an end, and I returned to the London office to see what else was in the offing.

The oil-related work was at one of its least active troughs, and I was asked to take on a contract at Sellafield, building a storage facility for radioactive waste – a contract that was estimated to run for about three years – and after considering that the alternative was a return to the London head office I agreed to go to Sellafield.

CHAPTER 170
Sellafield

Sellafield, or Seascale as it was previously known, lies on the west coast of Cumbria, about opposite the Isle of Man. It was used during the war as an ordnance factory, and was later developed into a facility for dealing with nuclear waste.

As on all nuclear installations the security arrangements were fairly strict, with armed police on the gate, and a high wire mesh fence all round the premises. I had to make special arrangements in advance of my arrival on site, and when I presented myself at the main gate had to sign all manner of documents and await the arrival of an escort from our site before I was allowed in. One of the staff came over and I was taken to the site, which was in a very early state of development. The area had been levelled, and a thin skin of weak concrete had been spread all over. Our premises comprised a couple of wooden huts and three Portakabin-type caravans, with neither water, electricity, nor sanitation, but a promise from London that further accommodation would be on its way.

By this time it was about one o'clock, and we went over to the canteen for some lunch. I cannot remember seeing such a disgusting place on any site I had ever visited, and the food was totally awful.

Having had such difficulty in getting into the site I was not prepared to go outside again for lunch, so I had a couple of Mars bars and a bottle of lukewarm lemonade to keep the pangs at bay, and resolved never to eat there again.

How can people who run major companies dealing with dangerous materials operate such a wretched canteen?

Out on the site we were agreeing the main survey lines and levels that would govern our construction, and setting up the reporting and accounting procedures with the client.

One of our first requirements was to have all members of the site crew indoctrinated into the site safety procedures, and for the senior staff to become acquainted with the special requirements of site safety, and the various permits to work systems that had been established to reduce the chances of someone digging up a main cable and putting the lights out from Barrow to Carlisle.

The authority conducted site safety courses every week in one of their lecture rooms, and this induction went through the explanation of the various hazards possible to be encountered on the site, the precautions to be taken in general, and the routine for behaviour in the event of an emergency.

The basic idea was that there would be assembly areas all round the site, and in the event of an emergency being declared the siren would be sounded, and everyone was directed to get inside the nearest designated building, report his presence, and be issued with a respirator. The senior designated person would then make a headcount, and report his tally to the safety office and wait further instructions.

The safety office would assess the situation and instruct the senior person whether or not to issue iodine tablets, and arrange for transport to come along and ferry the people off site.

The idea was that, if required, a bus would come along, the people in the assembly area building would put on their respirators and go aboard in an orderly fashion, the bus would take them to an established reception station where they would be looked after, and the bus would return for the next consignment.

I could imagine a situation where the passengers would encourage the driver to keep on going for a considerable distance, and he might be reluctant to return for the next load.

At the end of the course we were invited to examine the equipment available, try on the respirators, and marvel at the sensitivity of the detecting apparatus.

The instructor set off a very mild alarm signal in the room by holding his watch near the detecting head, and invited others to try this trick. When I put my old watch near the head, bells began to sound all through the building, and there had to be a bit of excited telephoning to explain that this was just a demonstration and all was well. I suppose that, by rights, I should have been going about all lit up like a Christmas tree, having been wearing that watch for all of 20 years, but no one seems to have noticed.

This was the time of the great blossoming of quality assurance, and we all had to write up procedures, and manuals, and sign our lives away over all sorts of obscure details. This reminded me of my earlier time in the nuclear power station industry, where it was rumoured that one station in America had to consider the possibility of a railway train running off the tracks opposite the station, and huge earthworks were demanded by the safety authority to protect against this possibility, and the layout of a number of twin-reactor stations in this country had to be such that, if the largest known aeroplane in service was to fall with one engine into the centre of one reactor, it would not be possible for another engine to land on the second reactor.

There must be people about who spend their time dreaming up these compound-disaster scenarios, and they must be very choked off when some simple engineering solution is found to deal with them. Unfortunately, such simple engineering solutions normally add to the cost, and can run into long delays before being accepted, or the scenario destroyed and some other consideration dreamt up.

Our building was to handle, pack and store relatively low-level waste, and had remote controlled handling, packing and sorting equipment installed in a safe area, behind thick concrete walls.

The packed material had then to be loaded out onto an internal railway system, and parked in a large storage area, which had sufficient capacity for the next 50 years.

During this construction period I lived in lodgings nearby, travelling home to Helensburgh each weekend, and became totally fed up with all this travelling.

One feature of such large construction sites is the pointless meetings that must be attended, mostly with subjects quite unconnected to the construction work in hand, and one method I found to relieve the boredom, and concentrate the mind, was to arrange my seat with my back to the window and focus the sunlight through one lens of my spectacles onto the knee of the person sitting alongside. This was an exercise requiring great steadiness of hand, and during all my time spent at such fruitless meetings, although I was never able to raise smoke, I did manage to cause some quite violent reactions.

Apart from all this frivolity the work proceeded more or less according to plan, and came to a satisfactory completion, with all the gadgets operating successfully, and several hundredweight of paper duly signed off to prove it.

At the end of this contract I was asked to move down to the head office in London but as I had managed to evade such a fate for the previous 30-odd years, and was losing my passion for weekend travelling, I politely declined, and opted for early retirement instead.

A decision that was, I think, one of the best I have ever made.

ABOUT THE AUTHOR

Born in 1929, the author attended the High School of Glasgow before studying as an articled apprentice civil engineer. Then came two years' national service, followed by three years as an assistant engineer, before he achieved full responsibility as an engineer in his own right. His other interests include sailing, maintaining his 50-year-old racing yacht, and walking the dog. He has also invented and patented some useful devices, among them a small hand-held sextant for measuring tacking angles ("all done by mirrors") and apparatus to test steel wire rope.